QUALITY AND STABILITY OF FROZEN FOODS

Time–Temperature Tolerance and its Significance

QUALITY AND STABILITY OF FROZEN FOODS

Time–Temperature Tolerance and its Significance

EDITED BY

WALLACE B. VAN ARSDEL
Assistant Director, Emeritus

MICHAEL J. COPLEY
Director

ROBERT L. OLSON
Assistant Director

Western Utilization Research and Development Division, Agricultural Research Service, U.S. Department of Agriculture, Albany, California

WILEY-INTERSCIENCE

A DIVISION OF JOHN WILEY & SONS
NEW YORK · LONDON · SYDNEY · TORONTO

The paper used in this book has pH of 6.5 or higher. It has been used because the best information now available indicates that this will contribute to its longevity.

10 9 8 7 6 5 4 3 2 1

Library of Congress Catalog Card Number 76–79148
SBN 471 897906

Printed in the United States of America

PREFACE

This volume is an outgrowth of extensive investigations conducted some years ago by the U.S. Department of Agriculture, encouraged and aided by farseeing leaders of the frozen food industry of the United States. The studies were designed to extend and strengthen the scientific, technological, and engineering foundations of this great industry, which utilizes many perishable agricultural commodities. The publications based on the work were known by the phrase used in the subtitle of this book, The Time–Temperature Tolerance of Frozen Foods.

Not only were the T–TT publications themselves voluminous, but they have been followed by extensive reports of important work in Canada and Europe as well as elsewhere in the U.S. These newer publications cover both a wider range of perishable food commodities and the rapid changes and developments in the technology of frozen food production and marketing. The consequence is that this field of knowledge is now in much need of a critical review and summary. The central objective of this volume is to meet that need.

To a quite unusual degree, this rapidly growing industry is technically minded. As is pointed out in the opening chapter of the volume, frozen foods have been, from the beginning, a scientific or technological development. Their present-day production and marketing call for the close attention of trained personnel and constant reliance upon scientific guidance. The Editors believe that the plants and laboratories of the frozen food industry and its allies and suppliers will furnish a large proportion of the users of this volume. Educators who need to offer a broad view of modern food industry to upper-division students will find this survey useful. The nonspecialist educated person will, in fact, find that only a few sections of the book are highly technical and many of the chapters will interest readers who wish only to keep abreast of current affairs by having an inside look at a dynamic industry.

An introductory chapter outlines the circumstances in the frozen food industry which led to the time–temperature tolerance investigations, and traces some of the more important consequences, in terms both of the *quality* of the product and of the *stability* of that quality during the subsequent marketing experience. The next two chapters describe and discuss the two classes of change which may occur in frozen foods to alter their initial quality in the course of time, namely, chemical reactions and physical effects, as one group of changes, and the effects of the presence and growth of microorganisms in the food as the other group.

Then follow six chapters devoted to more detailed examination of important classes of perishable foods preserved by freezing—fruits and fruit juices, vegetables, poultry and egg products, bakery products, meats, and seafoods, particularly the characteristics which are important for high quality and the ability to maintain high quality throughout the experiences of storage and distribution.

The last four chapters examine the concept and describe procedures of quality management and control in frozen food production and marketing; the many significant advances in the handling, storage, and distribution techniques which have aided the growth of the industry are described; the record of accomplishment of the industry in improving the conditions of handling and distribution of frozen foods is presented through summaries of recent industry surveys; and finally, the adoption of both voluntary and official standards of good practice in the frozen food industry, in Europe as well as in the United States, is summarized.

The editors of this volume are deeply grateful to the scientists and technologists who have so generously given of their time and special knowledge to make the book authoritative. Special thanks are due to the groups and organizations in the frozen food industry which bore the initial expenses of planning the book, securing a suitable publisher, recruiting well-qualified authors, and getting manuscripts into acceptable final form, as more fully recounted in the Foreword. We also acknowledge with thanks the kind permission granted by the following firms and publishers to use photographs of equipment and illustrations and tables which were previously published elsewhere: Academic Press, Inc., New York, N.Y., Alterman Transport Lines, Inc., Miami, Florida, American Association of Cereal Chemists, St. Paul, Minn., American Chemical Society, Washington, D.C., Clark Equipment Co., Battle Creek, Mich., Continental Freezers of Illinois, Chicago, Ill., Frozen Food Express, Dallas, Tex., Fruit Growers Express Co., Washington, D.C., Institute of Food Technologists, Chicago, Ill., International Association of Milk, Food, and Environmental Sanitarians, Shelbyville, Ind., Michigan Agricultural Experiment Station, East Lansing, Mich., Quincy Market Cold Storage and Warehouse

Co., Boston, Mass., Reinhold Publishing Co., New York, N.Y., Royal Society, London, Sea-Land Service, Inc., Elizabeth, N.J. (Subsidiary of McLean Industries, Inc.), Terminal Ice and Cold Storage Co., Portland, Ore., U.S. Department of Agriculture, Washington, D.C., U.S. Public Health Service, Washington, D.C.

W. B. Van Arsdel
M. J. Copley
R. L. Olson
EDITORS.

FOREWORD

One may assert confidently that no facet of the American food economy has been more completely a product of scientific and technological research and development than the frozen food industry. It is fully justifying the faith of its pioneers, less than a half-century ago, that this completely new method of distributing highly perishable foods could be carried out practically on the largest scale. Its growth to that stature, though, progressed despite some temporary setbacks only because industry and government have extended and deepened our exact knowledge of commodity behavior whenever the need for the information became apparent.

The U.S. Department of Agriculture, which had been deeply concerned with frozen food research for many years and was closely in touch with developments in the industry, became aware in the period about 1947–1948 of an urgent need for further investigations, especially into the maintenance of the high quality of commercially packed frozen foods under the conditions of temperature and time they would experience in nationwide distribution. Authoritative leaders of the frozen food industry in the United States recommended to the Department a thorough research study of the entire situation, and pledged full industry cooperation in such a study. As more fully recounted in the introductory chapter of this volume, the Department designated the Western Regional Research Laboratory (now the Western Utilization Research and Development Division of the Agricultural Research Service) at Albany, California, to undertake this responsibility. In the course of the next several years a great body of new knowledge about frozen foods came out of the laboratories and pilot plants of the Department. Technical results appeared especially in a series of articles in *Food Technology,* under the general heading, "The Time–Temperature Tolerance of Frozen Foods"—T–TT for short. Stories in trade journals, talks before meetings of industry associations, and other forms of educational publicity were all used to assure a wide hearing for the message.

These T–TT studies by the Department of Agriculture, although by far the most searching investigation of the quality and stability aspects of frozen foods that had ever been carried out, were somewhat limited in the number of different kinds of frozen food that were considered. Other Department laboratories later broadened the studies, and have continued to carry out important work in the frozen food field up through the present time.

Significant research and development work on frozen foods was, of course, being done in many other governmental, university, and industrial laboratories during this period. The number of foods being successfully frozen and the extent of public acceptance both grew steadily. Publication of the T–TT results stimulated other important studies in the underlying sciences and technologies at a number of places in North America and Europe.

Lessons taught by the time–temperature tolerance publications were taken with the greatest seriousness by responsible organizations in the frozen food industry, and were eventually translated into substantial physical improvements in transportation equipment and in handling practices. Efforts by industry leaders and by governmental regulatory bodies culminated in broadly supported rules of good practice in processing, handling, and storing frozen foods.

In short, the time–temperature tolerance investigation, extensive as it was, was only one phase in a continuing process of growth and improvement in the frozen food industry.

Many people have expressed the feeling that this whole subject of the maintenance of high quality in a frozen food has grown so large and complex as to be quite forbidding to the average intelligent person working in the industry or seriously interested in it. A book such as the present one, it is thought, can serve to introduce that person to the subject. Both the digested and summarized time–temperature tolerance results and those of important related work by other investigators can be brought into focus here. More people will see how the results can be applied in still other ways. As a reference volume, the book will provide ready information to workers in the related industries, food technologists who are not frozen food specialists, and educators.

Sponsoring the publication of the book, and assisting financially in late 1966 with its initial expenses, were six organizations prominent in the frozen food industry: National Frozen Food Association, National Association of Frozen Food Packers, National Association of Refrigerated Warehouses, The Refrigeration Research Foundation, Birds Eye Division of General Foods Corporation, and Trans-American Refrigerated Services, Inc.

This volume is, then, a commendable example of realistic cooperation between industry leaders and scientists conducting research on frozen foods for governmental, university, and private organizations. Its sponsors hope and expect that the information thus made widely available will assist materially to push back the great areas of ignorance about the properties and behavior of our food commodities, and thus reinforce the continuing efforts of far-seeing people who have worked to protect the widespread public acceptance now enjoyed by frozen foods both in the United States and abroad.

H. C. Diehl, D.Sc.

Colorado Springs, Colo.
July, 1968

CONTRIBUTORS

F. Bramsnaes *Food Technology Laboratory, Lyngby, Denmark*

Charles H. Byrne *Birds Eye Division, Research Laboratories, General Foods Corporation, White Plains, New York*

Michael J. Copley *Western Utilization Research and Development Division, Agricultural Research Service, U.S. Department of Agriculture, Albany, California*

Henry N. Couden *Safeway Brands Buying Division, Safeway Stores, Inc., Oakland, California*

L. E. Dawson *Food Science Department, Michigan State University, East Lansing, Michigan*

H. C. Diehl *Colorado Springs, Colorado*

William C. Dietrich *Vegetable Laboratory, Western Utilization Research and Development Division, Agricultural Research Service, U.S. Department of Agriculture, Albany, California*

K. G. Dykstra *Birds Eye Division, General Foods Corporation, White Plains, New York*

R. Paul Elliott *Microbiology Group, Technical Services Division, Consumer and Marketing Service, U.S. Department of Agriculture, Washington, D.C.*

D. G. Guadagni *Vegetable Laboratory, Western Utilization Research and Development Division, Agricultural Research Service, U.S. Department of Agriculture, Albany, California*

K. M. Hayes *Department of Food Science and Technology, University of Massachusetts, Amherst, Massachusetts*

Eugene F. Jansen *Plant Enzyme Pioneering Research Laboratory, Western Utilization Research and Development Division, Agricultural Research Service, U.S. Department of Agriculture, Albany, California*

Mogens Jul *Protein Advisory Group, United Nations, New York, New York*

H. David Michener *Vegetable Laboratory, Western Utilization Research and Development Division, Agricultural Research Service, U.S. Department of Agriculture, Albany, California*

Robert L. Olson *Western Utilization Research and Development Division, Agricultural Research Service, U.S. Department of Agriculture, Albany, California*

James W. Pence *Cereals Laboratory, Western Utilization Research and Development Division, Agricultural Research Service, U.S. Department of Agriculture, Albany, California*

F. M. Sawyer, *Department of Food Science and Technology, University of Massachusetts, Amherst, Massachusetts*

Wallace B. Van Arsdel *Western Utilization Research and Development Division, Agricultural Research Service, U.S. Department of Agriculture, Albany, California*

CONTENTS

Chapter 1

INTRODUCTION

W. B. Van Arsdel, M. J. Copley, and R. L. Olson
Western Utilization Research and Development Division, Agricultural Research Service, U.S. Department of Agriculture, Albany, California

I. COMMERCIAL FROZEN FOODS ARE A RECENT PHENOMENON

Frozen foods are now so familiar a part of our lives that we find it hard to realize how recent a phenomenon they really are. We now take them for granted. However, the middle-aged person today can easily remember the first can of frozen orange juice concentrate he ever saw, and the first plump, bright green peas put on the dining table in the dead of winter. We may already have forgotten how miraculous it seemed then that such things—and others even more exotic—could be bought in the nearest market.

To be sure, we had long been familiar with some significant forerunners. Ice cream and sherbet were no novelties. The general public was hardly aware of the large-scale movement of frozen meat, poultry, fish, eggs, and "cold-pack" berries of several kinds, and the term "cold storage" connoted merchandise that was *not fresh* and therefore, possibly inferior.

What we know as the "frozen food industry" really emerged into the public eye when "fresh-frozen" vegetables, fruits, and seafoods appeared in the markets in the early 1930's (1).

A. The Scientific and Technologic Base

The commercial freezing preservation of perishable foods came as the culmination of extensive scientific and technologic background work in a number of different institutions and then finally a determined move ahead into large-scale production and distribution. More than other major food-preservation methods, it had been built upon research from the beginning. Rapid and full publication of experimental results had characterized the development in both public and private institutions, so there were never any important elements of secrecy and few restrictive patents. Much of the early study had been carried out by Pennington (3,4), Diehl (5,6), and other workers in the U.S. Department of Agriculture (for example, see refs. (6) and(7)), and several state Agricultural Experiment Stations (8–14). By this time the published knowledge had become sufficiently extensive and detailed to warrant publication in hard-cover books (15–18). The early days of the industry were described in review articles by some of the principal actors in the drama, for example, Birdseye (19,20); see also the account by Williams (21).

B. A Technically Oriented Industry

The serious commercial developments were naturally designed around the best results of the small-scale experimental trials. In many cases the men who had been involved in the research went on to direct segments of the commercial development. To a marked degree, the commitment to technically sound procedures has continued to characterize the industry—not an easy philosophy to follow in the keenly competitive struggle for the limited cabinet display space.

The several national, regional, and state associations of companies within the industry have done much to strengthen this attitude, through extensive "cuttings" of comparative samples, educational training courses and use of training materials, industry-sponsored resarch on problems of general concern, and the like. Preparation of this volume, as has been explained in the Foreword, has been made possible through that kind of industry collaboration.

C. The Meaning of "Quality" and "Stability"

Of the two key words in the title of this volume, only "stability" is quite readily supplied with satisfactory quantitative meanings. "Quality," on the other hand, is a notoriously vague and elusive thing. Everyone is quite sure he knows what he means when he uses the word, and yet if he tries to pin down the meaning it turns out to be uncommonly slippery.

As an introduction to the collected papers of an AAAS Symposium entitled "Food Quality" (22), Irving and Hoover quote several of the definitions of "quality" which were applied by different contributors to the symposium (23–25). They range from the simple, pragmatic test, "that which the public likes best," to the composite of properties or attributes which are thought to be most important to the economic success of a food commodity.

No attempt has been made by the editors of this volume to impose a single, closely defined meaning of "quality" upon the chapter authors. Consequently, the reader will find a considerable degree of variance in concept and approach—enough, no doubt, to leave room for some argument between exponents of differing views. This, the editors believe, is a reflection of the complex truth, rather than a fault.

The difficulties met in defining the concept "quality" are apparent in all discussions of means of measurement or specification. They extend even to the terms used by different investigators. For example, a number of the authors of this volume have made use of the fact that if samples of a particular food product are subjected for a period of time to conditions under which gradual changes in the properties of the product are bound to occur, and if a trained panel of observers notes the flavor or appearance of successive samples, there will surely come a time at which 70–75% or more of the observers will agree that the product is now distinguishably *different* in flavor or appearance from a sample of the same kind of product that has been rigorously kept at a very low temperature so as to be sure that any change in it has been negligibly small. This is a highly important principle, because it provides our closest approach to a quantitative measure of a *sensory* phenomenon. We badly need that dependable way of correlating the easily made objective tests (such as ascorbic acid loss or chlorophyll degradation) with the corresponding sensory changes. Examples of its use will be found in Chapters 4, 5, and 8–10 of the following text. But the difficulties of definition we have referred to are illustrated by the variations in usage in those chapters:

In Chapters 8 (Jul) and 10 (Van Arsdel) this key quantity is called "just noticeable difference" (JND); in Chapter 9 (Bramsnaes) it is "just noticeable change" (JNC); in Chapter 5 (Olson and Dietrich) it is just perceptible difference"; while Guadagni, (Chapter 4), although he uses the same experimental technique, gives no special name to the difference in sensation but only defines the *length of time* during which *no definite quality change* has occurred as the product's "stable period" or, in short, its "stability." Here again the editors believe it is desirable to allow the shade of difference to remain, testifying to the complexity of the facts.

"Stability," on the other hand, has the very simple meaning of "steadiness," or the ability to endure without substantial change. This seeming simplicity, though, is mostly illusory, because of the practical difficulty of making measurements (objective or subjective) of some attribute of a food product with the accuracy which is necessary if we are to determine whether or not some change had, in fact, actually occurred in it. This situation, with its consequences, leads to much of the discussion in Chapters 4, 8, and 10 of this volume.

In thus emphasizing the desirability of good *stability* in frozen foods we do not intend in any way to play down the vital importance of high *initial quality* in the freshly packed goods. This is the tangible factor which is directly within the control of the manufacturer, through his selection of raw materials and his meticulous attention to every detail of the processing and packing operation. Without an assurance of excellent quality in the newly frozen product, high stability of the packed material would have little interest to anyone.

II. THE GRADUAL EXPANSION AND BROADENING OF THE FROZEN FOOD INDUSTRY

As we remarked earlier, meats, fish, eggs, and some other perishable foods had been stored and shipped in the frozen state for many years before the present "frozen-food industry" became important. The whole marketing concept of "fresh frozen" or "quick-frozen" foods was a novel development, especially when it was applied nationwide to consumer markets for perishables. The logic of the situation led step step to the inauguration of radical improvements in many related branches of the manufacturing and distributing industries.

A. Related Developments

1. In Refrigeration

A great surge in the installation of household freezers or refrigerator–freezer combinations in American homes followed the large-scale marketing of retail packs of frozen vegetables, fruits, and seafoods, and in turn the ready availability of dependable frozen-food storage space in the home accelerated the growth in purchase and everyday use of the full gamut of frozen foods. The appliance industry estimates that by 1966 the total zero-degree (0°F., or −18°C.) food-storage capacity in U.S. homes totaled well over 5 *billion* pounds (26).

There can be no doubt that households and institutions in northern Europe are now in the midst of a similar great expansion in the ownership and use of food-freezing equipment (27), although the period of rapid growth occurred a number of years later. European equipment manufacturers have developed some of the most advanced freezing systems, and have sold many of them in the United States.

A striking example of the way seemingly unrelated developments contributed to the expansion of frozen food production just after World War II is furnished by the utilization of the fluorocarbon refrigerants. The whole field of fluorine chemistry had received enormous stimulation by wartime necessities, and among the "spinoff" benefits was the economical large-scale production of a whole series of fluorocarbon derivatives. Some of these turned out to have nearly ideal properties for a refrigerant in household freezers and refrigerators (28); they were odorless and nontoxic, and permitted good thermoynamic efficiency. A large proportion of the appliances manufactured for some years have used refrigerant R12, which is dichlorodifluoromethane, CCl_2F_2. Development of this and other refrigerants of its group dovetailed neatly with the creation of a mass market for household freezing units.

2. In Transportation

The refrigerated rail transport of perishable foods is now more than 100 years old, according to McKee (29); the modern ice-bunker refrigerator car which has evolved from the crude beginnings is now the standard means for bringing fresh fruits and vegetables great distances from western and southern growing areas to the population centers of the east and north.

However, the ice-salt bunker car cannot effectively maintain an interior temperature below about 15°F. (−9°C.). Consequently, after the war, when the long-distance tranportation of frozen foods became highly important, several kinds of mechanically refrigerated cars or "reefers" were tried out on the nation's railways. They were so successful that they now (1967) number about 18,500, or more than 28% of the total number of refrigerated cars. A typical present-day car costs about $30,000, so the total expenditure by the railroads to convert their equipment for satisfactory frozen food operation has been very large indeed.

Intermediate-distance transport of frozen foods is increasingly handled by over-the-road refrigerated truck-trailers, and of course short-haul and local distribution is done almost exclusively by refrigerated trucks. Redit (30) says there are more than 270,000 trucks, 75,000 trailers, and 5,000 containers in refrigerated service in the United States, about 50% of them capable of maintaining a temperature of 0°F. (−18°C.) or lower. Just as in the case of rail transport, the exacting requirements for the handling of frozen foods made very large capital expenditures by the trucking companies obligatory.

3. *In Packaging*

Mass distribution of frozen foods was just one of several new developments of the past quarter century which happened to coincide with a remarkable proliferation of new and better ways to package consumer goods. A satisfactory frozen food package is the product of highly sophisticated developments in such manufacturing fields as metal foils and flexible plastic films; at the same time, in today's self-service food markets the package also has an important selling function as well as protecting the food (31). It might be noted that the first in the numerous family of moisture-resistant cellophanes, forerunners of many other types of plastic film, was developed in response to the need for a frozen food package.

III. THE MAGNITUDE AND BREADTH OF THE FROZEN FOOD INDUSTRY

A. The Growth Curve

Production statistics for the frozen food industry are quite incomplete and uneven in quality, but all estimates agree on the phenomenal growth exhibited by the industry during the past 30–35 years. The estimated value of combined retail and institutional sales in 1940 was only $108 million, whereas by 1965 this value had grown to nearly $6 *billion* (32), an average rate of increase of about 17% per year.

1. Temporary Setbacks and Their Causes

However, the history of the industry was by no means an unvarying succession of record-breaking years; there were significant setbacks whose circumstances will repay study.

The two most conspicuous declines in total production (occurring from 1956 and 1957 to 1958, and from 1962 to 1963 and 1964) were both simply consequences of disastrous freezes in Florida and the resulting precipitous declines in production of frozen orange juice concentrate. However, the major decrease from a peak in 1946, not fully erased for several years thereafter, signified a completely different kind of difficulty—a loss of markets. Enochian (32) ascribes that experience to the effects of a derationing boom at the end of the war, accompanied by a hastily improvised and ill-prepared rush to produce frozen foods; the careless manufacture resulted in very poor quality of many of them—badly separated, or even gelled, orange juice concentrate, inedible precooked specialties, and so on. The presence of inferior products in the frozen food cabinets did serious damage, lasting for several years, to consumer acceptance of all frozen foods. Total production did not again reach a new record high until 1949.

2. The Major Influence of New Product Introductions

Study of the production statistics confirms the general impression that first one group of products, then another, has taken the lead in rate of growth and in public notice.

a. Frozen Orange Juice Concentrate. The first of these leaders was frozen orange juice concentrate, introduced on a large scale in 1948, and immediately creating a sensation. The product succeeded enormously in a field which had been mostly abandoned to the heat-processed fruit juices; new and better processes of juice extraction, low-temperature concentration, flavor improvement by add-back of fresh juice, and rapid freezing had created a product consumers were anxious to have. However, in national distribution serious troubles were not long in appearing and for a time the continued success of frozen concentrate seemed in doubt. Eventually the difficulties were traced to production practices and processing methods, and to neglect of properly low temperature (0°F. or below) during the entire distribution sequence. Extensive research and development were required before the difficulties were overcome.

b. Precooked Frozen Foods. A few fully cooked foods had been frozen since the early days of the industry, but all the emphasis was placed on the novel and attractive ability of freezing to preserve the *fresh* appearance and flavor of perishables. Many food packers had experimented with various precooked, frozen food products, especially gourmet and nationality foods. However, it was not until 1946, with the ending of price controls, that real interest began to develop. About 40 million pounds of precooked, frozen foods were packed that year (33). However, within a single year the bubble had burst. Martin (33) observes,

> Dozens of new packers had frozen anything that would freeze, without any scientific experimentation to ascertain proper methods for freezing, without any regard for quality, without any notion of proper packaging, pricing, . . . or marketing. The result was a near disaster. . . .

By 1947, consumers had so completely abandoned the precooked products that only 5 million pounds were sold that year.

Recovery from the debacle was slow. Such products as French-fried potatoes, meat and fruit pies, and some nationality foods led a steady and soundly based growth. Frozen dinners and platters began to appear. Breaded shrimp and fish sticks reopened the field of seafoods. Then the available selection of precooked or prepared item really began to proliferate over the entire spectrum of acceptable foods, with a big emphasis on those that were well adapted to institutional use, ranging from in-flight meals, through dishes for plant and school cafeterias, and hospitals, to high-class restaurants. By 1954 or 1955 total production in this group of frozen foods was beginning to lead in growth rate, and by 1966 the prepared frozen foods constituted about 20% of the tonnage of frozen foods produced and led all the other categories in dollar value, with more than $1.5 *billion* (33).

c. Frozen Potato Products. The early recovery of French-fried potatoes from the 1947 slump has been mentioned. However, growth of the output of frozen potato products has been well sustained, rather than spectacular. Production was about 3 million pounds in 1947 (32); by 1966 the output had grown to 1.46 *billion* pounds, almost equaling the total production of all other frozen vegetables.

d. Retail Pack of Frozen Meats and Poultry. It seems most likely that a big surge in frozen food markets will come with general consumer acceptance of frozen meats and young poultry. Whatever technical problems there may have been with the frozen storage and distribution of these products appear to have been solved. Many housewives, in fact, although they purchase "fresh" meat, turkeys, and broilers, buy them in excess of their immediate needs, and keep them in their home freezers. Several efforts in the past to build strong markets for the frozen products have been unsuccessful, due to causes unrelated to the intrinsic usefulness and quality of the products; nevertheless, the economies of frozen storage and distribution are large, and the potential markets are tremendous. In years to come meats and poultry may well become one of the largest product categories in the frozen food industry.

e. The Institutional Pack. While the present very large and rapidly growing "institutional pack" of frozen foods is not truly made up of *new products,* it deserves special mention because its producers and distributors have carried out vigorous exploration of new and different markets. Most of the products are designed for convenience and rapidity of preparation, as well as accurate portion control, so as to facilitate cost analysis. With the growth of food consumption away from home, and the increasing cost and decreasing skill of kitchen labor, factory preparation of foods can not only reduce cost, but also assure tighter control of quality. The institutional pack of frozen foods is currently (1968) growing at a rate of more than 15% per year, compared with about 5% for all frozen foods.

IV. THE TIME–TEMPERATURE TOLERANCE INVESTIGATIONS

A. The Genesis and Course of the T-TT Work

The growing accumulation of difficulties within the frozen-food industry a number of years ago, tracing back to some of the unfortunate experiences we have recounted, led Diehl, with the vigorous support of responsible people in the industry early in 1948, to call for and help organize a thoroughgoing, deeply probing investigation of the scientific and technological bases for the commercial practices of the industry. The full cooperation of the U.S. Department of Agriculture was readily obtained, because the Department had itself been carrying on research on frozen foods for many years, growing out of studies which led to establishment of a U.S. Frozen Pack Laboratory in Seattle in 1930. Major facilities of the Department, in laboratories at Albany and Pasadena, California, were made available for the needed investigation, and a broadly competent staff of scientists, food technologists, and engineers was assembled. Specialized, experimental, cold-storage equipment was designed, procured, and installed. In consultation with industry leaders, a comprehensive series of investigations of the quality and stability of commercially processed and packed frozen fruit, vegetable, and poultry products was initiated, drawing its materials from cooperating packing plants throughout the United States during two to three packing seasons.

The general design of the experiments is described in what turned out to be the first of 24 technical papers which were published in *Food Technology,* beginning in January, 1957 (34) which described many of the experimental results. The series was given the general title "The Time–Temperature Tolerance of Frozen Foods" after considerable discussion among the staff, because at that time the opinion was held that while an ideal situation might well be presented if a freshly frozen product could be held ever thereafter in an unchanging environment at very low temperature, perhaps 0°F. (−18°C.), it was only realistic to suppose that in practice somewhat higher temperatures would be encountered for some period of time before the product was finally consumed. The practical question was, *how much* of such less-than-ideal treatment would a product take without sustaining substantial damage to its quality? In other words, what is the *tolerance* of a frozen food to adverse condtions, measured in terms of time and temperature combinations? The feeling was, in general, that most of the serious trouble which had been encountered in the recent past by various segments of the frozen-food industry was the result of overestimation of the degree of tolerance to mistreatment really possessed by frozen foods, i.e., overestimation of the *stability* of these foods at temperatures below, but perhaps not far below, the freezing point of water.

The "T-TT" series of publications, while somewhat limited in breadth (covering only certain fruit, vegetable, and poultry products and a few "precooked" items), provided an enormous amount of previously lacking quantitative information on the characteristics and behavior of commercial packs of the commodities studied. The wide variablity of single varieties of most fruits and vegatables was one of the unexpected observations; statistical treatment of the experimental results was usually required.

The T-TT series does not, of course, give complete coverage of all the important information, especially that obtained in subsequent years, about the quality and stability of frozen foods. Many investigators have described other important research. The freezing and storage characteristics of frozen bakery products, for example, were broadly studied a few years later (see Chapter 7).

1. *Relation to the Present Volume*

Many of the chapters in the present book summarize results obtained in the T-TT investigations (especially Chapters 4 and 5). However, we also describe much other reseach and development work on the improvement of quality and stability in frozen foods, not only in laboratories of the U.S. Department of Agriculture (for example, see Chapter 7, on frozen bakery products), but also in a state university (see Chapter 6, on frozen poultry and egg products), and in Europe (see Chapter 8, on meats, and Chapter 9, on seafoods. Other extensive research, known to have been stimulated, and perhaps in part suggested, by the T-TT work, is under way in the United States, Canada, and Europe.

B. Equipment, Processes, and Practices in the Frozen Food Industry Have Been Materially Improved

The wide dissemination of the results and conclusions from the T-TT investigations, by means of industry conferences, publication of popular bulletins and training materials, addresses at national and regional meetings, unremitting publicity by the frozen-food trade associations, and pressure applied by many convinced plant managements, have resulted in a continuing drive to upgrade capabilities and performance in all elements of the industry.

Chapter 12 of this volume describes many of the developments which have taken place. Improved stability of the frozen products is, of course, sought wherever possible by such changes in packaging or processing procedures as will weaken the causes of rapid quality loss. Better stability is also assured, though, if the normal warehouse storage temperature is lowered and all unnecessary exposures of the frozen goods to higher temperatures during shipment or handling are eliminated. How stable, however, is "stable enough"? In the earlier days of the industry, the high degree of dependence on each year's new harvests of fruits and vegetables automatically set the industry's sights on sufficient stability to keep the stored products marketable for a year. Since warehouse storage at a lower temperature usually costs more, the standard storage temperature would normally be just low enough to give the major stored products an expected "storage life" of a year, with a reasonable additional margin for possible carryovers. This was, in fact, the genesis of the traditional rule for storage temperatures: "0°F. or lower."

More complete knowledge of "time–temperature tolerance" has made it plain that while strict adherence to that rule would at least forestall any such disastrous losses of quality as have sometimes occurred in the past, it would not by any means guarantee a useful "storage life" of a year or more for all frozen commodities. This "life" at 0°F. (−18°C.) for some seafoods and precooked foods, for example, is no more than 2 or 3 months (see Chapter 10). European practice, in fact, insists that meats and seafoods generally should be held at $-10°F.$ ($-23°C.$) or lower (see Chapters 8 and 9).

The question arises at this point, how much of a loss in quality is permissible in a frozen product that is "stable enough"? The answer is evidently wrapped up with a whole series of difficult problems concerning the measurement or specification of "quality" on the one hand, and the psychology of sales appeal in the marketing of frozen foods on the other. These problems are discussed to some extent in several of the chapters in this volume, but more especially in Chapter 4 (fruits and juices), and Chapter 11 ("quality management").

1. Investments in Better Facilities

As pointed out earlier in this chapter, capital expenditures by U.S. railroads, trucking companies, and refrigerated warehouse companies for refrigerated equipment adapted to frozen-food transportation have been very large indeed—many hundreds of millions of dollars. A substantial part of this investment was undertaken in direct response to the public release of the main findings of the time–temperature tolerance investigations, which had left no doubt about the inadequacy of ice-bunker rail cars and makeshift refrigerated trucks for frozen-food handling. Similarly, improved facilities were installed by centrally refrigerated warehouses, wholesale distributors and, especially in the very large frozen-food sections of recently built supermarkets, the food retailers. The net result was a marked improvement in the conditions for distributing and marketing frozen foods, as compared with conditions that were being found some 10–15 years earlier (see Chapter 13).

2. Codes of Good Practice and Legal Regulations

Responsible producers and merchandisers of frozen foods have always been highly conscious of their obligations, as packers and sellers of *food* products, to assure the safety and wholesomeness of their goods. The early experimenters on frozen foods invariably based their work on the premise that freezing the food would not improve it, so they must always start with a high-quality raw material. For example, an important aspect of the very early research on the freezing preservation of vegetables was recognition of the necessity to show conclusively that food poisoning by *Cl. botulinum* was not a danger (35). In fact, commercial packs of frozen foods have had an excellent public record (see Chapter 3).

However, at about the time when the category of precooked or pre-prepared frozen dishes began to expand rapidly and scores of old-favorite, prepared dishes joined the frozen list (including many bland combinations like creamed fish or meat), a new wave of concern about the assurance of wholesomeness and safety made itself felt. The unease was centered especially on the precooked foods. It was, in fact, supported in a general way by the industry's own knowledge of occasional inexcusable mishandling of some frozen foods at one stage or another of their manufacture or distribution. The evident need, apparent to both leaders in the frozen-food industry and to public health authorities, was for far more detailed and more definite guidelines for satisfactory performance all the way through the complicated series of steps in the production and distribution of this new class of foodstuffs.

In the early 1950's the Association of Food and Drug Officials of the United States (AFDOUS) had taken note of the problem, and the National Association of Frozen Food Packers, cooperating with AFDOUS, had made a broad study of retail handling practices. Committees of AFDOUS studied freezing-plant operations, transportation, and warehousing. The published findings of the Time–Temperature Tolerance investigations were very influential and impressive upon regulatory officials and industrial leadership in their conclusive showing of the kinds and rates of quality damage which occur in frozen foods at various below-freezing temperatures and in forming standards of reference for high-quality (36). In 1962 AFDOUS adopted a general frozen-foods code; this, of course, did not have the force of a legal regulation, but it has been adopted in whole or in part by several states, and has had a large influence on regulations adopted by various agencies of the Federal Government and by several foreign countries.

During about the same period vigorous efforts were under way within the industry itself to establish sound guidelines for the rapidly changing situation. A leading frozen food packing firm, for instance, set up its own code of rules and practices for all of its handling of frozen fruits and vegetables (37). A committee of national associations dealing with all phases of frozen foods, from packer to retailer, known as the All-Industry Frozen Foods Coordinating Committee, was formed under the chairmanship of Dr. H. C. Diehl to draw up recommended operating procedures and standards of practice and, through education and special publicity, secure general voluntary compliance. In June, 1961, the committee issued a recommended set of "Voluntary Handling Practices" covering handling of frozen foods by the processor, warehouse equipment and handling practices, transportation equipment and practices, and distribution and retail facilities and practices (38). See also Chapter 14 for a review of the present situation on regulations and codes.

3. Educational and Training Programs

Many producing companies and a number of industry associations have instituted and continued training programs for their operating personnel and supervisors, because the average worker would have no conception of the subtle damage his minor neglect or inattention can cause. Educational posters, training courses, and incentive systems are in general use. These programs must be continuous because new workers are constantly entering the industries, and must be trained appropriately.

4. Present Level of Performance

All factors in the frozen-food industry can be justly proud of its accomplishments and be gratified by its present image in the minds of the general public. An objective appraisal of the performance of this $6 billion industry would, however, be enormously difficult to make; no one, so far as we know, has attempted the task. No information is published on many of the facts which should enter into such a study.

a. Continuing Effort Will Be Necessary. There can be no doubt that many of the weak spots that caused so much trouble at times during the industry's history have been eliminated. Processors, wholesalers, distributors, retailers, and housewives all have learned from experience how to deal with frozen foods.

That fact does not mean, however, that no problems remain. Some of these can be classed as consequences of the gaps in our scientific knowledge or as technological difficulties, as new products are produced, and they will be discussed in the next section. The more common kind of problem, familiar in every major industry, arises from human errors, attempts to reduce necessary costs, indifference, and laziness. On-the-job training and watchful supervision can do much to ease this, even if it can never be entirely eliminated.

V. THE UNSOLVED SCIENTIFIC AND TECHNOLOGICAL PROBLEMS

The proliferation of new frozen products, different packaging materials and techniques, and new production practices (for example, IQF freezing and bulk storage of frozen products for subsequent packaging) have kept the frozen food industry in a state of flux. Few guidelines are broad enough to cover all the unforeseen new developments as well as they did the older ones, and the large-scale manufacture of new or different items demanded by a marketing situation usually cannot wait for the result of two or three years of exhaustive testing. For an industry that is so completely dependent on the maintenance of near-perfection in its products, this is an uncomfortable and risky position. One of the persistent, unsolved problems of the industry centers right there.

A. Quality Changes During Frozen Storage

The underlying reason why extensive studies of changes of quality during storage have been thought necessary is that we do not know enough about the causes of deterioration to be able to predict accurately either its character or its rate (see Chapter 2 of this volume). As we have remarked earlier, holding a product at 0°F. (−18°C.) dependably preserves good quality for a year or more in some products, but not by any means in all.

1. Oxidative Deterioration.

Quality losses due to oxidation of components of the food include loss of ascorbic acid, discoloration or fading of color, and the onset of rancidity and allied off-flavors. Enzymatically accelerated oxidation is usually controlled by a heat treatment during processing, *i.e.,* "blanching" in steam or boiling water. Added ascorbic acid is sometimes used as an antioxidant, and certain other approved antioxidants are regularly used to stabilize the flavor of various compounded foods. However, in general the protection afforded by antioxidants is imperfect and transient. This is one of the areas in which more research is greatly needed.

2. Protein Denaturation

Freezing and frozen storage often have complex effects upon the molecular structure and corresponding properties of the protein foods such as meat, eggs, and seafoods, but the overall effect is almost always disadvantageous to the food quality, especially its texture. The protein loses more or less of its water-binding effectiveness. Meats lose some of their juiciness, fish muscle becomes more dry and crumbly, and cooked egg white becomes rubbery. In fish muscle the progressive change in texture upon freezing is most rapid just a few degrees below the freezing point, but further change still continues at the temperature of liquid nitrogen (39).

B. Poor Texture is an Unsolved Problem for Some Products

From the very beginning of the frozen food industry, altered or unsatisfactory texture of the thawed food (in comparison with the fresh product) has been an insurmountable obstacle to general acceptance of some frozen products—frozen lettuce, for example—and has limited the extent of use of several, especially strawberries. Vigorous experimentation within the past few years has awakened some new hopes. The freezing of tomato slices has been reported (40); very rapid freezing has also produced snap beans (41) and strawberries (42) which had good texture when they were thawed. However, the commercial success of these products, often imprecisely described as "cryogenically frozen," remains to be established. Relatively unsatisfactory texture of thawed products remains as a perennial problem upon which much more work will need to be done.

C. Difficulties Point to Needed Research

In the last few pages we have pointed out several of the continuing difficulties and problems which remain to bother the frozen-food industry from the scientific and technical points of view. These selected examples by no means exhaust the roll of projects which will amply repay time and money spent on them by very competent research teams. At numerous places in the following chapters the reader will find puzzles and unanswered questions; every such question should be looked upon as an opportunity, a door which only needs to be unlocked.

VI. THIS IS A GOOD TIME TO TAKE STOCK

To the authors of this chapter it seems that the industry has now reached some kind of critical point. It would not be surprising if the furious pace of new developments we have seen during the past 15 years or so were to give place gradually to a consolidation of the past gains, and a deepening of the roots the industry has put down in the food industry. This should be an appropriate time for thoughtful people in the industry to take stock of where they are and whither they are going. Such is the purpose of this book.

REFERENCES

(1) Anon., *Food Inds., 2,* 175 (1930).

(2) S. H. Fulton, *Cold Storage of Small Fruits,* Bull. 108, Bur. Plant Ind., U.S. Dept. Agr., 1907.

(3) M. E. Pennington, in *Yearbook of Agriculture,* J. A. Arnold, Ed., 1912, U.S. Dept. Agr., Government Printing Office, Washington, D.C., p. 285.

(4) M. E. Pennington, *U.S. Egg Poultry Mag., 47,* 554 (1941).

(5) H. C. Diehl, *Food Inds., 2,* 162 (1930).

(6) H. C. Diehl, J. R. Magness, G. R. Gross, and V. B. Bonney, *The Frozen-Pack Method of Preserving Berries in the Pacific Northwest,* Tech. Bull. 148, Bur. Plant Ind., U.S. Dept. Agr., 1930.

(7) J. S. Caldwell, J. M. Lutz, H. H. Moon, and C. W. Culpepper, *Comparative Studies of Varietal Suitability for Freezing Preservation of Peas, Green or Snap Beans, Lima Beans, and Sweet Corn Grown under Eastern Conditions,* Tech. Bull. 731, Bur. Plant Ind., U.S. Dept. Agr., 1940.

(8) M. A. Joslyn, *Preservation of Fruits and Vegetables by Freezing Storage,* Circ. 320, Calif. Agr. Expt. Sta., Berkeley, 1930.

(9) M. A. Joslyn and G. L. Marsh, *Changes Occurring During Freezing, Storage, and Thawing of Fruits and Vegetables,* Bull. 551, Calif. Agr. Expt. Sta., Berkeley, 1933.

(10) J. G. Woodroof, *Preserving Fruit by Freezing,* Bull. 163, Georgia Agr. Expt. Sta., Experiment, 1930.

(11) J. G. Woodroof, *Fruit Prod. J., 11,* No. 5, 138 (1932).

(12) E. H. Wiegand, *The Frozen-Pack Method of Preserving Berries,* Bull. 278, Oregon Agr. Expt. Sta., Corvallis, 1931.

(13) E. H. Wiegand, *Preservation of Fruits and Vegetables by Freezing,* Circ. 116, Oregon Agr. Expt. Sta., Corvallis, 1936.

(14) D. K. Tressler, *Ind. Eng. Chem., 24,* 682 (1932).

(15) D. K. Tressler and C. F. Evers, *The Freezing Preservation of Fruits, Fruit Juices, and Vegetables,* Avi Publishing, New York, 1936.

(16) D. K. Tressler and C. F. Evers, *The Freezing Preservation of Foods,* Avi Publishing, New York, 1943.

(17) W. V. Cruess, "Frozen-Pack Fruits and Vegetables," in *Commercial Fruit and Vegetable Products,* 2nd ed., McGraw-Hill, New York, 1938, p. 688.

(18) H. Carlton, *The Frozen Food Industry,* University of Tennessee Press, Knoxville, 1941.

(19) C. Birdseye and G. A. Fitzgerald, *Ind. Eng. Chem., 24,* 676 (1932).

(20) C. Birdseye, *Refrig. Eng., 61,* 1182 (1953).

(21) E. W. Williams, *Quick Frozen Foods, 26,* No. 1, 145 (1963).

(22) G. W. Irving, Jr. and S. R. Hoover, Eds., *Food Quality,* Amer. Assoc. for the Advancement of Science, Washington, D.C., 1965.

(23) O. J. Cotterill, "Evaluation and Measurement of Quality of Poultry and Eggs," in *Food Quality,* G. W. Irving, Jr. and S. R. Hoover, Eds., Amer. Assoc. for the Advancement of Science, Washington, D.C., 1965, p. 179.

(24) H. D. Naumann, "Evaluation and Measurement of Meat Quality," in *Food Quality,* G. W. Irving, Jr. and S. R. Hoover, Eds., Amer. Assoc. for the Advancement of Science, Washington, D.C., 1965, p. 239.

(25) A. M. Pearson, "The Influence of Nutrition and Management upon Meat Quality," in *Food Quality,* G. W. Irving, Jr. and S. R. Hoover, Eds., Amer. Assoc. for the Advancement of Science, Washington, D.C., 1965, p. 267.

(26) H. K. Schauffler, *1968 Frozen Food Factbook and Directory,* National Frozen Food Association, New York, 1968, p. 48.

(27) C. L. Rasmussen, "Frozen Foods in Western Europe," in *The Freezing Preservation of Foods,* Vol. 3, 4th ed., D. K. Tressler, W. B. Van Arsdel, and M. J. Copley, Eds., Avi Publishing, Westport, Connecticut, 1968.

(28) W. R. Woolrich, "Principles of Refrigeration," in *The Freezing Preservation of Foods,* Vol. 1, 4th ed., D. K. Tressler, W. B. Van Arsdel, and M. J. Copley, Eds., Avi Publishing, Westport, Connecticut, 1968, p. 31.

(29) R. F. McKee, "Rail Transportation of Frozen Food," in *The Freezing Preservation of Foods,* Vol. 1, 4th ed., D. K. Tressler, W. B. Van Arsdel, and M. J. Copley, Eds., Avi Publishing, Westport Connecticut, 1968, p. 259.

(30) W. H. Redit, "Refrigerated Trucks," in *The Freezing Preservation of Foods,* Vol. 1, 4th ed., D. K. Tressler, W. B. Van Arsdel, and M. J. Copley, Eds., Avi Publishing, Westport, Connecticut, 1968, p. 281.

(31) B. Feinberg and R. P. Hartzell, "Packaging Materials and the Packaging of Frozen Foods," in *The Freezing Preservation of Foods,* Vol. 2, 4th ed., D. K. Tressler, W. B. Van Arsdel, and M. J. Copley, Eds., Avi Publishing, Westport, Connecticut, 1968.

(32) R. V. Enochian, "The Rise, Present Importance, and Future of Frozen Fresh Food," in *The Freezing Preservation of Foods,* Vol. 3, 4th ed., D. K. Tressler, W. B. Van Arsdel, and M. J. Copley, Eds., Avi Publishing, Westport, Connecticut, 1968.

(33) S. Martin, "The Rise of Prepared and Precooked Frozen Foods," in *The Freezing Preservation of Foods,* Vol. 4, 4th ed., D. K. Tressler, W. B. Van Arsdel, and M. J. Copley, Eds., Avi Publishing, Westport, Conn., 1968.

(34) W. B. Van Arsdel, *Food Technol., 11,* 28 (1957).

(35) J. A. Berry, *Western Canner Packer, 30,* No. 4, 32 (1938).

(36) Anon., *Assoc. Food Drug Officials U.S., Quart. Bull., 26,* No. 1, 25 (1962).

(37) K. G. Dykstra, *Refrig. Eng., 64,* No. 9, 54 (1956).

(38) Anon., *Quick Frozen Foods, 23,* No. 11, 50 (1961).

(39) R. M. Love and M. K. Elerian, *Intern. Congr. Refrig. Proc.,* 11th, Munich, IV-1, *2,* 887 (1963).

(40) Anon., *Food Processing, 25,* No. 1, 72 (1964).

(41) E. R. Wolford and M. S. Brown, *Food Technol., 19,* No. 7, 109 (1965).

(42) R. C. Webster, E. J. Benson, and W. H. Lucas, *Quick Frozen Foods, 25,* No. 5, 35 (1962).

Chapter 2

QUALITY-RELATED CHEMICAL AND PHYSICAL CHANGES IN FROZEN FOODS

Eugene F. Jansen
Plant Enzyme Pioneering Research Laboratory, Western Regional Research Laboratory, U.S. Department of Agriculture, Albany, California

I. INTRODUCTION

The purpose of this chapter is to describe the principal types of chemical and physical change that influence the quality of frozen foods. This includes changes that occur prior to freezing as well as those caused by freezing, frozen storage, and thawing. Subsequent chapters will describe and discuss in detail the changes which occur in important plant and animal food products. The basis of discussion in this chapter will be type of change, rather than commodity, because the same kind of change can often be observed taking place in very different foods.

Plants and animals from which food is derived are composed of living cells that are highly organized entities, containing large populations of many kinds of subcellular entities—for example, the nucleus, mitochondria, ribosomes, chloroplasts (in green plants), Golgi bodies, membrane systems, and enzymes. In fact, the cell may be pictured as a community of several separate, organized systems, living symbiotically with one another. One such system comprises the enzyme-formation system of the nucleus (DNA) and ribosomes. Another contains the chloroplasts and their DNA enzyme-generating system. Still another comprises the mitochondria which, with their own protein-generating machinery, are responsible for the cell's respiration. Thus the cell contains many separate, semiautonomous but interdependent subsystems.

It is easy to see why the disturbance of one part of this interlinked system should turn organization into chaos. The mere bruising of living tissue is enough to produce disorder and ruin.

The subcellular unit which houses the hydrolytic enzymes is worthy of special attention. de Duve (1) called these units in animal tissue "lysosomes"; Balz (2) named those in plant tissue "sperosomes." These small particles break down and dissolve complex nutrients taken up by the cell and thus act as its digestive tract.

Many cell constituents must remain in their elaborately complex forms in order to play their parts in the life processes of the cell. The hydrolytic enzymes could break down and thus inactivate them were it not for the protection afforded by a selective membrane which encloses each lysosome or sperosome. Disruption of this membrane in a plant or animal tissue would be sufficient to cause anarchy and the death of the cells involved.

Considering the complexity of the cell, many changes could occur on freezing and in the frozen storage of uncooked or unheated commodities. The unleashing of the hydrolytic enzymes mentioned above is one possibility. The mitochondria in supplying cellular energy do so stepwise through a complicated series of enzymes. An interruption at any step would cause an accumulation of certain substances, some of which might impart off-flavor. The suggestion that off-flavor formation in unblanched frozen vegetables might be due to the lack of an enzyme rather than the presence of a "bad" enzyme was made many years ago.

Reduction in temperature and the change of state from water to ice can affect the activities of certain enzymes more than others, as shown by Lineweaver in 1939 (3). For example, proteinase activity is much more reduced by lowering the temperature, and particularly by the change of state upon freezing, than is lipase activity. Because of this disproportionate effect of temperature the end products of certain enzyme reactions might accumulate in frozen foods, particularly when the food is subjected to adverse times and temperatures, whereas the end products from other reactions would be unmeasurable.

The locale and occurrence of isozymes should also be considered. Isozymes are enzymes performing the same catalysis but possessing different chemical and physical properties. Work in the author's laboratory has demonstrated in plant-tissue cultures that several peroxidase isozymes are present, each in its own locale within the cell (4). Each isozyme may have different specificity and stability. Should one or more of the isozymes be responsible for adverse changes in frozen vegetables, then the measurement of total peroxidase and of the loss therein would be of little meaning. It should be emphasized that our present catalase and peroxidase tests for adequacy of heat treatment of plant products intended for freezing is purely empirical. No one claims that these enzymes are directly involved in off-color and off-flavor formation.

Not all changes that occur in foods prior to freezing, during frozen storage, and during thawing are catalyzed by enzymes. As a matter of fact, the majority of changes are either physical changes or nonenzymic chemical changes.

The changes in the state of aggregation of the starches due to freezing and frozen storage are very important in such products as cream sauces (*e.g.*, in chicken à la king) and bread and other baked products. Another type of simple change in the state of aggregation is that which occurs in the freezing of egg yolks. Here the lipoprotein (lecithovitellin) separates in frozen storage and aggregates, with the result that after thawing the yolk becomes a stiff paste. These changes will be discussed in more detail later.

The concentration of components which occurs during relatively slow freezing is another type of change. As freezing proceeds, the water-soluble components concentrate in the remaining aqueous phase. This can result in some undesirable changes, such as the denaturation of a protein in certain frozen fish.

Still other types of change which ocur during frozen storage are those resulting from nonenzymatic oxidative changes, particularly when the foods are held under adverse time–temperature conditions. The complicated formation of rancid off-flavor in frozen meats, poultry, and fish is one example and the changes in vitamin C which occur in vegetables and fruits is another.

Simple reactions such as hydrolysis may also occur. An example of this change, which can occur in vegetables, is the conversion of the bright green pigment, chlorophyll, to the olive drab derivative, pheophytin.

Crystallization during frozen storage is another type of adverse change which may occur. Sugar may crystallize on the surface of high-sugar fruit spreads in frozen storage to give the appearance of mold growth on the surface of the spread. This crystallization of a sucrose hydrate proceeds faster at $-10°$F. ($-23°$C.) than at either lower or higher temperatures, but it can be prevented by the addition of some invert sugar to replace part of the sucrose (5). The milk sugar, lactose, will sometimes crystallize in ice cream to give the product a "sandy" property. Ice crystals will grow in ice cream and ice milk stored for an extended period, particularly under adverse temperature conditions, and will give the product an unfavorable texture.

Local dehydration due to the migration or sublimation of water is another type of adverse change in foods which can occur in frozen storage. This localized dehydration is called "freezer burn" and permits changes to occur which would not occur, or at least not as rapidly, in the hydrated material. For example, oxidative rancid off-flavor formation occurs most rapidly in frozen turkeys or chicken at the area of freezer burn, where ice does not prevent access of oxygen to the fat. See also Chapter 10, Section IV.D.1. For the most part, freezer burn can be eliminated by proper packaging. This is certainly true for poultry products. Nevertheless, vegetables suffering from severe freezer burn can still be found in the normal channels of distribution. Some of the specimens of deteriorated product occasionally found in retail markets have evidently experienced more severely adverse conditions than an experimenter would be likely to set up deliberately.

It is apparent that the physical and chemical changes that can and do occur in frozen foods are many. Proper processing and handling will limit the changes, but still more basic information is needed about the physical and chemical systems involved in both desirable and adverse changes that occur at various stages in the life of a frozen food product, if we are to deliver highest quality frozen foods to the consumer.

II. CHANGES IN TEXTURE

Texture may be defined as the "manner of structure" or "structural quality." Desirable texture varies from food to food. In vegetables, it is the proper turgor and freedom from tough fibers; in cooked meats, poultry, and fish, it is the proper degree of tenderness; in fruit, it may be turgor; in cream sauces, it is uniformity of suspension of thickening agent; and in orange juice and certain other juices, it is uniformity of suspension of the cloud. Conversely, examples of texture defects are: limpness and sloughing in cooked green beans, toughness in cooked meats and poultry, separation of the water phase in cream sauces, limpness in strawberries, and separation of cloud in reconstituted orange juice concentrate.

A. Texture and Polysaccharides

Starch granules as deposited in the plant cell consist of two components, amylose and amylopectin, in microcrystalline form. The pro-

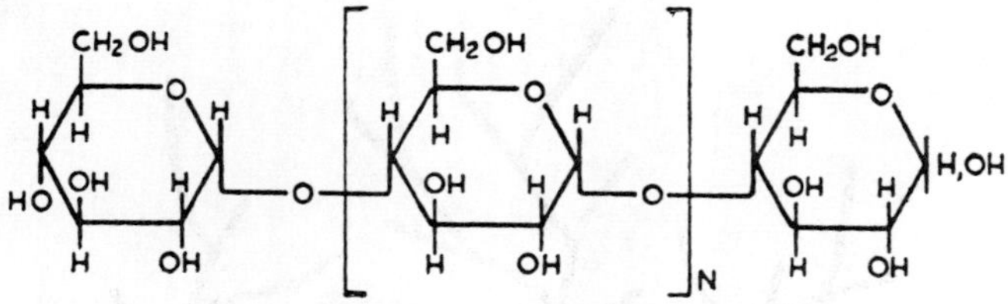

Fig 2.1. Structure of amylose molecule.

portions of each vary from plant to plant. Amylose consists of glucose units in 1,4-glycosidic linkage, as illustrated in Fig. 2.1. Amylose is a large, linear polymer containing approximately 4000 glucose units. On the other hand, amylopectin consists of glucose units not only in 1,4-glycosidic but also 1,6-glycosidic linkage. A segment of an amylopectin molecule is presented in Fig. 2.2. The 1,6-linkage results in branching of the chain. Thus, amylopectin is a highly branched, extremely large polymer, several times larger than amylose. A schematic representation of the whole amylopectin molecule is shown in Fig. 2.3. When starch granules are heated in water the microcrystalline structure is destroyed and the starches dissolve. Because of the linear structure of amylose, when its solutions are cooled the molecules can, like logs in a log raft, again aggregate to form microcrystals, a phenomenon called "retrogradation." Conversely, like a large group of tumble weeds, the branching of amylopectin interferes with the alignment of molecules so that the crystallization of amylopectin from cold water solutions is greatly retarded. Retrogradation of either variety of starch results in a separation of starch from the aqueous phase, a phenomenon which can be reversed by heat.

Retrogradation of the starches is involved in the stability of frozen sauces (as used for such products as chicken à la king) and the staling (firming) of baked products such as bread. Hanson *et al.* (6) clearly

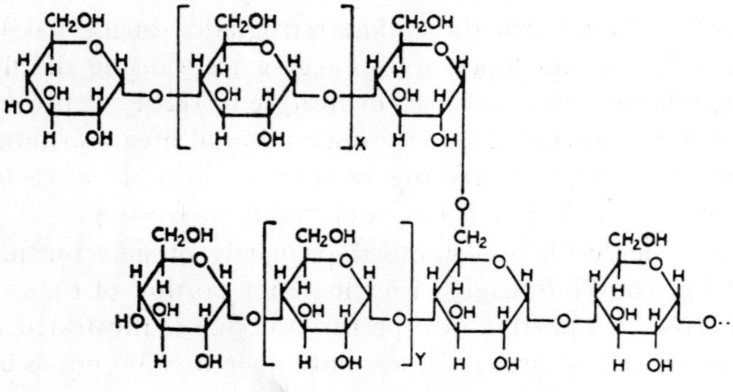

Fig. 2.2. Structure of a segment of amylopectin molecule.

Fig. 2.3. Schematic representation of amylopectin.

demonstrated that the use of amylopectin (for example, waxy rice flour) as a thickening agent produced a white sauce much more stable to frozen storage than that prepared from a mixture of the starches (*e.g.*, wheat flour). Sauces prepared with wheat flour and subjected to freezing and frozen storage at 0°F. (–18°C.) develop a curdled appearance and separation of liquid upon thawing that can only be reincorporated by heating *and* vigorous stirring. Sauces prepared with waxy rice flour when subjected to adverse times and temperatures will develop the curdled appearance, but it can be easily eliminated by merely warming.

A study of staling of bread by Schoch and French (7) revealed that firming, the principal staling defect (other than flavor changes), is due to the retrogradation of amylopectin. Through the use of water solubility tests they found that the amlose retrogrades during baking. As the bread stales the amylopectin also aggregates, causing the firming. The retrogradation can be reversed by heating to 122°F. Pence *et al.* (8) found that in the successful frozen preservation of bread, firming must be minimized by rapidly freezing as soon as possible after baking, storing at 0°F. (–18°C.) or below, and rapidly defrosting.

The pectin molecule is composed primarily of galacturonic acid units in 1,4-glycosidic linkage, with the major portion of the carboxyl groups esterified. A portion of a pectin molecule is illustrated in Fig. 2.4. Pectins are rather large polymers but apparently are not as large as starches.

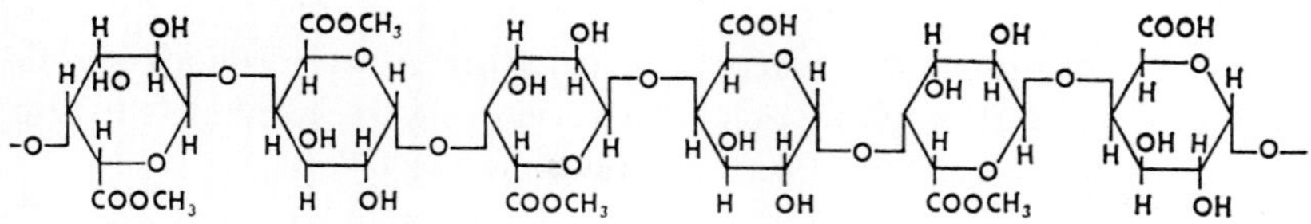

Fig. 2.4. Structure of a portion of pectin molecule.

Deesterification of pectin, *i.e.,* hydrolysis of the ester groups, results in the formation of pectinic acids and methanol. Dependent upon the degree of deesterification, the pectinic acids are water insoluble, particularly in the presence of calcium salts. When the concentration of soluble pectin is sufficiently great, deesterification will result in gel formation. The deesterification of pectin is catalyzed by the enzyme "pectinesterase," which is present in all fruits and vegetables preserved by freezing.

The principal quality defect in frozen citrus juice concentrates resulting from exposure to adverse times and temperatures is cloud loss after reconstitution, or even gellation in the concentrate (9,10). These defects result from the action of pectinesterase on the pectin present in the cloud and/or in solution in the concentrate. The resulting pectinic acid can then form insoluble salts or become insoluble *per se,* thus causing a gelation or a rapid settling of cloud in reconstituted juices. In the early days of frozen orange juice concentrate, when gentle squeezing procedures were being used, little cloud loss was encountered during commercial distribution. The problem became troublesome when, because of competitive pressure, increased yields of juice were required. The more vigorous extraction yielded more solid particles and, incidentally, more pectinesterase activity because the pectinesterase occurs in the solids of orange juice (11). The additional enzyme in the juice made a heat treatment step necessary to stabilize the concentrate sufficiently. Much effort has been devoted to determining the amount of heating required to produce a stable orange juice concentrate (12–14). In all this work the guide for assessment of a satisfactory heat treatment has been the measurement of residual pectinesterase and no cognizance was taken of the state of the pectin. Heat treatment not only destroys the enzyme but also solubilizes part of the pectin from the cloud.

Our results (15) demonstrate that pectinesterase is firmly bound to the suspended material of orange juice, that the enzyme is essentially inactive at the pH of the juice on the *in situ* pectin, that soluble pectin in combination with salts extracts part of the enzyme from the suspended matter, and that the enzyme so solubilized hydrolyzes soluble pectin to an extent that causes gelation. These results suggest that gelation and cloud loss of concentrates might be influenced in part by solubilization of pectin into the serum during processing and in part by the action of pectinesterase. Moreover, it is clear that the food technologist, in studies of the effects of heat treatment on enzymes, should determine not only the extent of destruction of the enzyme but also, where possible, the effects of the treatment on its substrate.

When frozen Montmorency cherries are subjected to temperatures above 0°F. (−18°C.) for an appreciable length of time, a pronounced increase in the firmness occurs (16). This increase persists even after the cherries are baked into a pie. Gee and McCready (17) have shown that the toughening is due to pectinesterase deesterification of the pectin and the formation of a calcium pectinate gel when the calcium present in the fruit tissue combines with the so-formed carboxyl groups. Accordingly, one step in the toughening of cherries is the same enzyme-catalyzed change that is involved in the cloud loss and gelation in frozen orange juice concentrate.

Glycogen, a polysaccharide which occurs in all animal tissue, has a highly branched structure similar to amylopectin. This carbohydrate is not concerned with structure as is pectin but rather serves as an energy–storage substance. Glycogen is the energy source for muscular work. After slaughter, when oxygen is no longer available, glycogen in meat and poultry continues to be metabolized but in this case anaerobically. In anaerobic glycolysis, glycogen is converted into lactic acid whereas aerobically it is converted to CO_2. The disappearance of adenosine triphosphate occurs when glycogen is nearly exhausted and the muscle then goes into rigor mortis (18). Subsequently, meat and poultry muscle become tender by other but unknown processes that occur during ageing. Curiously, however, if glycolysis is avoided then poultry muscle is tender without ageing as demonstrated by the work of de Fremery (19). He depleted the glycogen in chicken and turkeys by the subcutaneous injection of epinephrine prior to slaughter (see refs. 20 and 21) and found the meat to be tender without aging.

B. Texture and Proteins

There are four general types of change that occur in proteins: denaturation, hydrolysis, interaction with other constituents or other proteins, and simple precipitation. The occurrence of some of these reactions influences the texture of frozen foods. "Denaturation" is a term difficult to define exactly because it really refers merely to changes in the properties of a protein. Denaturation does produce a decreased solubility. Where the protein has biological activity, as in enzymes, denaturation brings about a destruction of this activity. Denaturation can be caused by any of several different treatments. Among these are heating, treatment at acid or alkaline pH values, and many more. Proteins show a wide difference in their sensitivity to any one method of denaturation. The blanching of vegetables prior to freezing increases their storage stability, at least in part, by virtue of the denaturation of proteins, especially the enzymes.

In this regard, the destruction of one of two enzymes, catalase or peroxidase, has been used as an index of adequacy of heat treatment in the blanching of vegetables. These two enzymes differ considerably in their susceptibility to heat, catalase being much more labile than peroxidase. This difference is at least partially due to the locale of the enzymes. Catalase occurs in the cytoplasm of the cells, whereas peroxidase for the most part is associated with the cell walls; such an association generally increases the heat stability of enzymes. It has been found that blanching just to the point of inactivation of all of the catalase, but not all of the peroxidase, is not sufficient to assure stability of most vegetables in frozen storage. Therefore, the stabilities of the enzymes probably involved in enzymatic off-flavor formation are closer to the stability of peroxidase.

Egg yolk upon freezing and thawing loses its normal fluidity and becomes a stiff paste. This phenomenon was investigated by Moran in 1925 (22). He found that the effect did not occur in eggs held at their freezing point (27°F.) but did occur in a relatively short time at temperatures below 21°F. He demonstrated that the effect was due to the separation of the lipid–protein complex, lecithovitellin, at the lower temperatures and failure of the complex to redissolve on thawing. The effect can be minimized by the addition of 10% glucose or fructose to the yolks prior to freezing (23). The pastiness or gelation of frozen and thawed yolk products can be lowered by salt or sugar before freezing (24). Without any additives gelation can be lowered by colloid milling prior to freezing (25). Thus, this texture change is due simply to the precipitation of a protein.

The principal structure of frozen meats and fish is the muscle. We may therefore start with a very general discussion of muscle proteins before we consider changes in muscle prior to and during frozen storage.

The sarcoplasm is the plasma which bathes all of the formed elements of the muscle, the myofibrils. Sarcoplasm contains several proteins, including the myoglobin (see Sect. III). It also contains the enzymes responsible for glycolysis, wherein glycogen energy is converted to ATP energy for muscular work (contraction), as discussed above. The sarcoplasm is the portion of the muscle from which drip arises upon thawing. The sarcoplasm also contains microsomes which are responsible for the synthesis of muscle proteins.

The myofibrils are the formed elements of the muscle and are respon-

ɛ contraction, relaxation, and rigor. Proteins make up 95% of the myofibrils, with three major protein com- n, 55%; actin, 25%; and tropomyosin, 15%. In muscle :ombination occurs between actin and myosin filaments ediation of the hydrolysis of ATP. Relaxation is the ıtraction, and represents a dissociation of actin and myosin, probably mediated through the formation of more ATP and relaxing factors by the sarcoplasm enzymes. Rigor mortis is the stiffening of muscles which occurs after death; it is due to a partial combination of actin and myosin. The muscle glycogen, which is responsible for maintaining the ATP at a normal level, is quite rapidly depleted and hence the ATP level falls; rigor then sets in. Muscles of animals which are subjected to severe exercise prior to death go into rigor more quickly than muscles of unexercised animals. This is due to depletion of glycogen reserve in the exercised muscles.

Post mortem tenderization of meat and poultry following rigor is by no means completely understood, in spite of the research attention which has been given to the phenomenon. However, some facts are known about tenderization and some educated guesses have been made about it. Some protein hydrolysis occurs during tenderization of beef (26). Dahl (27) has demonstrated that after 22 days of aging, certain amino acids have been converted to nonprotein form (tryptophan increases from 17 to 39 mg./100 g. of raw protein, and tyrosine from 19 to 100 mg). This can be due only to hydrolysis of proteins. Reed (26) states that the change is brought about by the proteolytic enzyme, cathepsin, inherent in muscle. Hydrolysis occurs in the muscle protein *per se* and not in the connective tissue (27). Connective tissue is made up of the special protein, collagen. Collagen, unlike any other animal protein, contains the amino acid hydroxyproline. During tenderization no nonprotein hydroxyproline is formed; hence, the protein hydrolysis occurs in the muscle proteins *per se* (27).

Accordingly, during the tenderization of beef muscle, protein hydrolysis occurs. However, the explanation of tenderization on this basis alone may be insufficient. Wierbicki *et al.* (28) have suggested, with some evidence based on protein solubility measurements, that the increase in tenderness of beef may also be related to the dissociation of actomyosin, which increases the protein extractability, and to a redistribution of the within muscle, thus causing increased hydration and tenderness.

The tenderization of poultry during ageing cannot be attributed to protein hydrolysis. The development of tenderness as a function of ageing time is illustrated in Fig. 2.5 (29). In this case, the chickens

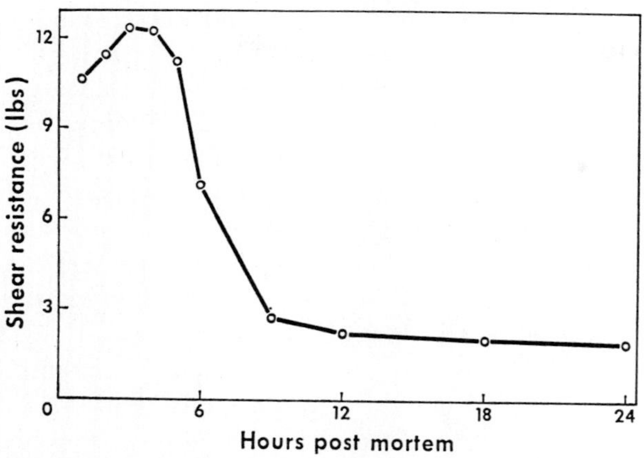

Fig. 2.5. Tenderness of chicken breast muscle (29), aged on carcass, then excised before cooking in boiling water.

were held in ice water. At room temperature or slightly below, poultry does not become as tender as in ice water (29). If the change were enzymatic, *i.e.*, if tenderization were simply protein hydrolysis, the higher temperature would accelerate the rate and yield a more tender bird. Poultry tenderization thus occurs by some as yet unknown mechanism. This, coupled with the fact that freezing *per se* may result in some tenderization (30), points to the need for further research in this field.

It is safe to say that no rule applies to all proteins except that they are all polymers composed of amino acids joined by peptide bonds. As a general rule, but not invariably, proteins are stable to freezing, *e.g.*, the freezing accompanying the freeze-drying of plasma. An enzyme chemist thinks first of freezing as a method of preserving his enzme preparations. However, some enzymes are damaged by freezing.

Myosin of certain fish, *e.g.*, cod, can become denatured when intact fish muscle is frozen and subsequently thawed. The denaturation of myosin in cod fish is accompanied by a toughening of the texture (31). The deterioration of herring during frozen storage, during defrosting, and during subsequent storage at refrigerator temperatures, is likewise due to the denaturation of its myosin (32). The myosin of tuna, on the other hand, is stable to freezing (33) as is that of beef and poultry muscle.

A decrease in the solubility of myosin in 5% sodium chloride is a criterion of its denaturation. Using this criterion, Snow (34) demonstrated that the rapid freezing of cod resulted in less myosin denaturation than did slow freezing. Using solutions of myosin as a model system to study the cause of denaturation, Snow found that almost complete denaturation occurred on freezing concentrated solutions, whereas clear dilute solutions showed no immediate denaturation. The denaturation was dependent upon pH, since solutions at pH 6.5–7.8 showed no change upon freezing, whereas solutions at higher and lower pH values exhibited rapid change on freezing . That denaturation of myosin by freezing is a change in state of aggregation was confirmed by Connell (35) who demonstrated by classical protein methods that the first observable change in myosin caused by freezing is an increase in aggregation. Later the loss of solubility is grossly evident. The aggregation proceeds more rapidly at higher salt concentrations.

Increases in salt and myosin concentrations resulting from the removal of water during ice formation, especially in slow freezing, undoubtedly play important roles in the changes observed in the texture of frozen fish of the cod type. However, the reason for the difference of cod myosin from that of tuna and meat is not clearly established.

C. Texture and Other Constituents

Delay in processing of peas and certain other vegetables not only results in the development of off-flavor (see below) but also an increase in toughness (36). Both of these deleterious effects, the chemistry of which is unknown and which may or may not be related, are reduced by rapid processing.

Individually quick-frozen green beans and, even more, package-frozen beans suffer from texture and flavor defects. As compared with the crisp texture of cooked fresh green beans, the cooked frozen beans are soft. In fact, they may suffer from a sloughing off of the outer layer during cooking. This defect, usually a result of overblanching, is undoubtedly caused by the translocation of water permitted by slow freezing. Wolford and Brown (37) have demonstrated that this defect can be prevented by freezing the blanched beans in liquid nitrogen. In the liquid nitrogen system used, beans were frozen in 10 sec. Later it was found that the texture was preserved if the beans were frozen to the center in any time shorter than 6 min. (38). Wolford and Brown suggest the use of "fluidized bed" freezers to preserve the texture of green beans and similar vegetables. A fluidized bed freezer is one which employs the principle of through-flow of cold air to effect freezing.

Thawed frozen strawberries, either whole berries or those sliced and with sugar added, are a far cry from fresh strawberries with respect to texture. As a matter of fact, the product frequently might be called strawberry sauce, or "mush." This must be regarded as a serious quality defect; it probably is responsible for the approximately 100-

million pound drop in frozen strawberry pack from 1956 to 1961 (39). This pronounced weakness of texture can be eliminated in part by freezing the berries by immersion in liquid nitrogen (40).

Certain other commodities, such as tomatoes, might be called the "hard to freeze" commodities, because on thawing their firm texture is gone and the product is soft and mushy, much like defrosted, conventionally frozen strawberries. Here again, freezing by immersion in liquid nitrogen results in a much more acceptable thawed product (41).

The changes in texture in the freezing, storage, and thawing of frozen foods are therefore numerous in kind. Only trial and error progress has been made in the elimination of texture quality defects. Accordingly, much research remains to be done, particularly from a basic research approach, in this field.

III. CHANGES IN APPEARANCE

Any change in the appearance from that natural to the commodity, during the processing and storage of frozen foods, will be regarded by the consuming public as an adverse change, even though it may not reflect any decrease in nutritional excellence. Such changes will, of course, result in poor sales. Among the first food products frozen were peas. Frozen peas had a tremendous appeal to the public because of their bright green color, in contrast to the olive color of canned peas. In reality, the growth of the frozen food industry was based on the fact that the frozen product was much more like the fresh commodity than the canned product was.

The color changes in frozen foods are of three types: (*a*) Fading or change in natural color constituents, (*b*) loss of color to the surrounding medium, and (*c*) development of off-color from originally noncolored constituents of the commodity.

The change of chlorophylls to the pheophytins is a typical example of the loss of true color with the formation of an off-color. The bright green color of vegetables is due to the photosynthetic pigments, chlorophyll a and chlorophyll b. These pigments are complex magnesium compounds and the magnesium is readiy removed and replaced by hydrogen. This occurs, for example, if the medium is made slightly acid. A rise in temperature accelerates the change. Removal of magnesium leaves the olive-colored pigment pheophytin a or b, hence the bright green color becomes decidedly duller and yellower.

The adverse change of chlorophylls to pheophytins occurs in frozen green vegetables such as peas (42), snap beans (43), and spinach (44) when they are held at temperatures much above 0°F. (−18°C.) or even are held for extended periods at 0°F. This quality defect can some extent before and during the blanching process (45). It ... be minimized by use of the proper combination of time and temperature in blanching (46).

A similar type of change occurs on the surface of stored frozen beef. However, here the change is due to oxidation whereas that involving chlorophyll was simple hydrolysis. Myoglobin, a pigment very similar to the hemoglobin of blood, occurs in muscle. In the living animal its color is purplish red. In cut fresh meat it is oxygenated to the red oxymyoglobin. This pigment imparts the "bloom" to fresh meat. When the ferrous iron of oxymyoglobin is oxidized a red-brown pigment, metmyoglobin, is formed. It is this change that finally gives beef an off-color at the surface during frozen storage, even at low temperatures (47).

Examples of the second type of color change, *i.e.,* loss of color to the surrounding medium, occur in frozen raspberries (48) and red, sour, pitted cherries packed in sugar syrup (49) where the red color diffuses from the fruit into the syrup. This change occurs only very slowly at 0°F. (−18°C.) storage but is greatly accelerated at temperatures above 0°F. After approximately 25 days at 20°F. (−7°C.) the concentration of color in the syrup is equal to that in the fruit.

The third and most often experienced type of color change is the formation of off-colors without the loss of natural color. One example is the darkening of the bones of frozen poultry upon cooking. Brant and Stewart (50) clearly demonstrated the cause of the darkening. Freezing and thawing breaks down the red blood cells of the bone marrow. The hemoglobin thus liberated penetrates the walls of bones and causes discoloration on cooking. When bone-marrow cells were hemolyzed without freezing, no darkening resulted. Thus, freezing and thawing not only liberated the hemoglobin in the bone marrow but also made it possible for the pigment to penetrate the bone.

The most important and most noticeable quality change in sliced, frozen peaches is the browning of slices exposed to the headspace atmosphere (51). The browning is due to the enzyme polyphenoloxidase, which catalyzes the following type of reaction:

$$\underset{\text{(many)}}{\text{phenols}} + O_2 \longrightarrow \underset{\text{(light color)}}{\text{quinones}} \longrightarrow \text{brown polymers}$$

The enzyme has been variously named (phenolase, plant tyrosinase, potato oxidase, and catechol oxidase). This is one of the few enzymes that reacts directly with the oxygen of the air. At the quinone stage the reaction can be reversed by reducing reagents such as ascorbic acid (vitamin C), which in turn is depleted. However, at the polymer stage reducing reagents are ineffective. Sliced peaches are commonly packed with ascorbic acid added to the syrup. Slices exposed to the headspace darken first because of the oxygen there, with the result that the ascorbic acid is depleted there first. The exclusion of oxygen will prevent the darkening and prolong the high–quality storage life of the product. The polyphenoloxidase color change also occurs in many other fruits, such as apple slices, bananas, and avocados. In the case of apple slices, sulfur dioxide is used as an antioxidant.

But not all the fruits that go off-color in frozen storage at the higher storage temperatures do so through the intervention of polyphenoloxidase. This enzyme is completely absent in strawberries. At temperatures above 0°F., the color change in strawberries is from bright red to a dull, reddish gray appearance; the rate of change increases exponentially as temperature rises (52). The mechanism of this change is unknown, but it may also be oxidative.

Brussels sprouts which are insufficiently blanched develop an undesirable pink coloration at the centers during frozen storage (53). To prevent this pink coloration by severe blanching is also undesirable, since that would cause an excessive loss of the green chlorophyll color. Therefore, blanching methods have been developed for Brussels sprouts which compromise between the two ill effects (54).

Some vegetable products, particularly when stored under adverse temperature conditions, develop a yellow or brown off-color of unknown origin. This is the case with cauliflower (55) where a water–acetone soluble dark color is formed at temperatures above 0°F. (−18°C.).

The examples given of the three types of color change which occur during the processing and storage of frozen foods illustrate the chemical diversity of the changes and the examples of the changes where the mechanisms are unknown illustrate the need for further research.

IV. CHANGES IN FLAVOR

Before discussing off-flavor formation it might be well to define flavor. The broadest definition in Webster's unabridged dictionary defines flavor as "that quality of anything which affects the taste or gratifies the palate." Under this definition all qualities of a food, *i.e.,* texture, appearance, odor, and taste, contribute to its flavor, and any changes therein from the desirable or normal would constitute off-flavor. However, in the present discussion of off-formation in frozen foods only the changes in odor and taste will be considered.

Some of the off-flavor formation occurring during the processing and storage of frozen foods, particularly when the handling and storage are carried out at adverse temperatures, will be discussed. Little is known about the chemistry of off-flavor formation except for the one type, rancidity. Accordingly, most of our attention will be given to the chemistry of rancidity development.

That vegetables must be blanched prior to freezing to prevent off-flavor formation has been known for many years (56). More recently Lee *et al.* (57) have demonstrated that rancidity development occurs in unblanched frozen vegetables. As a matter of fact, the rancidity develops in a relatively short time at 0°F. (−18°C.). These authors imply that rancidity may be the flavor change which occurs in blanched vegetables stored for an extended period at 0°F., or for shorter times at higher temperatures. However, the flavor change which occurs in the frozen storage of blanched vegetables, *e.g.*, peas (42), snap beans (43), spinach (44), and cauliflower (55), is not rancidity development. Moreover, Rhee and Watts (58) found that the amount of lipid oxidation occurring in frozen raw peas stored at 14°F. (−10°C.) was too small to produce rancid odors. Furthermore, these authors found that the enzyme system responsible for lipid oxidation, lipoxidase, is rapidly inactivated by a short blanching time. Accordingly, the nature of the off-flavor and its mode of formation in frozen vegetables remain unknown.

Another type of off-flavor which can form in frozen vegetables is the so-called "delay" off-flavor, a haylike flavor. This type of off-flavor was first found in connection with the canning industry and later became an important difficulty in the frozen food industry. Kertesz and Green (36) found that shelled peas held at room temperature prior to canning developed the delay off-flavor and a tougher texture. Holding the peas at 30°F. (−1°C.) retarded off-flavor formation. Kohman and Sanborn (59) subsequently found that the bruising of some vegetables, *e.g.*, peas, lima beans, and corn, prior to processing, produced the delay off-flavor. They found that bruising alters the respiration of vegetables; it causes a decrease in oxygen consumption, but an increase in carbon dioxide production. This anaerobic like metabolism may be connected with the off-flavor formation. Subsequently it was demonstrated that the delay off-flavor in peas is not due to microbial growth, and that unbruised peas (hand shelled) failed to develop this flavor (60). Accordingly, the highest quality frozen vegetables are those processed rapidly and in such a way as to minimize bruising, *e.g.*, peas vined at the plant.

Little is known about the chemistry of off-flavor formation in frozen fruits and vegetables, but the most usual kind of off-flavor in meat and fish, namely rancidity, can be discussed in detail because the chemical mechanism of its development in the lipids (including fats) has been studied intensively and by many people. Had other types of off-flavor formation received only a fraction as much effort we would undoubtedly know a great deal more about them today.

Fats undergo two types of change en route to development of off-flavor. The first of these is simply hydrolysis (by the enzyme lipase) of the fat into its component parts, *i.e.,* glycerol and free fatty acids as illustrated in Fig. 2.6. This type of rancidity, called hydrolytic rancidity, is important only in milk products, not in frozen meat, fish, or fowl. In milk products, free fatty acids impart an objectionable bitter flavor and the pungent, sharp, unpleasant, rancid odors that are characteristic of such volatile free fatty acids as butyric, caproic, and caprylic. The proper handling and pasteurization of milk destroys the lipase enzyme responsible for the fat hydrolysis (61). The formation of free fatty acids in frozen poultry is minor and of no significance with regard to flavor (62).

The most prevalent type of change that occurs in fats en route to off-flavor formation is an oxidative reaction. The mechanism involved in development of oxidative off-flavor is by no means a simple reaction, but rather is a group of reactions including chain reactions (Fig. 2.7).

The chain reactions involve formation and rapid further reaction of short-lived free radicals. The first phase of rancidity formation is characterized by an induction period during which oxidation is slow; this is followed by an accelerating rate of oxygen absorption, with concurrent development of the products indicated above. The first products of oxidation of the unsaturated fats, hydroperoxides, do not *per se* impart rancidity, although measurement of their content in foods has been widely used to give an index of rancidity (63). However, they are extremely reactive and, as indicated by Fig. 2.7, can lead to a whole series of compounds, most important of which are the fission products. Some of these cleavage products arise as a result of free radical forma-

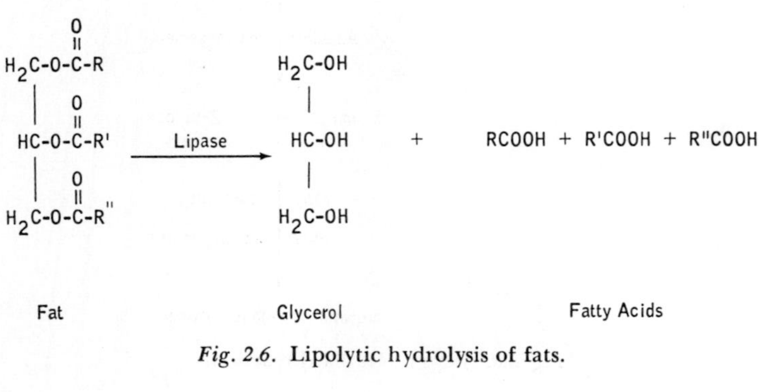

Fig. 2.6. Lipolytic hydrolysis of fats.

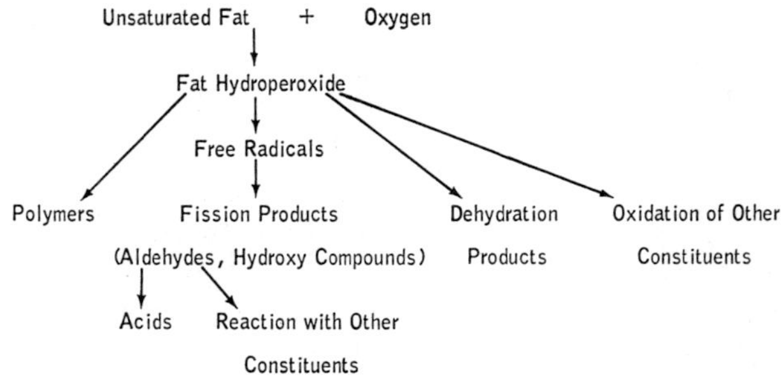

Fig. 2.7. Reactions involved in rancidity formation.

tion and decomposition. The rancid odor and flavor in foods is mainly due to various volatile aldehydes and compounds derived from aldehydes. The color reaction yielded by the aldehyde, malonic dialdehyde, with thiobarbituric acid sometimes been used as an objective indicator of rancidity. However, this particular aldehyde contributes little to rancid off-flavor.

The aldehydes react with amino acids to form colored compounds. This reaction is known as the Maillard reaction. One example of the mechanism involved is the series of reactions in Fig. 2.8. It is this reaction which is responsible for the "rusting" of stored frozen fish, an oxidative deterioration which may occur in addition to rancidity. Rusting consists of a light yellow to brown discoloration on the exposed surfaces of the fish.

Rancidity formation is still further complicated by the occurrence of rancidity accelerators in the frozen fish and meat. Examples of accelerators are organic iron catalysts like hemoglobin, and trace metal ions, *e.g.*, copper, iron, etc. Since the organic iron catalysts are inherently

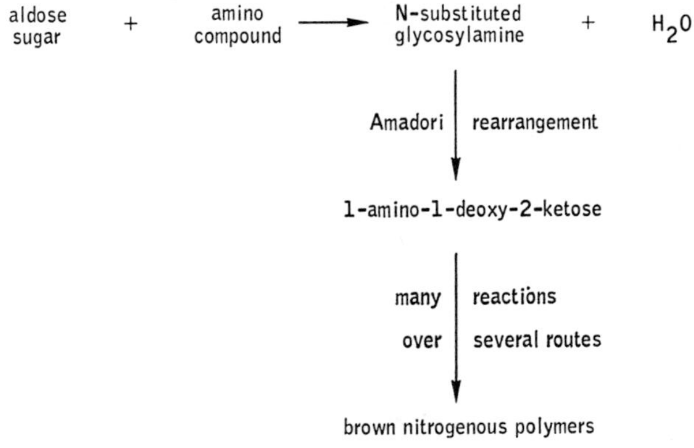

Fig. 2.8. Steps of Maillard reaction.

bound to proteins and since proteins are rendered insoluble by hea denaturation, cooked and uncooked meat would be expected to possess different stabilities with respect to rancidity in frozen storage. Watts and Peng (64) found that precooked frozen pork sausage is more stable than raw, presumably due to the protein denaturation in the former.

Rancidity development in frozen poultry is accelerated by "freezer burn" (65). Freezer burn of poultry (now seldom observed because present packaging prevents dehydration) is, in reality, surface drying. The fact that in the dried surface condition the fat is not protected from oxygen by a barrier of ice is undoubtedly a factor in the acceleration.

Certain naturally-occurring substances are antioxidants, and delay the onset of rancidity by interrupting the chain reaction. An example of such an antioxidant which occurs in meats and fish is tocopherol, vitamin E. Synthetic antioxidants have also been developed. Some of the latter are permitted in foods.

Retardation of rancidity in frozen meats and fish has been accomplished, then, either by excluding oxygen or by adding antioxidants. These measures are, of course, in addition to the proper low-temperature storage. Present packaging methods exclude oxygen from poultry. Tightly fitting plastic film used also eliminates freezer burn. Rusting and rancidity in fish can be delayed by covering the fish with a glaze of ice.

Supplementation of the diet of chickens and turkeys with tocopherol increased the birds' tocopherol content and hence the stability of their fat (66). However, at the present time this feed supplementation is not economical. Synthetic antioxidants are used in the ice used for glazing of fish (67).

The knowledge of the chemistry of rancidity formation, limited though it is, has led to practical methods of retarding development of this serious defect. Clearly a need exists for more extensive research on the chemistry of other kinds of off-flavor development in all frozen foods, so that their high-quality storage life may be extended. Indeed, the need exists for more research to define and characterize the true flavors of all frozen foods, both cooked and uncooked.

V. CHANGES IN NUTRITIONAL VALUE

The loss of and change in different vitamins during the preparation, freezing, and frozen storage of various foods varies greatly. For example, the vitamin C (ascorbic acid) in vegetables is quite labile, whereas that in cans of frozen orange juice concentrate is highly stable. As a matter of fact, frozen orange juice concentrate held under an adverse time–temperature condition will become unpalatable before any essential loss of vitamin C occurs (14).

The commercial operations such as washing, blanching, and cooling necessary for the preparation of vegetables for freezing are responsible for appreciable losses of vitamins. Most measurements have been made on vitamin C loss, but it is probable that what can be said about vitamin C would also apply to the water-soluble B vitamins present. In this regard, blanching is the worst offender. Tressler *et al.* (68) found that water blanching of lima beans resulted in a 20–40% loss in vitamin C. Jenkins *et al.* (69) reported a 30% loss of this vitamin in peas during the various steps of commercial preparation, with a loss of 10% occurring during the blanching process. The most rapid loss of ascorbic acid was found to occur during the cooling and washing operation. In boiling-water blanching the loss became greater as the time of contact was lengthened, whereas with steam blanching the loss became negligible after the first 60 sec. Similar results were obtained by Retzer *et al.* (70) with snap beans and cauliflower. Recently developed procedures have moved toward a short-time, high-temperature blanch, followed by rapid cooling, for better retention of the green color and ascorbic acid of green beans (46) and other vegetables. However, the necessary blanching of vegetables prior to freezing always results in some vitamin loss.

The next question is, "What effect does freezing rate have on vitamin retention?" In a study conducted by Lee *et al.* (71) with peas and snap beans, it was demonstrated that the rate of freezing had little or no effect on changes in ascorbic acid and the B vitamins, thiamine and riboflavin. The times necessary for freezing varied from minutes (realized by freezing with liquid nitrogen) to 50 hr. They employed three other rates of freezing within this range. Over the range studied, the rate of freezing of vegetables had little effect on vitamin retention. Similarly, the rate of freezing of beef did not measurably alter the B vitamins (72). No differences in thiamine, riboflavin, niacin, pantothenic acid, and pyridoxine were found with freezing times of 1.5, 7, and 24 hr.

The changes in vitamins during frozen storage, particularly under adverse conditions, are important. Vitamin C will be considered in detail. The vitamin, in the form of ascorbic acid, is used in the syrup added to peaches intended for freezing, in order to decrease the rate of enzymatic browning. Vitamin C is labile to oxidation by both enzymatic and nonenzymatic catalyzed reactions. The oxidation reaction

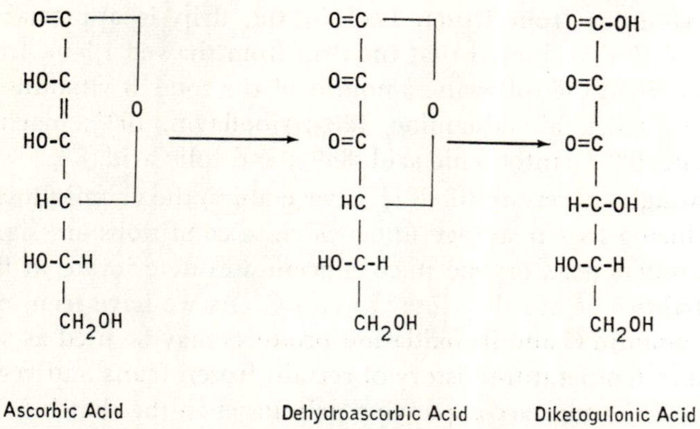

Fig. 2.9. Structure of ascorbic acid and its oxidation products.

proceeds as indicated in Fig. 2.9. The first product of oxidation, dehydroascorbic acid, possesses approximately three-fourths of the antiscorbutic activity of ascorbic acid, whereas its derived product, diketogulonic acid, is void of activity. This reaction proceeds in all frozen fruits and vegetables studied when oxygen is available, very slowly at 0°F. (−18°C.) but much more rapidly at higher temperatures. A method of estimating the total time–temperature experience of frozen fruits and vegetables, based on measurement of the components in this reaction, has been developed for nonhermetically sealed retail packages of strawberries (52), raspberries (48), and vegetables (45).

During the thawing of frozen fruits and vegetables, at least using recommended procedures, little, if any, significant loss in vitamins will occur. However, the picture is a little different for the thawing of meat and poultry. Substantial drip losses may occur in the thawing of meats, fish, and poultry. Drip loss is the liquid that exudes from the muscle during thawing. The amount of drip loss is dependent upon the relative amount of cut muscle surface. For example, the loss occurring during the thawing of whole eviscerated turkeys is only 3%, whereas with cutup chicken the loss is 10% (73). Part of the drip loss in thawing poultry is undoubtedly due to water incorporated during processing. In cutup chicken the drip contains appreciable amounts of the B-complex vitamins (thiamine, riboflavin, and niacin) in addition to other nutrients such as protein (73).

Drip losses from frozen meat are also significant. Drip occurs at the cut surface of the meat, and in cuts such as small beef steaks the loss can be approximately 10% (74). The composition of the drip is similar to that of the relatively free fraction of muscle fluid which can be expressed from the cut surfaces of the same unfrozen muscle (74). The loss of vitamins from frozen beef in the drip is appreciable (75). Pearson *et al.* (75) showed that the drip from thawed 1.5-in. frozen rib steak contained the following amounts of the total B vitamins present in the original steak: thiamine, 12%; riboflavin, 10%; niacin, 14%; pyridoxine, 9%; pantothenic acid, 33%; and folic acid, 8%.

Accordingly, in certain foods (*e.g.*, vegetables) the vitamin losses prior to and during frozen storage under adverse conditions are significant, whereas others (*e.g.*, orange juice concentrate) deteriorate in flavor or other qualities before they lose vitamin C. As we have seen, measurement of vitamin C and its oxidation products may be used as an index of the time–temperature history of certain frozen fruits and vegetables. The loss of vitamins accompanying drip losses in the thawing of meats and cutup poultry may be significant.

Accordingly vegetables suffer a loss in nutritional value during processing prior to freezing. Of all the foods studied, none lose significant amounts of essential nutrients during freezing. Fruits and vegetables, depending upon their individual character and upon the method of packaging, may lose vitamins in frozen storage. The loss of vitamins accompanying drip losses in the thawing of meats and cut-up poultry may be significant.

REFERENCES

(1) C. de Duve, *Sci. Am., 208,* No. 5, 64 (1963).
(2) H. P. Balz, *Planta, 70,* 207 (1966).
(3) H. Lineweaver, *J. Am. Chem. Soc., 61,* 403 (1939).
(4) D. W. DeJong, A. C. Olson, K. M. Hawker, and E. F. Jansen, *Plant Physiol., 43,* 841 (1968).
(5) J. E. Brekke and W. F. Talburt, *Food Technol., 4,* 383 (1950).
(6) H. L. Hanson, A. Campbell, and H. Lineweaver, *Food Technol., 5,* 432 (1951).
(7) T. J. Schoch and D. French, *Cereal Chem., 24,* 231 (1947).
(8) J. W. Pence, N. N. Standridge, T. M. Lubisich, D. K. Mecham, and H. S. Olcott, *Food Technol., 9,* 495 (1955).
(9) C. W. Du Bois and T. J. Kew, *Refrig. Eng., 59,* 772 (1951).
(10) A. H. Rouse, *Citrus Ind., 31,* 7 (1950).
(11) L. R. McDonnell, E. F. Jansen, and H. Lineweaver, *Arch. Biochem., 6,* 389 (1945).
(12) O. W. Bissett, M. K. Veldhuis, and N. B. Rushing, *Food Technol., 7,* 258 (1953).
(13) R. B. Guyer, W. M. Miller, O. W. Bissett, and M. K. Veldhuis, *Food Technol., 10,* 10 (1956).
(14) R. J. McColloch, R. G. Rice, M. B. Bandurski, and B. Gentili, *Food Technol., 11,* 444 (1957).
(15) E. F. Jansen, R. Jang, and J. Bonner, *Food Res., 25,* 64 (1960).

(16) D. G. Guadagni, C. C. Nimmo, and E. F. Jansen, *Food Technol., 12,* 36 (1958).

(17) M. Gee and R. M. McCready, *Food Res., 22,* 300 (1957).

(18) E. C. Bate-Smith and J. R. Bendall, *J. Physiol. (London), 106,* 177 (1947).

(19) D. de Fremery, *J. Agr. Food Chem., 14,* 214 (1966).

(20) C. F. Cori and G. T. Cori, *J. Biol. Chem., 79,* 309 (1928).

(21) C. Radouco-Thomas, C. Lataste-Dorolle, R. Zender, R. Busset, H. M. Meyer, and R. F. Mouton, *Food Res., 24,* 453 (1959).

(22) T. Moran, *Proc. Roy. Soc.* (London), *Ser. B, 98,* 436 (1925).

(23) O. M. Urbain and J. N. Miller, *Ind. Eng. Chem., 22,* 355 (1930).

(24) A. W. Thomas and M. I. Bailey, *Ind. Eng. Chem., 25,* 669 (1933).

(25) A. Lopez, C. R. Fellers, and W. D. Powrie, *J. Milk Food Technol., 17,* 334 (1954).

(26) G. Reed, *Enzymes in Food Processing,* Academic Press, New York, 1966, pp. 348–366.

(27) O. Dahl, *Nährung, 6,* 492 (1962).

(28) E. Wierbicki, L. E. Kunkle, V. R. Cahill, and F. E. Deatherage, *Food Technol., 8,* 506 (1954).

(29) D. de Fremery and I. V. Streeter, *J. Food. Sci.,* in press.

(30) O. G. Hankins and R. L. Hiner, *Refrig. Eng., 41,* 185 (1941).

(31) W. J. Dyer, *Food Res., 16,* 522 (1951).

(32) O. E. Nikkila and R. R. Linko, *Food Res., 19,* 200 (1954).

(33) C. S. Chung, E. G. Richards, and H. S. Olcott, *Biochem., 6,* 3154 (1967).

(34) J. M. Snow, *J. Fisheries Res. Board Can., 7,* 599 (1950).

(35) J. J. Connell, *Biochem. Biophys. Acta, 74,* 374 (1963).

(36) Z. I. Kertesz and E. L. Green, *J. Agr. Res., 45,* 361 (1932).

(37) E. R. Wolford and M. S. Brown, *Food Technol., 19,* 1133 (1965).

(38) H. K. Burr, M. S. Brown, and K. Robe, *Food Processing Marketing, 27,* No. 4, 116 (1966).

(39) *Frozen Food Pack Statistics,* Natl. Assn. of Frozen Food Packers, Washington, D.C., 1964.

(40) Anon., *Canner/Packer, 131,* No. 13, 31 (1962).

(41) Anon., *Food Eng., 36,* No. 1, 56 (1964).

(42) M. M. Boggs, W. C. Dietrich, M-D. Nutting, R. L. Olson, F. E. Lindquist, G. S. Bohart, H. J. Neumann, and H. J. Morris, *Food Technol., 14,* 181 (1960).

(43) W. C. Dietrich, M-D. Nutting, R. L. Olson, F. E. Lindquist, M. M. Boggs, G. S. Bohart, H. J. Neumann, and H. J. Morris, *Food Technol., 13,* 136 (1959).

(44) W. C. Dietrich, M. M. Boggs, M-D. Nutting, and N. E. Weinstein, *Food Technol., 14,* 522 (1960).

(45) W. C. Dietrich, F. E. Lindquist, J. C. Miers, G. S. Bohart, H. J. Neumann, and W. F. Talburt, *Food Technol., 11,* 109 (1957).

(46) W. C. Dietrich, R. L. Olson, M-D. Nutting, H. J. Neumann, and M. M. Boggs, *Food Technol., 13,* 258 (1959).

(47) J. M. Ramsbottom and C. H. Koonz, *Food Res., 6,* 571 (1941).

(48) D. G. Guadagni and C. C. Nimmo, *Food Technol., 11,* 604 (1957).

(49) D. G. Guadagni, C. C. Nimmo, and E. F. Jansen, *Food Technol., 12,* 36 (1958).

(50) A. W. Brant and G. F. Stewart, *Food Technol., 4,* 168 (1950).

(51) D. G. Guadagni, C. C. Nimmo, and E. F. Jansen, *Food Technol., 11,* 33 (1957).

(52) D. G. Guadagni, C. C. Nimmo, and E. F. Jansen, *Food Technol., 11,* 389 (1957).

(53) F. E. Lindquist, W. C. Dietrich, M. P. Masure, and M. M. Boggs, *Food Technol., 5,* 198 (1951).

(54) W. C. Dietrich and H. J. Neumann, *Food Technol., 19,* No. 7, 150 (1965).

(55) W. C. Dietrich, M-D. F. Nutting, M. M. Boggs, and N. E. Weinstein, *Food Technol., 16,* No. 10, 123 (1962).

(56) M. A. Joslyn and W. V. Cruess, *Fruit Prod. J., 8,* No. 7, 9 (1929); *8,* No. 8, 9 (1929).

(57) F. A. Lee, A. C. Wagenknecht, and J. C. Hening, *Food Res., 20,* 289 (1955).

(58) K. S. Rhee and B. M. Watts, *J. Food Sci., 31,* 675 (1966).

(59) E. F. Kohman and N. H. Sanborn, *Ind. Eng. Chem. 26,* 773 (1934).

(60) R. U. Makower and A. C. Ward, *Food Technol., 4,* 46 (1950).

(61) V. N. Krukovsky, *J. Agr. Food Chem., 9,* 439 (1961).

(62) W. H. Cook and W. H. White, *Food Res., 4,* 433 (1939).

(63) C. H. Lea, in *Lipids and Their Oxidation,* H. W. Schultz, E. A. Day, and R. O. Sinnhuber, Eds., Avi Publishing, Westport, Connecticut, 1962, pp. 3–28.

(64) B. M. Watts and D-H. Peng, *J. Home Econ., 39,* 88 (1947).

(65) W. H. Cook and W. H. White, *Can. J. Res., 18D,* 363 (1940).

(66) E. P. Mecchi, M. F. Pool, M. Nonaka. A. A. Klose, S. J. Marsden, and R. J. Lillie, *Poultry Sci., 35,* 1246 (1956).

(67) H. L. A. Tarr, *J. Fisheries Res. Board, Can., 7,* 237 (1948).

(68) D. K. Tressler, G. L. Mack, R. R. Jenkins and C. G. King, *Food Res., 2,* 175 (1937).

(69) R. R. Jenkins, D. K. Tressler, and G. A. Fitzgerald, *Food Res., 3,* 133 (1938).

(70) J. I. Retzer, F. O. Van Duyne, J. T. Chase, and J. I. Simpson, *Food Res., 10,* 518 (1945).

(71) F. A. Lee, W. A. Gortner, and J. Whitcomb, *Ind. Eng. Chem., 38,* 341 (1946).

(72) F. A. Lee, R. F. Brooks, A. M. Pearson, J. I. Miller, and F. Volz, *Food Res., 15,* 8 (1950).

(73) E. R. Larson, *J. Am. Dietet. Assoc., 32,* 716 (1956).

(74) W. A. Empey, *J. Soc. Chem. Ind., 52,* 230T (1933).

(75) A. M. Pearson, J. E. Burnside, H. M. Edwards, R. S. Glasscock, T. J. Cunha, and A. F. Novak, *Food Res., 16,* 85 (1951).

Chapter 3

Microbiological Conditions Affecting Frozen Food Quality

H. David Michener

Western Regional Research Laboratory, Agricultural Research Service, U.S. Department of Agriculture Albany, California

R. Paul Elliott

Microbiology Group, Technical Services Division, Consumer and Marketing Service, U.S. Department of Agriculture, Washington, D.C.

I. INTRODUCTION

Processors, distributors, and retailers of frozen foods should understand basic food microbiology. Microbial contamination and growth can cause loss of quality, frank decomposition, or even health hazard. The processor must maintain good sanitation, remove natural contamination if possible, handle the food promptly, and freeze it promptly. Warehousemen, distributors, and retailers must keep it from contamination, and maintain it at a sufficiently low temperature to prevent microbial growth and rapid quality loss.

The microorganisms that are most significant in the processing, storage, and distribution of frozen food can be divided into several groups. Those that cause deterioration in the flavor or appearance of foods are often called food-spoilage organisms. Yeasts and molds are common spoilage organisms on fruits and other acid or partially dried foods; bacteria are common on neutral or slightly acid foods such as meats, fish, and vegetables. When the spoilage organisms are enumerated, they are commonly referred to collectively as "total aerobic count." Two small but important groups of bacteria commonly found on food are related to human safety. The food poisoning bacteria produce toxins when they grow in the food or cause human infections when ingested. A small group of bacteria are used as indicators of possibly hazardous conditions. Sometimes called indicators of hazard or fecal-indicators, they are harmless in themselves, but their presence indicates possible contamination by man or animals, and therefore possible presence of human pathogens.

On the other hand, many microorganisms contribute to the flavor or stability of foods. Indeed, ripened cheeses, fermented beverages, pickles, and soy sauce could be considered partly decomposed; but we enjoy their strong flavors of "decomposition" without harm. Likewise some Orientals eat decomposed eggs or fish, and some primitive Indian tribes of Alaska prefer rotten salmon to the fresh. Yeast bread depends on microbial action to give the desired flavor and physical properties.

II. MICROBIOLOGY IN THE PROCESSING PLANT

A. Food Spoilage Microorganisms

1. Growth on the Product

Microorganisms cannot grow on frozen foods stored at sufficiently low temperatures (Sect. IV). However, during processing before freezing, contamination and time–temperature abuse can result in microbial growth and eventual spoilage. Even if no spoilage can be detected during processing, the product can better resist subsequent time–temperature abuse if it enters the freezer with a low microbial population rather than with a high one.

Since microorganisms have no digestive tracts, they excrete their enzymes of digestion into the surrounding menstruum. Thus most of the products of digestion are lost to the cell that produced the enzymes. To obtain some small benefit, the cell must produce an excess of enzymes. A relatively small weight of microorganisms will therefore cause a disproportionately rapid rate of breakdown of the surrounding medium—in this case, of a food. Within limits to be discussed later, freezing interrupts this process.

Food spoilage microorganisms come into the processing plant with the food. Foods of plant origin are contaminated primarily with microorganisms from soil and water, but sometimes from animals. Foods of animal origin come to be processed with the microflora of the animals from which they were derived. Many kinds of microorganisms can grow on foods while they are being processed and up to the time they are frozen. Thus, a reduction of processing time reduces the microbial growth that can occur and therefore the final microbial population on the product.

When microbial growth occurs, the types of organisms that will predominate in the total microbial flora depend on many factors. A heavier inoculum of one species will often give it an advantage in early stages that other species cannot overcome. The pH, temperature, nutrients, availability of water and oxygen, etc., all play roles in favoring one type over another (Reviewed by Elliott and Michener (1)). These factors may determine whether the abused product merely spoils, or whether food poisoning bacteria originally present in small numbers increase to hazardous levels. For a particular product being processed under particular conditions, one or a few species of microorganism usually predominate. Species of *Leuconostoc* and *Streptococcus* predominate among the spoilage bacteria on vegetables, but each tends to have its characteristic flora. *Pseudomonas* species are the usual spoilage bacteria on meats and fish at refrigerator temperatures.

2. *Blanching and Precooking*

Most vegetables are blanched in steam or hot water to inactivate their enzymes. This reduces their microbial population to very low levels. Bacterial spores, however, are commonly so heat-resistant that they survive this treatment. Blanching also renders vegetables more susceptible to bacterial attack. After blanching, the product is normally recontaminated with the microorganisms from equipment surfaces. The microbial load when it enters the freezer then depends on such factors as the duration and temperature of post-blanch processing, the amount of handling it receives, and the cleanliness of the plant. The possibility that processes may be developed in the future which would allow the product to enter the freezer with little or no recontamination is a matter for some concern, since temperature abuse might permit growth of food poisoning bacteria if they happen to be present. (See Sect. VI. A. 3.)

Many meat and fish foods, ingredients of prepared dinners, soups, etc., are precooked before they are frozen. The heat generally kills most or all of the pre-existing microflora except for bacterial spores.

After this, any microorganisms that have access to the product (including sporeformers that survived the heat) can establish themselves and grow until the product is frozen. As with blanched vegetables, it is possible that new manufacturing techniques would be introduced that would prevent or enormously reduce recontamination after precooking. (See Sect. VI. A. 3.)

3. Plant Sanitation

A major problem presented by these spoilage organisms is that they multiply not only in the product but also on the processing equipment. As any food is processed (especially if it is juicy or sticky), it leaves a residue of juice and food particles on the equipment it touches. This residue is an excellent growth medium for the spoilage organisms. If cleanups are infrequent, or if a food residue is allowed to accumulate on hard-to-clean parts of the equipment, its microbial population reaches enormous proportions. This, in turn, contaminates more of the food.

As an example, an accumulation of corn on a feed pipe to a cleaner in a corn freezing plant (Fig. 3.1) contained 17 million bacteria/g.

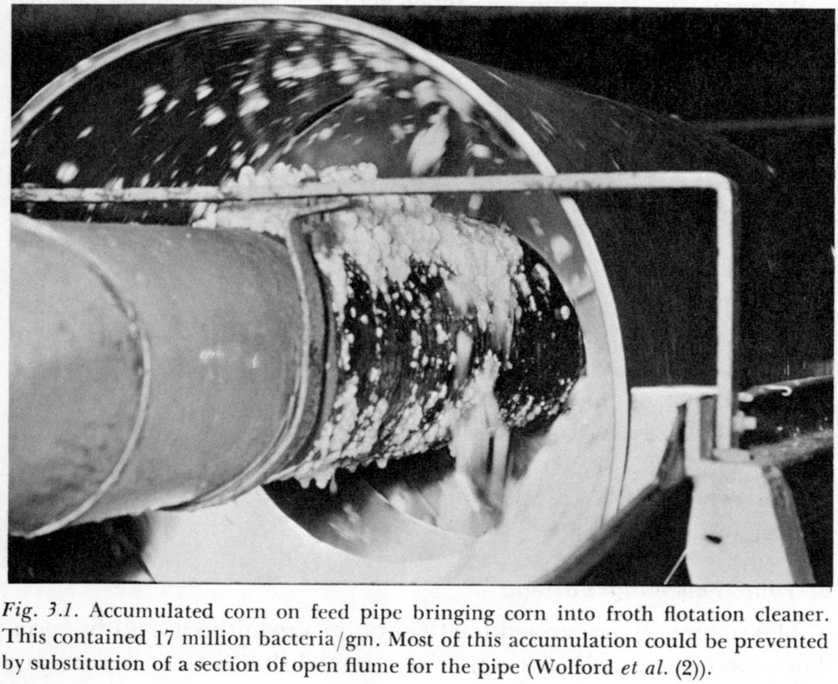

Fig. 3.1. Accumulated corn on feed pipe bringing corn into froth flotation cleaner. This contained 17 million bacteria/gm. Most of this accumulation could be prevented by substitution of a section of open flume for the pipe (Wolford *et al.* (2)).

(Wolford *et al.* (2)). From time to time chunks of the accumulated corn could be seen to fall back into the product stream, thus contaminating not only the product but the succeeding equipment along the line. Corn also accumulated at the distributor, from which it dropped onto the belt in the freezing tunnel. This was a difficult location to clean because wash water would fall into the freezing tunnel, and the authors recommended a water-tight door which could be closed during the clean-up. Such accumulations were observed in numerous other locations. Other products present similar problems, although corn may be particularly difficult because of its sticky juice.

Even without product accumulation, some processes increase the bacterial count in the product. A slightly worn belt caused a 100- to 200-fold increase in counts on lima beans and peas, probably because it contained minute cracks or holes that harbored rapidly growing bacteria. Belt contamination can often be controlled by continuous spraying of the returning belt with chlorinated water (Wolford *et al.* (2); Splittstoesser *et al.* (3)). Addition of chlorine to the entire water supply ("in-plant chlorination") reduces the buildup of microbial populations on equipment. The chlorine residual usually recommended is 1–6 p.p.m.

Bacterial counts can more easily be kept at a low level if the product is blanched near the end of the processing line (2), because then there is less time for regrowth of the microbial population and there are fewer operations which can cause recontamination. When possible, operations requiring much handling of the product should precede the blanch, because of the probability of contamination from the hands of the operator. However, good sanitation is needed even before the blanch, and law enforcement officials object if heating (or other bactericidal procedure) is used as a substitute for recognized, good sanitary practice.

When careful visual inspection discloses inadequate or infrequent cleanup of equipment, bacteriological samples will often disclose gross contamination of the product. On the other hand, contamination of the product directly from human sources rarely results in measurable increases in total bacterial count, but rather in the appearance of specific types such as *Escherichia coli, Salmonella,* or *Staphylococcus,* usually in small numbers.

B. Food Poisoning Bacteria

The commonly recognized food poisoning organisms are all bacteria. They can create hazardous conditions in food that is to be frozen, just as they can in any other food. The fact that they have not usually done so is dependent on some special considerations that will be discussed later (Sect. VI. A. 3). However, they frequently survive freezing (Sect. III).

Certain food poisoning bacteria are generally harmless in themselves but excrete toxins into food in which they grow. Others grow in food and become harmful when they are eaten, after which they cause infection and disease. Many other pathogenic or infection-producing bacteria can be carried on food although they rarely or never grow there. No sharp distinction can be made between food poisoning and pathogenic (infection-producing) bacteria, because these groups overlap; but by common usage the food poisoning organisms are usually those discussed below.

1. Staphylococcus

One of the most common of the food poisoning bacteria is *Staphylococcus aureus*. Certain strains secrete a toxin into the food in which they grow. Characteristically, gastrointestinal distress which may be severe but rarely if ever fatal appears from 2–6 hr. after the toxic food is eaten (Dack (4)). The toxigenic strains are nearly always "coagulase positive" (*i.e.*, they coagulate blood plasma) although not all coagulase-positive strains produce toxin. Tests for the toxin have been developed but are difficult and time-consuming. Most bacteriologists use the coagulase test to indicate potentiality for toxin formation.

Since growth to hundreds of thousands or even millions per gram (depending on the strain) is required before sufficient toxin is formed to cause illness, the presence of a few hundred staphylococci per gram is not in itself hazardous to the consumer. The danger lies in the possibility of time–temperature abuse that provides suitable conditions for growth in foods. Any moist, nonacid food is likely to offer such conditions, particularly if the normal microflora is destroyed. The harmless spoilage bacteria may prevent *Staphylococcus* growth and toxin production, especially at low temperatures (See Sect. VI. A. 3).

Staphylococci are easily killed by heat. In foods that have been cooked, or otherwise treated so that all staphylococci are destroyed, their occurrence at a later time indicates the likelihood that human or other animal contact has occurred. Up to 50% of human beings carry coagulase-positive staphylococci in the mucous membranes of the nasal passages. All of us carry them on our skin at times, especially when we have pimples or boils. For this reason, foods that have been handled are likely to have a few of the organisms. Unless and until human contact can be eliminated, we can expect a few of these potentially hazardous organisms to be present on many foods. However, a high incidence may reflect an unwarranted degree of human contact.

Once the toxin is formed, it is not as readily destroyed by heat as are the microorganisms themselves; indeed, a large outbreak of *Staphylococcus* intoxication occurred from dried milk in which no viable staphylococci could be found. They grew to large numbers in the fluid milk, and died in the drying process, leaving the toxin. On the other hand, the relatively high temperatures used in the retorting of canned foods would probably destroy any preformed toxin.

2. Clostridium botulinum

The most feared of all the food poisonings and food infections is botulism, for up to 65% of all affected persons die (Dack (4); Gordon and Murrell (5)). Fortunately, only 5–50 persons are afflicted by this dread disease yearly in the United States, whereas many thousands become ill from the less serious food-borne diseases.

Like *Staphylococcus* poisoning, botulism is caused by a toxin formed during the growth of the organism. It is the most potent poison known; 10^{-4} μg. will kill a mouse, and one ounce could kill 200 million people (5). Unlike *Staphylococcus* toxin, it is heat labile and does not withstand boiling.

Also, unlike *Staphylococcus, C. botulinum* forms spores with sufficient heat resistance to survive the heat processes normally used in preparation of frozen vegetables or precooked frozen foods. The spores can be eaten without harmful effect; it is when they germinate and grow that they produce toxin.

Growth usually occurs in neutral or only slightly acid foods; and outbreaks of botulism have come most frequently from vegetables, less often from meats and fish, and only very rarely from milk products (5). Growth rarely if ever occurs below pH 4.5, and the occasional outbreaks of botulism from fruit have been attributed to growth of *C. botulinum* after another contaminating organism has altered the pH of the product.

C. botulinum is an anaerobe and is thus able to grow in hermetically sealed containers. It can also grow in other situations where oxygen fails to penetrate to all parts of a food, either because of packaging or because of the nature and consistency of the food itself. Toxin has formed in frozen-thawed peas in cardboard containers (Perry *et al.* (6)), in a chicken pie (U.S. Public Health Service (7)), in loosely wrapped smoked fish (Kautter (8)), and in sausages (Dack (4); Gordon and Murrell (5)). Growth is favored if a food is heated enough to destroy the normal spoilage flora, leaving the heat-resistant Clostridium spores to grow with little or no competition (6). The minimum temperature permitting growth of *C. botulinum* strains is discussed in Section IV. B.

Of the six types known, three (A, B, and E) are those commonly associated with human disease. Spores of Types A and B ordinarily occur in soil and hence commonly on vegetables, less often on meats and poultry. Spores of Type E are much less heat resistant than Types A and B, so that normal cooking inactivates them. They usually occur in marine and fresh water muds; hence mostly on fish. Recently marine strains of Types B and F have been discovered, which have many of the characteristics of Type E, including relatively high sensitivity to heat (9).

Although the raw product for freezing can contain *C. botulinum* spores, these normally have neither enough time nor the right conditions to grow during processing. There is no practical way to insure their total removal except heating at 250°F. (121°C.) or above as is done in canning. One can imagine conditions of gross mishandling that would permit them to grow, but in practice they are not a hazard unless frozen food is mishandled after thawing. This possibility is discussed in Section VI. A.

3. Clostridium perfringens (welchii)

An undetermined number of food poisonings occur each year from *Clostridium perfringens.* The spores are widely distributed in feces, garden soil, and house dust. In fact, they are so ubiquitous that they can hardly be avoided in nonsterile foods. Like *C. botulinum,* the spores are not harmful, but if they germinate and grow to a large population on a food, ingestion will produce a mild, short-lived intoxication, characterized by abdominal cramps and diarrhea, and rarely vomiting (Hobbs (10)).

C. perfringens grows best at temperatures somewhat above room temperature. It grows well on meat. Poisoning generally occurs when spores survive cooking in a deep pot or other large container of meat, then germinate and grow when the meat remains warm for several hours (10).

4. Salmonella

Salmonellosis is an infection, rather than an intoxication. Its symptoms resemble those of staphylococcal food poisoning except that they take longer to appear after the offending food is eaten and the illness is often more prolonged. It is caused by any of several species of *Salmonella.* Although they produce infections, they are classed as food poisoning organisms because they are frequently carried on foods. (Typhoid fever is a much more serious form of salmonellosis, formerly common but now controlled through supervision of milk and water supplies and isolation of anyone having the disease.)

In the normal healthy adult, illness occurs only after ingestion of many millions of *Salmonella* cells; however, in the very young or the debilitated, far fewer cells can produce serious, sometimes fatal, infection. The records of the U.S. Public Health Service (11) show 69 deaths from salmonellosis in 1966, giving a mortality rate of 0.39%. Fig. 3.2 shows the predilection of this disease for the very young. Furthermore, in the young and the very old of all species affected, the microorganisms may enter the blood stream, causing frequent fatalities. In normal adults, the infection is generally limited to the digestive tract, sometimes producing asymptomatic carriers.

The reported incidence of salmonellosis in human beings has increased in the United States from about 4,000 cases per year in 1942 to over 20,000 per year in 1966 (Fig. 3.3). Much of the increase can be attributed to a progressively greater interest in investigating and reporting outbreaks. Nevertheless, numerous cases are thought to go unreported. For example, in the investigation of three separate cases of salmonellosis, it was discovered that there had been numerous other unreported cases associated with each: 8, 7, and 238, respectively, in the three outbreaks.

In addition, the National Communicable Disease Center of the United States Public Health Service in Atlanta has received 5,000–7,000 *Salmonella* isolates per year from sources other than human disease. A better idea of the incidence of the organisms in foods can be obtained from objective surveys, summarized in Table 3.1. In some instances, the

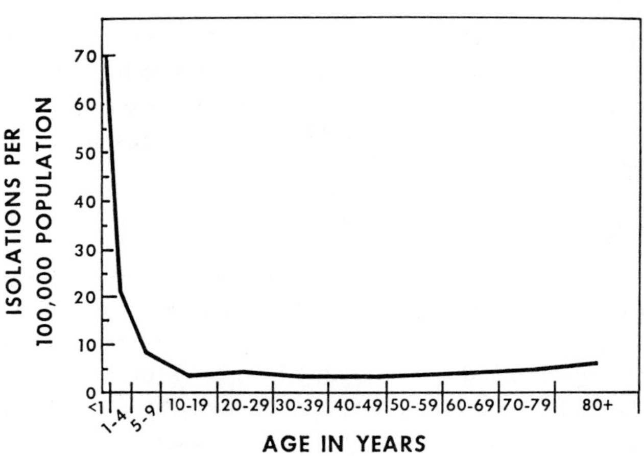

Fig. 3.2. Rate of human isolations of salmonellae by age group, 1966 (11).

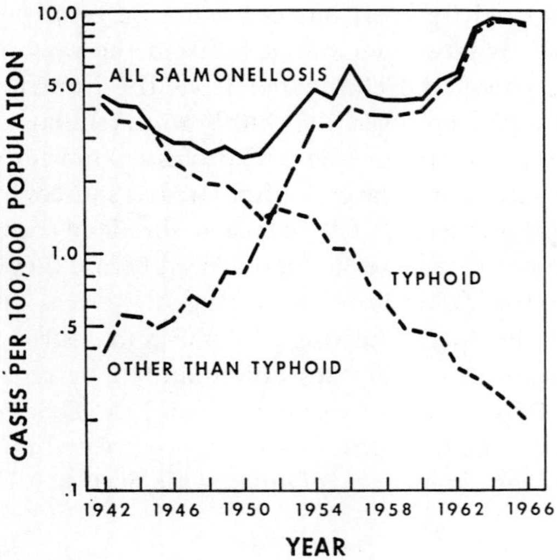

Fig. 3.3. Reported incidence of human salmonellosis in the United States, 1942–1966 (11).

food industry involved instituted control measures after such data were obtained.

Salmonellae are typically from animal fecal sources, and are therefore commonly found on meats and other animal products, such as eggs. However, they may also be transmitted by human carriers, through food, or otherwise. They can grow in a wide variety of foods, and have been found, for example, in raw meat, eggs, dried yeast, coconut meats, animal feeds, powdered milk, cochineal dye, and cottonseed flour. They

Table 3.1. Incidence of Salmonella in Some Products of Animal Origin

Product	Number of published reports summarized [a]	Total number of samples analyzed	Percentage of positive samples	Percentage of positives in separate reports, lowest and highest
Shell eggs	2	21,755	0.02	0.02 to 0.6
Frozen and dried eggs	7	19,157	16	3 to 35
Milk, dried milk	2	4,665	0.9	0 to 1.1
Red meat	11	5,794	13	0.2 to 38
Poultry meat	10	13,703	7	0.7 to 50

[a] Published in the past 20 years. Reports covering less than 100 samples are excluded from this summary.

are easily destroyed by heat, but can often be reintroduced into the cooked product by cross contamination from the raw.

Salmonellae should be eliminated from the human and domestic animal food supply; however, the task is a tremendous one because of the ubiquity of the organisms. Several products known to be sometimes contaminated will soon be largely eliminated as sources by special treatments employed on them. A few years ago, shredded coconut was found contaminated; now the meats are pasteurized before they are shipped to the United States. Some firms producing dried yeast and cottonseed flour have installed special heating procedures to destroy the organisms. Liquid, frozen, or dried eggs are now required by regulations of the United States Department of Agriculture and the U.S. Food and Drug Administration to be pasteurized or otherwise treated to destroy salmonellae. For recent reviews on *Salmonella* see Section VIII.

5. *Other Food-Borne Bacterial Diseases*

About 50% of the outbreaks reported to the U.S. Public Health Service are of unknown etiology, and no one knows how many other outbreaks remain uninvestigated and unreported. No doubt many of these are caused by bacterial agents already described. However, because few laboratories look for other agents, no one knows the extent of their importance. The bacterial agents less commonly implicated in food-borne disease usually produce abdominal cramps and diarrhea, rarely vomiting, similar to *Clostridium perfringens* poisoning. The alpha-type streptococci, *Bacillus cereus, Bacillus subtilis,* some strains of *Escherichia coli,* and perhaps others fall into this category. The Japanese have reported extensive outbreaks from raw seafood, with symptoms almost identical to those of *Salmonella;* but the causative agent is *Vibrio parahaemolyticus. Shigella,* a genus related to *Salmonella,* can be transmitted from human carriers through food.

6. *Food-borne Viruses*

Infectious hepatitis, poliomyelitis, and other viral diseases can be transmitted by foods contaminated by human carriers. But viruses cannot multiply outside of the living host cell; therefore they are unlikely to be present in high concentrations in foods. This makes their isolation difficult. Methods are being developed (Lynt (12)), but are not yet in general use.

7. *Food-Borne Parasites*

Several food-borne parasites can cause human disease when ingested. However, they die promptly on storage at or near 0°F (−18°C). This applies to *Cysticercus bovis* (13), *Trichinella spiralis* (14), *Endamoeba histolytica* (15), and *Toxoplasma gondii* (16). The first of these is the larval stage of the beef tapeworm; the others cause, respectively, trichinosis, amoebic dysentery, and toxoplasmosis.

C. Possible Elimination of Food Poisoning Bacteria

Some food poisoning bacteria are naturally present on raw agricultural products. It will probably never be possible to prevent completely the occurrence of spores of clostridia in such foods. Staphylococci will invariably be present in small numbers on raw meats because of their close association with the living animals. Salmonellosis affects nearly all of the higher animals, both domestic and wild, so that complete control will be difficult. Tests for the presence of pathogens are unreliable because the number of bacteria may be small and they may be unevenly distributed in the product. One must assume, therefore, that pathogens are present on raw agricultural products.

Food poisoning bacteria are found from time to time in new situations, so that foods previously considered safe have been found to be capable of causing disease. The following foods have been so categorized in the past 20 years: liquid and frozen eggs, *Salmonella;* shredded coconut, *Salmonella;* dried yeast, *Salmonella;* dried milk, *Staphylococcus, Salmonella;* Cheddar and Monterey jack cheese, *Staphylococcus;* canned tuna, *C. botulinum* Type E; smoked fish, *C. botulinum* Type E, *Salmonella;* processed cheese, *C. botulinum* Type B; chicken pie, *C. botulinum* Type B; cochineal dye (Carmine red), *Salmonella;* soya milk, *Salmonella;* cottonseed flour, *Salmonella;* pepperoni sausage, *Staphylococcus.* Most of the problems associated with these products have been or are being corrected by industry and health authorities.

Nonsporeforming bacteria such as *Staphylococcus* and *Salmonella* can be eliminated by comparatively mild heat treatments. This is commonly called pasteurization. For example, an effective treatment for egg magma is 3½ min. at 140–145°F. (60–63°C.). Pasteurization of milk is nearly universal in this country. Nevertheless, indiscriminate use of heat treatments, in the absence of complete microbiological investigation, may lead to undesirable results. First, the amount of heat required increases with the microbial load on the product, so that a low count on the raw product and good sanitation remain important even when the final product is heated. Second, a mild heat treatment can greatly reduce or eliminate the normal spoilage organisms, leaving only bacterial spores such as those of *C. botulinum.* Subsequent temperature abuse of a nonacid food might allow this organism to grow and render the product extremely hazardous (See Sec. VI. B).

D. Indicator Organisms

The presence of coliforms generally, of *Escherichia coli* specifically, or of the enterococci *(Streptococcus faecalis)* are taken as evidence of fecal contamination because these bacteria are present in enormous numbers in the gut of man and other animals. Although these organisms are normally harmless in themselves, it is assumed that if they are present in a food, there is danger that pathogenic organisms will also be present. Consequently the requirement is frequently made that these indicator bacteria (usually coliforms or *E. coli*) be absent from a specified sample of food or have a count below a certain numerical level (See Sec. VI. B).

Unfortunately, some of the indicator organisms have been found to be widely distributed in nature, so that their presence in a food is not conclusive evidence of fecal contamination. Certain coliforms are common in soil and therefore likely to be present on any fruits or vegetables which are not heated during processing. Although *E. coli* has more significance than other coliforms, its source should be determined. Its presence in small numbers is unavoidable, for example, on vegetables from heavily manured soils or in some animal products such as eggs, poultry, or red meats. The enterococci present a similar situation as they have been found in the interior of vegetables opened aseptically and thus cannot represent recent fecal pollution. On the other hand, when indicator organisms are found on foods which have been heated enough to kill the original nonsporeforming bacterial flora, their presence indicates recent contamination by the food handler or by contact with equipment which was previously used to handle the unheated product.

Another complication in the use of indicator organisms is that they may grow on food. Thus an initially small population can grow on a food sufficiently to cause it to be rejected as contaminated. This cannot happen as long as the food is frozen (Sec. IV. C).

E. coli and, to a lesser degree, other coliforms do not survive frozen storage as well as many other bacteria (Sec. III. C), so that the size of their population may be greatly reduced on food that is examined after frozen storage. Since salmonellae also die off relatively rapidly during frozen storage, it has been argued that when coliforms fail to survive, there is less danger from salmonellae. On the other hand, a food that is produced under insanitary conditions may appear to be in good bacteriological condition (or to comply with legal or industry standards) after a period of frozen storage, if judged on the basis of its coliform count.

III. SURVIVAL OF MICROORGANISMS IN FROZEN FOODS

Microbial populations almost always survive freezing although they are commonly somewhat reduced in size. It follows that if a product has a high microbial population before it is frozen, it will usually remain in this condition after frozen storage.

A. The Effect of Freezing and Thawing

Even when a microbial population is merely frozen and thawed, without intervening frozen storage, there is usually a reduction in the number of live organisms. The reduction is usually slight to moderate, but occasionally survival is less than 1%. That killing is an effect of freezing and not merely of low temperature is shown by the slight effect of supercooling (97% survival), as compared with freezing at the same temperature (2% survival), and by the fact that repeated freeze–thaw cycles are more lethal than continuous frozen storage.

Freezing and thawing has occasionally been reported to cause an apparent increase in the bacterial count on frozen foods. This is usually attributed to mechanical breaking up of clumps of bacteria so that the product gives a higher plate count. Such increases in apparent population are usually less than twofold; but a 50-fold increase in count has been observed on peas. In this case a microscopic examination showed that there were clumps of bacteria in the unfrozen material, but not after freezing and thawing.

B. The Effect of Frozen Storage

1. The Harmless Spoilage Organisms

When foods are frozen and then stored frozen for considerable periods of time, the microbial count typically drops sharply at or near the time of freezing, and continues to drop slowly during storage. Often it levels off after a few weeks or months. The total counts on several food products before freezing and after various periods of frozen storage are shown in Fig. 3.4.

Survival of mixed populations of spoilage organisms varies greatly with the circumstances. It is usually between 1 and 75% on nonacid foods. Survival rates are often lower in acid foods (fruits, fruit products), in which 0.1–10% of the original total count frequently survive, although survival rates above 50% have been reported (Michener and Elliott (21)). It rarely if ever happens that no survivors are reported from a mixed population, and even though some spoilage

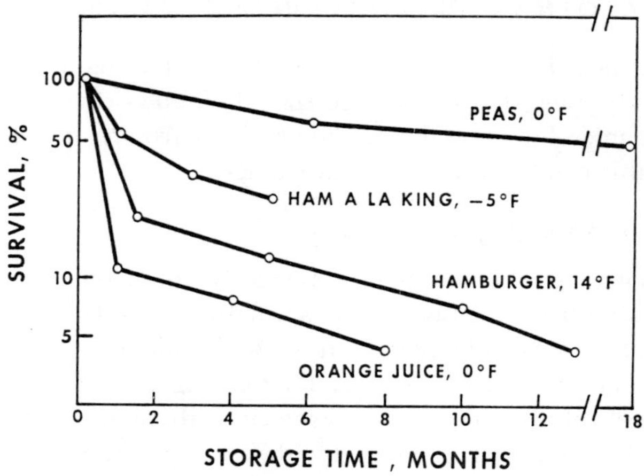

Fig. 3.4. Effect of freezing and frozen storage on total microbial counts on several foods: peas (17), ham à la king (18), hamburger (19), and orange juice (20). The initial counts were made prior to freezing.

bacteria in a population die on prolonged storage, enough survive so that a food usually spoils as fast after thawing as if it had never been frozen (Sec. V. C).

Bacterial spores almost always survive frozen storage. Several authors have reported survival rates approximating 100%. Young, rapidly growing cultures of spore-forming species are, of course, mostly vegetative cells, and their survival rates may be much lower. Spore-forming species do not normally predominate in the microbial flora of frozen foods, and they contribute significantly to food spoilage only if their nonsporeforming competitors are killed by heat or otherwise.

Microbial populations can survive for years in frozen food (Table 3.2. These data cover a wide range of foods and organisms, and include acid foods in which survival is normally poor.

2. *Food Poisoning Organisms*

Spores of the clostridia are extremely resistant to prolonged frozen storage (Table 3.2) and several authors have reported 100% survival of spores of *C. botulinum* and *C. perfringens.* On the other hand, vegetative cells of *C. perfringens* are unusually sensitive to freezing and even chilling without freezing will destroy most of them. This becomes a problem for microbiologists in transporting suspected foods to the laboratory.

Survival of staphylococci during frozen storage is roughly the same

Table 3.2. Prolonged Survival of Microorganisms in Frozen Storage [a]

Organism or group	Food or other substrate	Surviving organisms recovered, years
Total count	Poultry	3
Total count	Egg	6
Total count	Vegetables	11
Total count	Cherries	2
Yeast	Grape juice	4
Sporeformers	Cheddar cheese	7
Streptococcus	Cheddar cheese	7
E. coli	Ice (distilled water)	2.3
Salmonella enteritidis	Ice cream	7
Clostridium spores [b]	Culture medium	20

[a] See the review by Michener and Elliott (21) for references.
[b] Spores of PA 3679, a close relative of *C. botulinum*.

as that of the food spoilage organisms. Thus survivors are to be expected in a food initially contaminated with staphylococci.

The salmonellae are relatively sensitive to frozen storage, but survivors can usually be found in artificially inoculated packs. Their survival rate varies with the species (Fig. 3.5) and other factors as described

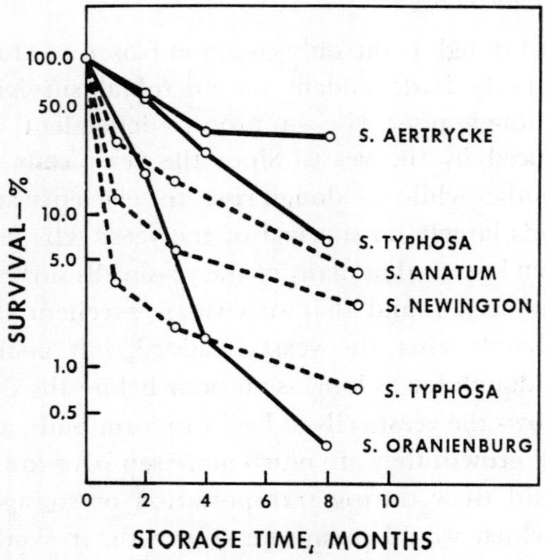

Fig. 3.5. Survival of several species of *Salmonella* during freezing and frozen storage. (———) Beef, 16°F. (—9°C.) (22); (- - -) Chicken chow mein, —14°F. (—26°C.) (23).

below. Occasionally there have been reports of complete destruction of salmonellae on frozen storage in acid foods, but since this is difficult to establish with certainty and highly dependent on conditions, it must be assumed in practice that if food is once contaminated with this organism it will remain so after frozen storage.

The food-borne viruses survive freezing in many foods. However, since methods for their isolation are inadequate, we must depend on epidemiological evidence. Laboratory tests with viruses in pure culture have shown their survival for many years at very low temperatures.

3. Indicator Organisms

E. coli is relatively sensitive to frozen storage, and is therefore considered as an indicator of recent pollution. Its rate of freeze-death is usually similar to that of *Salmonella* and related enteric pathogens. However, many factors can effect its death rate and prolonged freeze-survival is known to occur (Table 3.2). Other coliforms are also sensitive to frozen storage, but often less so than *E. coli.* As with *Salmonella,* coliforms sometimes fail to survive frozen storage, especially on acid foods.

The enterococci are more resistant to frozen storage; in fact, they are so persistent that they lose significance as indicators of recent pollution. Also, since they are often natural inhabitants of plant material, they are of doubtful significance as indicator organisms.

4. Yeast in Bread Dough

Frozen bread dough is the only common frozen food in which maintenance of quality is dependent on microbial survival. After it is thawed, the dough must rise—a process dependent on the carbon dioxide produced by the yeast. Since the yeast cells do not greatly increase in number while the dough rises, the efficiency and speed of the process depends largely on survival of the yeast cells initially present. It has long been known that many of the yeast cells survive freezing, but recently it has been found that survival is excellent if the dough is frozen immediately after the yeast is added, but poor if yeast cells remain in the dough for as long as an hour before the dough is frozen. This delay allows the yeast cells to begin to form buds, and in this condition of rapid growth they are much more sensitive to freezing. Thus, if dough should thaw during transportation or storage and reach a temperature which would permit yeast growth, it would be seriously damaged by refreezing.

C. Effect of Environmental Conditions on Survival

1. Effect of pH

In the pH range found in foods, the percentage of microorganisms which survive freezing is usually greatest at neutral or slightly alkaline pH values. For example, survival of *Salmonella* in ground meat for 21 days at −4°F. (−20°C.) varied from 0.12% in meat adjusted to pH 4.5 to 20% in meat at pH 7.5 (Georgala and Hurst (24)). In buffer, the yeast *Saccharomyces cerevisiae* showed the following survival rates after 2–3 hr. at −11°F. (−24°C.) (Stille (25)):

pH	Per cent survival
4.5	22
5.9	51
7.0	68
8.0	73
9.2	76

Similar, though smaller, effects of pH on survival have been noted by other authors.

2. Composition of the Substrate

Survival rate for a particular time–temperature situation is not inherently characteristic of each microorganism but depends on the composition of the surrounding substrate. For example, survival of a yeast on orange juice was increased from below 0.1 to 54% by the addition of 50% sucrose. Such protection is also given by glycerol, by milk, and by other colloidal materials. Cations and anions have specific effects, and survival differs in the presence of different sugars. Thus, a number of unrelated substances affect survival, and it can reasonably be assumed that other food ingredients do likewise (See the review by Mazur (26)).

3. Temperature

In general, microorganisms do not survive as well at temperatures slightly below freezing as they do at lower temperatures. For example, survival of *E. coli* increased over 10,000-fold as the temperature was reduced from +28°F. to −4°F. (−2° to −20°C.) (Fig. 3.6). Few organisms are as sensitive as *E. coli*, but in a number of foods, mixed populations have been shown to survive better at lower temperatures (Fig. 3.7). There are similar data for beef (31), vegetables (32), and other products. *Salmonella typhimurium, Staphylococcus aureus,* and

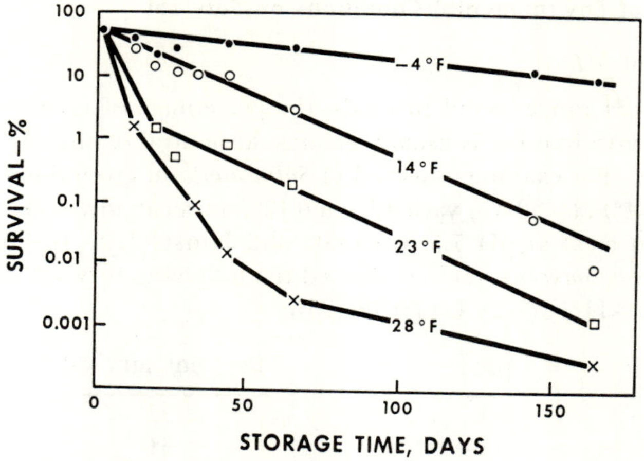

Fig. 3.6. Survival of *E. coli* in frozen aqueous suspension at several temperatures (Haines (27)). All lots were frozen rapidly at —94°F. (—81°C.) and then stored at the temperature indicated.

Streptococcus faecalis, in buffer and in several food ingredients, behaved similarly.

The introduction of liquid nitrogen freezing raises the question of survival of microorganisms in the neighborhood of −321°F. (−196°C.). Reports that they do survive at very low temperatures have been accumulating since the turn of the century. Most of these are unrelated to

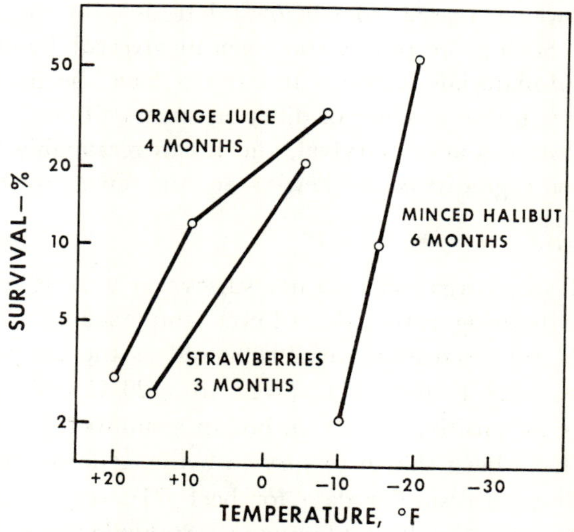

Fig. 3.7. Effect of storage temperature on survival of the natural microbial population of foods: orange juice (28), strawberries (29), halibut (30).

foods, but a mixed aerobic population has been completely recovered from vegetables frozen at −319°F. (−195°C.), and three lactic streptococci in skim milk showed 76–89% survival for 6 months at −321°F. (−196°C.). Nearly complete survival in this temperature range has been reported for fungus spores, yeast, *Staphylococcus aureus,* and *E. coli.* Cultures of many other organisms, including pathogenic bacteria and many fungi, withstand this temperature. It can be concluded that microorganisms withstand freezing at the temperature of liquid nitrogen at least as well, and perhaps better, than they withstand freezing at the customary freezer temperatures.

4. Survival Rate of Different Organisms

Even among closely related organisms, the survival rate depends on the organism, as shown by two separate sets of data (Fig. 3.5), each for three species of *Salmonella.* In general, there are wide differences in survival among different organisms (Engley (33); Michener and Elliott (21)).

D. Freezing Injury to Microorganisms

As a result of frozen storage, microorganisms may suffer such injury that a routine plate count fails to detect them. They become more fastidious; that is, they fail to grow on certain plating media that were previously quite acceptable to them, but retain the ability to grow if certain nutrients are added which they are no longer able to synthesize (Mazur (26)). Therefore an enriched plating medium is likely to give a higher count of injured cells than a simple one, even though the two media give equal counts of uninjured cells. For this reason, routine plate counts on frozen foods are not necessarily comparable to counts on the same foods before freezing.

E. Conclusions Relative to Survival

Microbial populations almost always survive freezing, frozen storage, and thawing. Some have survived several years of frozen storage. In general, not all of the individuals in a population survive. The proportion that does survive varies greatly, but is commonly in the range of 1–75%. Survival is poorer in acid than in nonacid foods. Generally the coliforms do not survive as well as the other organisms of significance in foods. In populations of bacterial spores, survival is often approximately 100%.

Since a considerable part of the microbial population usually survives, it follows that if a food has a high microbial count before it is frozen, the count will probably remain relatively high after frozen storage. This applies not only to the total count but to food poisoning, fecal indicator, pathogenic, or other groups of organisms having special significance in foods.

IV. MINIMUM TEMPERATURES FOR MICROBIAL GROWTH

If a food having a mixed flora of microorganisms is gradually brought to a lower and lower temperature, each species of microorganism will reach a temperature below which it will be unable to grow. A great many microorganisms have minimum growth temperatures above freezing. In general, these are the mesophiles. They grow best from room temperature up to body temperature. The common food poisoning bacteria are mesophiles, as are the organisms used as indicators of fecal contamination. The pathogenic, or disease-producing organisms are also mesophiles.

At temperatures somewhat below freezing, numerous organisms are still able to grow. These are usually called psychrophiles. A few can grow only at refrigerator or freezer temperatures, but most of those found on food can grow luxuriantly at temperatures up to room temperature, and a few grow well up to 98.6°F. (37°C.) or even 113°F. (45°C.). The psychrophiles are not dangerous, but they cause deterioration of the odor, taste, color, and texture of foods. They do this even to frozen foods, unless the temperature is low enough to prevent growth.

Since psychrophiles and mesophiles can both grow at room temperature, they grow on processing equipment and often contaminate food as it is being processed as discussed in Section II. A. Further growth may occur on the product if it is not frozen immediately.

In food held somewhat below freezing, the mesophiles cannot grow but the psychrophiles may grow and eventually become very numerous. These belong to a long list of genera but probably the most common are *Pseudomonas* species, followed by bacteria of the genera *Achromobacter, Flavobacterium, Alcaligenes, Escherichia,* and *Aerobacter* (Witter (35)). Psychrophilic molds often belong to the genera *Cladosporium, Sporotrichum,* and *Penicillium* (Michener and Elliott (34)).

If a frozen food thaws and remains at room temperature for a few hours, mesophilic organisms that have survived freezing can also grow and a different flora may result. The flora of thawed frozen vegetables commonly includes *Leuconostoc* species but many genera may be present (Smart (36)). Molds and yeasts are likely to grow on fruit, and meat and fish frequently contain numerous pseudomonads after thawing.

* Condensed from the review by Michener and Elliott (34).

Food poisoning organisms can sometimes grow under these conditions (Sect. IV. B).

A. Psychrophiles

A number of authors have made recommendations as to the highest temperature acceptable for storage of frozen food without danger of microbial action. All of these recommendations fall between 16 and 10°F. (−9 and −12°C.) (Table 3.3).

Beginning in about 1930, the literature contains numerous reports of low temperatures at which microorganisms grew on frozen foods. The lowest, for each of several classes of foods, are as follows: meat, 18°F. (−8°C.); cured meat, 14°F. (−10°C.); fish, 12°F. (−11°C.); fruit, 10°F. (−12°C.); vegetables, 10°F. (−12°C.). We found no report of growth on frozen egg although it doubtless can occur. Growth was reported at 14°F. (−10°C.) on ice cream, but literature on dairy products was not exhaustively reviewed. Growth has been reported down to 10°F. (−12°C.) on artificial media prepared for culture of microorganisms.

A compilation of about 150 reports of minimum growth temperatures of psychrophiles shows most observed minima in the range of 32 to 18°F. (0 to −8°C.). Molds grow more commonly than bacteria in the lower part of this range. A few of these reports (about 10%) showed growth in the range of 18–10°F. (−8 to −12°C.). Growth at subfreezing temperatures is relatively slow but it can neverthless proceed to the point where the product is "spoiled" microbiologically (Sect. V. C).

There are four other reports, by three authors, of growth at temperatures from 0°F. (−18°C.) down to −29°F. (−34°C.). None of these have been confirmed by later work. One was published before modern refrigeration was available, so that one questions whether the reported temperature could have been maintained. The best evidence that

Table 3.3. Maximum Recommended Storage Temperatures for Complete Suppression of Microbial Spoilage

Product	Maximum recommended storage temperature		Number of authors making this recommendation
	°F.	°C.	
Meat	14 to 16	−10 to −9	5
Fish	10 to 14	−12 to −10	3
Vegetables	14	−10	2
Butter	12 to 16	−11 to −9	1
Frozen food, generally	14 to 15	−10 to −9	6

vth does not generally occur at this temperature is that billions of ; of frozen food have been stored at or near 0°F. (−18°C.), without reported microbial spoilage. Without more experiments it is impossible to confirm or reject these few reports of growth at very low temperatures, but from a practical standpoint it is unlikely that microbial growth occurs in this temperature range.

Although growth apparently ceases below about 10°F. (−12°C.), microbial enzymes elaborated during growth can persist and cause deterioration even after the temperature is reduced sufficiently to prevent further microbial growth (Elliott and Michener (1)).

B. Food Poisoning Organisms

Food poisoning bacteria are not psychrophilic and have not been reported to grow below freezing. *Clostrodium botulinum* Type E grew and produced toxin at 38°F. (3.3°C.) in 38 days (Schmidt *et al.* (37)) and at 37.4°F. (3.0°C.) in three months but not at 33.8°F. (1.0°C.) in a year (Abrahamson *et al.* (38)). Types B and F of marine origin grew and produced toxin in 39–61 days at 38°F. (3.3°C.) (Eklund *et al.* (9)). The lowest temperatures at which growth or toxin production have been reported for the common food poisoning bacteria are shown in Table 3.4. As shown graphically in Fig. 3.8, food poisoning bacteria have not grown below 37.4°F. (3.0°C.), while in the range 37.4°F. (3°C.) to about 10°F. (−12°C.), microbial spoilage can be caused only by the psychrophiles.

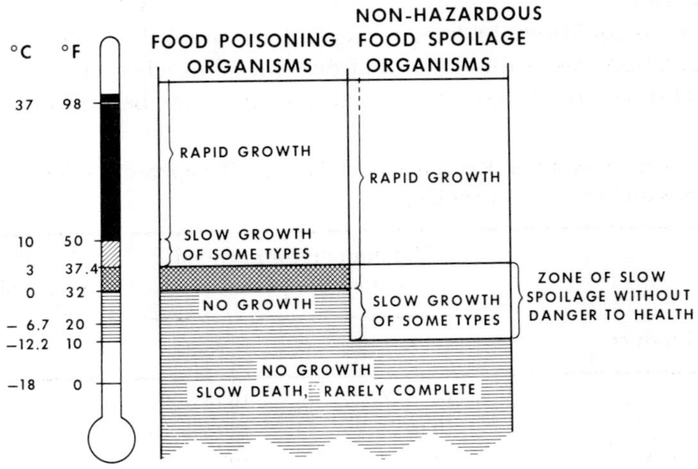

Fig. 3.8. The temperature range of growth of food poisoning and food-spoilage microorganisms.

Table 3.4. Lowest Temperatures Reported for Growth and Toxin Production by Food Poisoning Bacteria

	Lowest reported temperature				
	Growth		Toxin production		
	°F	°C	°F	°C	Reference [a]
Staphylococcus aureus	44	6.7	44	6.7	34
Salmonella species	44	6.7	— [b]		34
Clostridium botulinum					
Type A	50	10	50	10	34
Type B	50	10	50	10	34
Type B (marine origin)	38	3.3	38	3.3	9
Type C			50	10	34
Type E	37.4	3	37.4	3	38
Type F (marine origin)	38	3.3	38	3.3	9
Clostridium perfringens	50 to 68 [c]	15-20 [c]			10

[a] Where applicable, reference is made to the review by Michener and Elliott (34).

[b] *Salmonella* causes an infection but does not produce a toxin.

[c] Slight growth of one out of nine toxigenic strains of *Clostridium perfringens* has been observed at 43°F. (6°C.), but there was no further growth after the culture was transferred (39).

Thus there is no danger that these organisms will grow in food that is kept frozen. Even at normal refrigerator temperatures, they grow slowly or not at all, depending on the organism. But if food poisoning organisms have rendered a food hazardous before it is frozen, it will remain so after it is thawed.

The possibility should not be ignored that toxigenic molds might be able to grow under some circumstances on frozen foods. *Fusarium sporotrichioides,* for example, has not been found on frozen foods but can grow on grain at temperatures down to 10°F. (−12°C.) and it produces toxin optimally at temperatures from 35 to 39°F. (2 to 4°C.) (Forgacs (40)). On the other hand, the enormous volume and general safety of our frozen foods suggest the absence of such organisms. Even the possibility that such a mold could grow on frozen food is eliminated by the use of the recommended storage temperatures, which will prevent all microbial growth.

C. Indicator Organisms

The indicator organisms are also mesophiles and ordinarily will not grow on frozen foods. A single strain of *E. coli* grew at 28°F. (−2°C.)

and there are reports of another coliform (*Aerobacter*) and one of the enterococci which grew at 32°F. (0°C.) in milk. In general, however, these organisms grow only at 41°F. (5°C.) or higher.

E. coli and the coliforms generally are among the most sensitive organisms to freezing and frozen storage (Sect. III. B). Thus, while they will occasionally grow at or slightly below freezing, their population is much more likely to be lower on a frozen food than it was on the same food before freezing.

D. On the Accuracy of Minimum Growth Temperature Data

Most of this information exists as reports that microbial growth was observed at certain temperatures under specified conditions, on certain foods or other substrates. Often lower temperatures are given at which no growth was observed. Microbial growth at low temperatures is characterized by very long lag periods (Sect. V. B). It is never possible to say with certainty that growth would not have occurred if the experiment had been continued for a longer period of time. For example, growth of psychrophiles on fish at 23°F. (−5°C.) was observed only after 350 days; and growth of a fungus at 18°F. (−8°C.) only after 414 days, but other investigators have terminated their experiments after much shorter storage periods and reported absence of growth.

Also, the minimum temperature for growth of an organism is affected by the environment in which it finds itself. In general, conditions or substances which tend to retard growth at ordinary temperatures will have a more pronounced effect at temperatures approaching its minimum. As a result, in the presence of a substance which retards growth, or on a suboptimal substrate, the minimum growth temperature for a particular organism will not be as low as it would under conditions that are otherwise optimal. Stated differently, organisms are more fastidious at temperatures near their minima. It is therefore impossible to determine the real minimum for an organism unless other conditions are optimal. It is thus not surprising that there is considerable variation among published reports of minimum growth temperatures, even for the same organism.

E. Effect of Environmental Conditions on Minimum Growth Temperatures

1. *Incomplete Freezing of Foods*

Although foods appear to be completely frozen when cooled slightly below 32°F. (0°C.), they actually contain considerable amounts of liquid water until very low temperatures are reached. For example

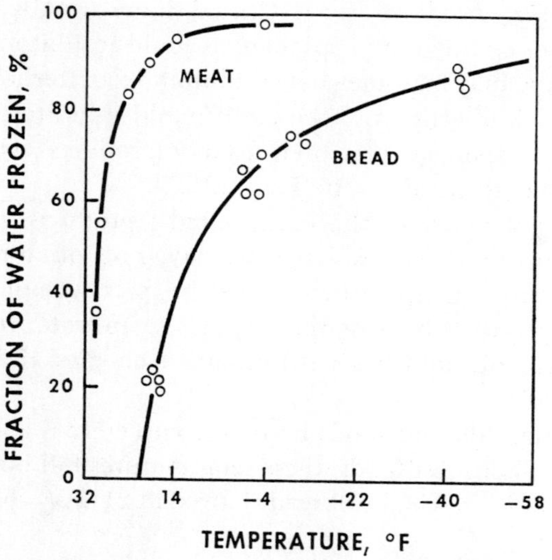

Fig. 3.9. Fraction of water frozen at various temperatures in bread (41) and meat (42) (from (34)).

(Fig. 3.9), the water in meat was only 82% frozen at 23°F. (−5°C.) and 98% at −4°F. (−20°C.). Fish gave almost identical results, with complete freezing only at −94°F. (−70°C.). In bread the water was approximately 20% frozen at 18°F. (−8°C.), and 87% at −42°F. (−41°C.). Vegetables and fruit became completely frozen in the range +3° to −6°F. (−16° to −21°C.), but frozen orange juice contains a viscous fraction which becomes gummy at −18°F. (−28°C.) and plastic at −139°F. (−95°C.). Thus, foods are not completely frozen in the temperature range in which we find growth of microorganisms.

As a food is cooled, crystals of nearly pure ice form at the expense of the solution which exists around and in the particles of food. The solution thus becomes more and more concentrated as freezing progresses. Most micoorganisms probably remain in the liquid portion, and grow there until the temperature is too low for growth or the solute concentration too high. The partial freezing and resultant dehydration is not, in itself, sufficient to explain why each organism has a rather definite minimum. Possible explanations include imbalance of metabolic processes with resultant cell damage, inhibition of growth-limiting master reactions, and mechanical effects of freezing of the substrate.

Microbial spoilage has not been reported in frozen breads, cakes, etc. Even above freezing, these products are too dry for growth of bacteria. A bread containing 36% moisture begins to freeze at 21°F. (−6°C.), below which some molds but relatively few bacteria can grow. Furthermore, bakery products in the United States usually contain sorbic acid or sodium or calcium propionate as mold inhibitors. At low temperatures, these may be somewhat inhibitory to bacteria as well. Thus, the relative lack of water and the use of mold inhibitors probably prevent microbial spoilage of frozen bakery products even if they are stored at temperatures above 10°F. (−12°C.).

This may not apply to the ready-to-bake products, as they have a higher moisture content. However, we have not found reports of their spoilage at freezing temperatures. Likewise, pies are not comparable to other baked products because the crust takes up water from the filling. Surface growth of mold has been observed on pies after 9 months at 20°F. (−6.6°C.).

On warming, different foods become completely thawed at different temperatures. For most foods, these temperatures fall between 31.8 and 25°F. (−0.1 and −4°C.), but bread thawed at 21°F. (−6°C.) and mince pie at 15°F. (−9°C.).

2. *Insufficient Cooling Rate*

Spoilage in foods can occur during the freezing process if the product temperature is not reduced rapidly enough below the minimum growth temperature of the microorganisms present. In large packages, the center can spoil as a result of microbial growth before it reaches a temperature low enough to prevent growth. This has happened, for example, in berries frozen in barrels, in egg magma in 30-lb. cans and in spinach frozen in large blocks. Salmonellae have grown on internal surfaces of large blocks of boned meat during freezing, because the center remained warm. The use of smaller containers, a lower freezer temperature, or an air blast freezer accelerates heat transfer and tends to prevent this.

3. *Effect of Food Ingredients*

It is evident from Sect. IV. D that the minimum growth temperature for an organism depends on the composition of its substrate. It follows that the composition of a food will influence the minimum growth temperature of its microflora. For example, bacterial growth has not been reported below 25°F. (−4°C.) on fruit. This is doubtless because of the low pH, for bacteria grow at far lower temperatures on foods that are neutral or only slightly acid. Molds, which are more acid-tolerant, can grow at lower temperatures than bacteria on frozen fruit.

The sugar commonly added to frozen fruit may also increase the minimum growth temperature for its flora. However, with berries in barrels, the added sugar changed the consistency of the product, making it more liquid and accelerating the loss of heat, so that it came more quickly to a temperature sufficiently low to prevent growth.

It was discovered early in the work on frozen fruits that molds do not grow if oxygen is excluded. They do not grow in the interior of barrels of frozen berries or in sealed containers of fruit. Berries in sealed containers did not spoil below 28°F. (−2°C.), because molds were inhibited by lack of oxygen, bacteria by the acid, and yeasts by the low temperature. Some other yeasts, however, grow at much lower temperatures.

V. MICROBIAL GROWTH AT LOW TEMPERATURES*

Although frozen foods are normally stored in a temperature range where no microbial growth occurs, they are occasionally brought for significant periods of time into the range which permits psychrophilic growth. This can take place by accident or be caused by equipment breakdown. Poultry is often thawed and stored at refrigerator temperature before it is sold. The product temperature in a retail frozen food cabinet can rise to above 20°F. (−6.6°C.) for a brief period during the daily defrost cycle (43). Over a prolonged storage period, this provides a time–temperature experience that may permit mold growth. Although the practice is not recommended, institutional–pack frozen potatoes for French frying are sometimes stored in restaurant refrigerators, and consumer–pack frozen food is stored in home refrigerators. For these reasons, we should consider the characteristics of microbial growth in the temperature range in which psychrophilic spoilage takes place.

A. The Growth Curve

When bacteria are introduced into an environment where they can grow (reproduce), their numbers remain relatively constant during the period when they are becoming accustomed to their new surroundings. This is called the lag phase. In Fig. 3.10, the lag phase at 32°F. (0°C.) lasted about 5 days. Then the reproduction enters the logarithmic growth phase, during which the logarithms of the numbers form a straight line when plotted against time. If the growth is permitted to continue for a considerable period, the numbers will cease to increase (the resting phase), then decrease (the death phase). In food microbiology, only the lag and logarithmic phases are important, because foods are spoiled by the time bacterial growth has begun to decelerate to the resting phase.

* Condensed from the review by Elliott and Michener (1).

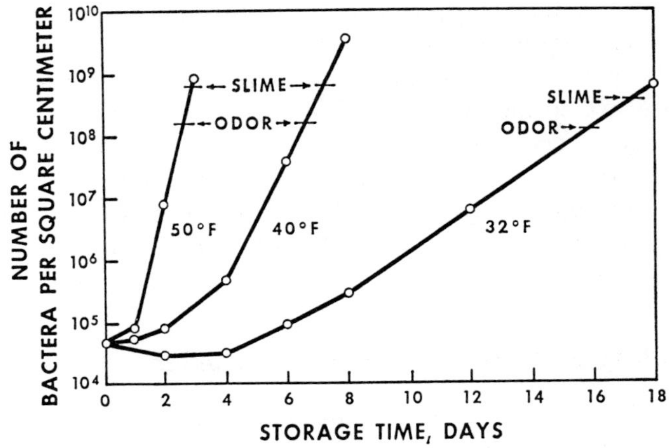

Fig. 3.10. Effect of temperature on growth of bacteria on chicken meat (Ayres, Ogilvy, and Stewart (44)).

B. Effect of Temperature

The effect of temperature on the length of the lag phase and on the slope of the logarithmic phase is immediately evident in Fig. 3.10. Note that the effect of the 8° difference between 32 and 40°F. (0 and 4°C.) is greater than that of the 10° difference between 40 and 50°F (4 and 10°C.). If we plot the shelf life, that is, the time it takes to reach the odor point, against temperature, the enhanced effect of low temperature is shown in a more dramatic way (Fig. 3.11). This is a composite of data from several groups of investigators, and shows that a given temperature reduction extends shelf life much more in a low-temperature range than in a high one. Indeed, fresh chilled foods such as fish or chickens can be held at temperatures slightly below 32°F. (0°C.) without freezing them. This practice, now a commercial reality, extends shelf life even longer. Freezing at appropriately low temperatures, of course, stops all microbial growth.

C. Low-Temperature Spoilage

As a means of detecting spoilage, organoleptic tests (odor, appearance, sometimes taste) are basic. No chemical or bacteriological test can do more than confirm the organoleptic. Many investigators have determined that at the time odor of decomposition appears or surface slime forms, the bacterial level is usually 10^6 to 10^8 per g. or per $cm.^2$, although detectable loss of flavor and odor occasionally occurs at 10^4 or 10^5 per g. (Elliott and Michener (45)). Off-flavor and odor apparently result from growth of micoorganisms on the food, since the simple ad-

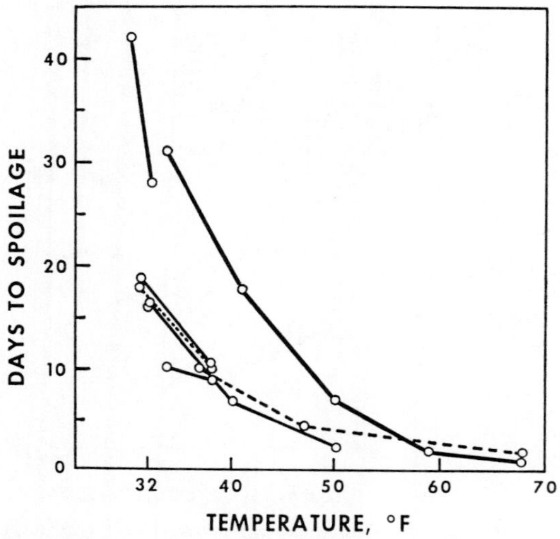

Fig. 3.11. Effect of storage temperature on shelf life of poultry meat. Each graph is from a different set of published data. Meat and fish gave similar results. For references see Elliott and Michener (1), Fig. 23.

dition of large microbial populations to foods just before freezing had little effect on flavor (Dalton and Strong (46)).

A high level of bacteria at the beginning of growth decreases the length of the lag period of the growth curve. As a result, a food product handled in an insanitary manner, then held at a temperature permitting growth, will decompose more quickly than a food product prepared in a sanitary manner and held at the same temperature. Fig. 3.12 shows, for raw fish, that there is approximately a straight line relationship between the logarithm of the initial bacterial count and the time for spoilage to occur. This principle applies to red meats, chicken, and other foods as well.

The nature of the inoculum, pH, water activity, composition, temperature, and oxidation–reduction potential, all affect the rate of microbial growth. These factors influence the growth of various strains of microorganisms differently; therefore in a mixed flora, one strain may be favored over another, and will come to predominate at the time spoilage is evident. Table 3.5 shows that psychrophilic bacteria of the pseudomonas group become dominant during spoilage of ground beef at 45°F. (7°C.). The same change occurs during low-temperature spoilage of many protein foods. On the other hand, molds and yeasts grow better than bacteria in acid conditions; therefore they become predominant in fruits during spoilage. Molds can tolerate lower levels of

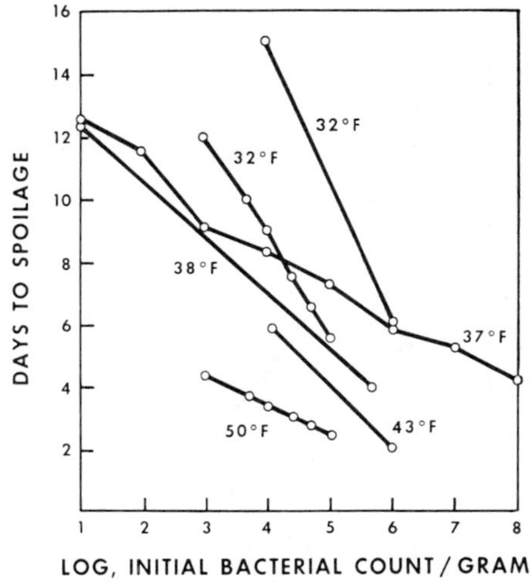

Fig. 3.12. Shelf life of fish as affected by initial bacterial count. For references see Elliott and Michener (1), Fig. 32.

water activity than yeasts or bacteria; therefore they predominate in spoilage of breads and partially dried foods.

After a frozen food thaws, microbial growth that began before the product was frozen will continue as before. Any time–temperature experience that allows a significant increase in microbial population brings the product that much closer to eventual spoilage. Several investigators have compared the spoilage rate of frozen-thawed foods

Table 3.5. Changes in Bacterial Flora during Spoilage of Ground Beef at 45°F. (7°C.)[a]

Microorganism	Proportion of total isolates after	
	0 days, %	14 days, %
Pseudomonas or *Achromobacter,* or both	4	84
Bacillus	28	6
Microbacterium	0	5
Micrococcus	26	5
Yeasts	42	0
Total	100	100

[a] From Rogers and McCleskey (47).

with that of the same foods never frozen, and have found that the rates are usually the same (reviewed by Elliott and Michener (1)).

The relationships of growth to spoilage have been described for foods at refrigerator temperatures but they hold for frozen foods whenever the temperature is high enough to permit microbial growth. Frozen peas, for example, can spoil from bacterial growth at 25°F. (−4°C.) (Fig. 3.13), although the lag period is long and growth slow. Decrease in microbial population during the lag phase, as shown here, is common at low temperatures and is probably due to death of some components of the population before others can begin to grow.

As explained elsewhere (Chapters 4–9), deterioration of a nonmicrobial nature usually takes place at temperatures too low to permit microbial spoilage. Thus, foods often suffer quality loss before they receive time–temperature abuse sufficient to permit microbial spoilage.

VI. MAINTAINING SAFETY IN FROZEN FOOD

The United States has a safe food supply—perhaps the safest in the world, and deaths from unsafe foods are rare. Nevertheless, as we have pointed out in the foregoing sections, foods destined for freezing can contain bacteria capable of causing food poisoning or human infection. We have shown that these bacteria usually survive frozen storage and that they will grow if the temperature rises above their minimum for growth, which is 38–50°F. (3–10°C.) or more, depending on the organism (Fig. 3.8).

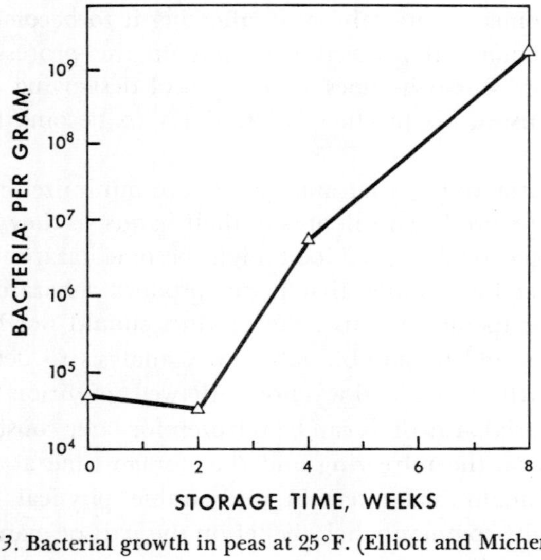

Fig. 3.13. Bacterial growth in peas at 25°F. (Elliott and Michener (1)).

Proper handling of a food before freezing can keep the incidence of harmful bacteria to a minimum and insure that the food is safe when it is frozen. In order to keep it safe during storage and transportation and up to the time when the consumer eats it, however, it is essential that it be handled in such a way as to prevent the growth of any harmful bacteria which it might contain.

A. Safety During Storage, Distribution, Sale, and Home Use

1. Comparison with Other Types of Foods

None of the standard types of preservation are completely without hazard. Although the principles of preservation by canning, drying, and freezing are well known, occasional food poisonings continue to occur from foods so preserved—even from commercial products. Each has its own points of vulnerability. The canned foods must be adequately sealed and processed, the dried foods kept dry, and the chilled and frozen foods kept at suitable temperatures. Frozen foods are as safe as chilled foods; indeed they might be considered safer, because microorganisms cannot grow in them at the usual low temperatures of frozen storage

2. Objectives in Handling Frozen Foods

The processor should do all he can to prevent or minimize contamination with food poisoning and disease-producing organisms, and to keep already processed foods from becoming contaminated with them. He usually has no control over the product after it leaves the processing plant, but if someone later abuses it, allowing it to become hazardous, part of the blame will invariably be put on the processor, whether justified or not. But if he does a good job of destroying or removing harmful organisms, his product is less likely to become hazardous if abused.

The distributor and retailer must prevent or minimize time–temperature abuse to the product while it is in their hands, as they have no way to know whether the processed food might become hazardous if abused. They must therefore assume that it can produce a hazard. In case of serious time–temperature abuse, the product should not be marketed until it is subjected to suitable laboratory analysis to detect possible microbial growth. If the food was not in thawed condition long enough to permit bacterial growth, it can be refrozen for later consumption. In such an instance, the refreezing and the storage time at high freezer temperatures might make certain undesirable physical or chemical changes. But the commonly held belief on the part of many consumers that refreezing is hazardous is fallacious. It is rather the time–temperature abuse above about 37°F. (3°C.) that can be hazardous.

Probably frozen food is most likely to suffer time–temperature abuse at the consumer level. Of course, most U.S. housewives will use care to avoid this, and most housewives have refrigerators and use them. But there are always those who are careless or ignorant, and in some countries where U.S. products are shipped, household refrigeration is uncommon.

Since new products are constantly being developed, the microbiologist and the food technologist must be alert to possible hazards associated with new products or processes. The Food Protection Committee (National Research Council (48)) described potential and real hazards from technological developments that do not take microbiology into consideration.

3. Possible Hazards of Time–Temperature Abuse

Many foods, if stored at temperatures above 50 or 60°F. (10 or 15°C.), can support growth of harmful bacteria. Yet billions of packages of frozen food have been marketed. Of these, surely some have been mistreated and then eaten. Nevertheless, frozen food is generally considered very safe and illness has been attributed to it only very rarely.

That the harmful organisms do not commonly grow in commercial frozen foods if they are mistreated is due partly to the presence in relatively large numbers of the harmless spoilage flora. Also, some components of the spoilage flora are psychrophilic, so that as a food thaws and warms up there is already an actively growing population before the temperature is high enough to permit growth of food poisoning and disease-producing organisms. Thus there is normally a large and growing spoilage flora which competes with any harmful organisms that may be present.

Several authors have shown that staphylococci do not produce toxin at temperatures below about 80°F. (27°C.) when grown in competition with the natural spoilage flora on various foods. Some similar evidence exists for growth of *Salmonella*. *C. botulinum* does not produce toxin on frozen-thawed vegetables in the presence of their natural spoilage flora because the pH becomes unfavorable (Perry *et al.* (6)). Thus there appears to be a wide range of circumstances in which harmful bacteria are suppressed by the competition of the natural spoilage flora. Probably this greatly reduces the chance of food poisoning when foods are accidentally mishandled and then eaten. However, the spoilage flora varies with the product, its source, the method of processing, etc. It is almost infinitely variable, so that it should not be assumed that it will always suppress growth of harmful bacteria. For example, lactic acid bacteria sometimes stimulate growth of staphylococci. Introduction of new products or processing methods or even slight changes in the existing methods may introduce altered microflora which will not behave as expected with respect to food poisoning or disease-producing bacteria. Some relationships of spoilage to food poisoning organisms were discussed by Michener and Elliott (34).

Rapid growth of the spoilage flora will normally render a food so objectionable in odor or appearance that it will not be eaten. However, this is not always an effective deterrent. Even putrid foods are sometimes eaten, and foods are occasionally hazardous even without evidence of putrefaction.

Heating to a temperature near boiling will eliminate practically all the spoilage flora from a food, but leave the heat-resistant *Clostridium* spores, which can then grow and produce toxin in many nonacid foods. This can readily be accomplished experimentally in freeze-thawed vegetables (Perry *et al.* (6)). With chicken á la king, a heat treatment recommended by one worker as a means of reducing the total count was shown by others to reduce the spoilage flora to a point where *C. botulinum* spores, if present in the product, could grow and produce toxin. Thus, a heat treatment used to reduce the spoilage flora, supposedly making the food more acceptable and giving it a longer shelf life after thawing, actually could allow it to become highly dangerous if subjected to temperature abuse. Blanching of vegetables would have the same effect except that it is normally done at an earlier stage of processing, so that recontamination with spoilage organisms occurs before they are packaged (Michener *et al.* (32)).

Consumer mishandling of foods intended to be heated in a closed package or shell can result in a similar problem. An outbreak of botulism occurred when a frozen chicken pie was baked, allowed to remain in the slightly warm oven, and eaten the next day (U.S. Public Health Service (7)). A similar possibility exists in the case of products designed for cooking in a sealed package. Cooking kills all of the nonsporeforming spoilage flora but not the heat-resistant spores. If eaten immediately as intended, the product is entirely safe, but if it happens through accident or ignorance to be held for several hours or days at room temperature, any *C. botulinum* spores which it contains may grow and render the product extremely toxic.

Botulism could be caused in this way only as a result of a succession of rather unlikely circumstances: *(1)* The cooked product must be saved rather than being eaten as was intended. *(2)* It must contain *C. botulinum* spores. *(3)* It must support growth of *C. botulinum*. *(4)* Other microorganisms must be killed or nearly all killed. *(5)* At least part of the product must be anaerobic. *(6)* It must be stored at or near room temperature for one to several days. *(7)* Someone must eat it. Although it is unlikely that many people would become victims of botulism in this way, such an outbreak would be unfortunate not only for the victim but for the luckless processor who would be involved. Therefore it is suggested that frozen prepared dinners, boil-in-bag vegetables, meat pies, and any other "heat-and-eat" nonacid foods be labeled to instruct consumers to refrigerate any packages not consumed soon after heating.

B. Microbiological Quality of Frozen Foods

Safety is of course one aspect of the microbiological quality of frozen foods. Poor quality can be the result of a high population of harmless spoilage microorganisms in the food before it was frozen (Sect. II. A) or of time–temperature abuse sufficient to permit microbial growth in the processed food (Sects. IV and V). The harmless spoilage organisms can grow at temperatures as low as +10°F. (−12°C.), and they cause off-flavors at populations of 10^6 to 10^8 per g. or per $cm.^2$, but occasionally at lower populations.

To protect the quality of frozen foods, it may be suggested or required that microbial counts must not exceed a predetermined limit. Such limits are termed microbiological standards or microbiological criteria. They often refer to the total count (usually limited to 50,000 or 100,000/g.), less often to coliforms, *E. coli, Salmonella,* or staphylococci, by giving a maximum count or requiring their absence from a sample of a certain size. They are commonly applied by one processor to products purchased from another. They may also be applied by law to products offered for sale, as is done almost universally in this country with milk. Microbiological standards for foods have been legally established in Massachusetts and in Canada, but these laws are not at present enforced. The various laws, industrial practices, and published suggestions regarding maximum permissible microbial populations have been reviewed (Elliott and Michener (45)).

The Food Protection Committee (48) recommended that before microbiological standards are established, (1) analytical methods should be agreed upon and *(2)* a survey should be made for each food in question to determine what size population levels of various groups of organisms are found under good commercial practice.

C. Analytical Methods for Frozen Foods

The literature contains numerous analytical methods for determination of the microbiological quality of various foods. Several excellent basic sources of information are: American Public Health Association, dairy products and frozen desserts (49), shellfish (50), water (51), foods generally (52); Association of Food and Drug Officials of the U.S. (AFDOUS), precooked frozen foods (53); Association of Official Analytical Chemists, frozen precooked or prepared foods (54); Food and Drug Administration, foods generally (55); Frazier (56) and Frazier *et al.* (57), food microbiology, including analytical methods; Lewis and Angelotti (58), food poisoning and indicator organisms; Galton *et al.* (59), *Salmonella.*

The National Academy of Sciences National Research Council Committee on Food Microbiology is now endeavoring to bring several agencies together to develop U.S. Standard Methods. Likewise, the International Committee on Microbiological Specifications for Foods of the International Association of Microbiological Societies is preparing a book entitled *The Significance of Microorganisms in Foods and Methods for their Enumeration.* This is an attempt at international standard methods. International collaborative studies of these methods are underway.

Interpretation of analytical results may be difficult because of variation among the samples analyzed. Regardless of the location, samples of a processed food will vary because they are taken from a continuously varying stream of product as it passes over the line. That rapid fluctuations can occur can be appreciated from Fig. 3.1 (see Sect. II. A. 3), showing clumps of high count corn which fall at intervals into the product stream. Such conditions can be identified and eliminated. However, Michener *et al.* (32) found small lots of frozen peas in which the highest total count was up to 24 times the average total count for the lot, and other vegetables in which the high count was up to three times the average. They also found large differences (up to sixfold) between the two halves of some packages. Messinger (60) found large variations among supposedly identical samples analyzed at different laboratories or at different times in the same laboratory. When Hartman and Huntsberger (61) took frozen chicken pies off the processing line at half-hour intervals, the total count sometimes increased over 10-fold within a half hour and dropped back within the following half hour. They concluded that if all portions of the product are to meet a proposed standard, the counts must be maintained generally at a level far below that standard. Because of package-to-package variation, AFDOUS (53) recommended that 10 units be analyzed, and that their geometric mean be used for any determination of acceptability.

VII. SUMMARY

The microbiological quality of frozen food can be no better than that of the food before it is frozen. When food enters the processing plant it contains microorganisms that are harmless but can cause it to spoil, and it may unavoidably contain food-poisoning bacteria. It may be additionally contaminated during processing by the equipment and by the people who handle it, and microorganisms can grow on it if it remains too long in the plant before freezing. The processor should, as far as possible, remove or minimize contamination and prevent microbial growth on the food.

Whatever microorganisms are on a food when it is frozen are likely to remain and grow again if the food is thawed. Freezing may reduce the total microbial population but it rarely eliminates it.

No microorganisms grow below about 10°F (−12°C.) Between 10 and 37°F., (−12 and 3°C.) only psychrophilic food spoilage organisms will grow. They are harmless but they may destroy the color, texture, and palatability of a food even though it is frozen. In the 37–50°F. (3°–10°C.) range, some food-poisoning organisms can grow slowly and could possibly render a food hazardous. With increasing temperatures above 50°F (10°C.), food-poisoning organisms grow more and more rapidly. In the temperature range of about 50–115°F. (10–46°C.), they can quickly render a food hazardous.

At a temperature only slightly above its minimum growth temperature, any microbial population grows very slowly and experiences a very long lag period before growth begins. Consequently spoilage or development of hazard is delayed. Any increase in temperature rapidly accelerates these processes.

If sanitation is poor in the processing line, there will be more spoilage organisms in the food and it will spoil more rapidly upon thawing; also, it may fail to meet established microbiological standards (criteria). Furthermore, it is more likely to contain food-poisoning, disease-producing, or indicator organisms.

In a food that is allowed to thaw and warm to ambient temperature, growth of food poisoning or disease-producing organisms may be suppressed by the spoilage organisms. A food is also rendered unpalatable by the spoilage organisms so that, safe or not, it will not usually be eaten. If a food is heated enough to kill the common spoilage organisms, bacterial spores may remain, including those of *Clostridium botulinum*. A food so heated could become toxic if held out of refrigeration. For this reason, it is suggested that frozen "heat-

and-eat" foods should be labeled to instruct consumers to refrigerate any packages not eaten soon after heating.

Thawing and refreezing a food is not hazardous, although it will probably cause some quality loss of a nonmicrobial nature. But if the time–temperature experience of the food above about 37°F (3°C.) is sufficient to permit growth of food-poisoning bacteria, then a serious hazard is introduced. Sufficient time–temperature experience between about 10 and 37°F. (−12 and 3°C.) may permit microbiological spoilage but it will cause no health hazard.

Frozen foods have an excellent safety record. However, the industry should examine the microbiology of each new and revised processing and packaging method to avoid the danger of introducing unsuspected hazards to the consumer.

VIII. REVIEWS ON THE MICROBIOLOGY OF FROZEN FOODS

The published literature in this field is extensive. In this chapter we have referred mainly to the more important publications and to a few of the very recent ones. Several reviews are therefore listed which will provide more information and give literature references: food microbiology (56,62), food-plant sanitation (63), effects of freezing on biological materials generally (64), microbiology of frozen or chilled foods (35,65–68), food poisoning generally (4,58) staphylococci (53), salmonellae (53,69–71), botulism (4,5,72) indicator organisms (53), microbial survival in frozen foods (21,26), minimum temperature for microbial growth (34), microbial growth at low temperatures (1), microbiological safety of foods (48), microbiological standards (criteria) (45,48), analytical methods (Sect. VI. C).

REFERENCES

(1) R. P. Elliott and H. D. Michener, *Factors Affecting the Growth of Psychrophilic Microorganisms in Foods,* U.S. Dept. Agr. Tech. Bull. 1320, 1965.

(2) E. R. Wolford, A. D. King, and H. D. Michener, *J. Milk Food Technol., 28,* 183 (1965).

(3) D. F. Splittstoesser, W. P. Wettergreen, and C. S. Pederson, *Food Technol., 15,* 329 (1961).

(4) G. M. Dack, *Food Poisoning,* 3rd ed., University of Chicago Press, Chicago, Illinois, 1956.

(5) R. A. Gordon and W. G. Murrell, *Food Preserv. Quart., 27,* 6 (1967).

(6) H. Perry, C. T. Townsend, A. A. Andersen, and J. A. Berry, *Food Technol., 2,* 180 (1948).

(7) U.S. National Office of Vital Statistics, Morbidity and Mortality Weekly Report, *9,* No. 27, 2 (1960).

(8) D. A. Kautter, *J. Food Sci., 29,* 843 (1964).

(9) M. W. Eklund, F. T. Poysky, and D. I. Wieler, *Appl. Microbiol., 15,* 1316 (1967).

(10) B. C. Hobbs, *J. Appl. Bacteriol., 28,* 74 (1965).

(11) U.S. National Communicable Disease Center, "Salmonella Surveillance Annual Summary 1966," October 9, 1967.
(12) R. K. Lynt, Jr., *Appl. Microbiol., 14,* 218 (1966).
(13) B. H. Ranson, *J. Parasitol., 1,* 5 (1914).
(14) S. E. Gould and L. J. Kaasa, *Am. J. Hyg., 49,* 17 (1949).
(15) N. D. Levine, *Protozoan Parasites of Domestic Animals and of Man,* Burgess Publishing, Minneapolis, Minnesota, 1961.
(16) B. C. Hobbs, Monthly Bull. Ministry Health, Public Health Lab. Service, *24,* 123, 145 (1965).
(17) G. J. Hucker, R. F. Brooks, A. J. Emery, *Food Technol., 6,* 147 (1952).
(18) B. E. Proctor and A. W. Phillips, Jr., *Refrig. Eng., 53,* 30 (1947); *Frozen Food Ind., 3,* No. 1, 16 (1947).
(19) J. Weinzirl and E. B. Newton, *Am. J. Public Health, 5,* 833 (1915).
(20) E. R. Wolford and J. A. Berry, *Food Res., 13,* 172 (1948).
(21) H. D. Michener and R. P. Elliott, *Advan. Food Res.,* in preparation for Vol. 18 (1969).
(22) S. E. Hartsell, *Am. J. Public Health, 41,* 1072 (1951).
(23) M. F. Gunderson and K. D. Rose, *Food Res., 13,* 254 (1948).
(24) D. L. Georgala and A. Hurst, *J. Appl. Bacteriol., 26,* 346 (1963).
(25) B. Stille, *Arch. Mikrobiol., 14,* 554 (1950).
(26) P. Mazur, "Physical and Chemical Basis of Injury in Single-Celled Micro-Organisms Subjected to Freezing and Thawing" in *Cryobiology,* H. T. Meryman, Ed., Academic Press, New York, 1966, p. 213.
(27) R. B. Haines, *Proc. Roy. Soc. (London), Ser. B, 124,* 451 (1938).
(28) E. L. Moore, R. L. Huggart, and E. C. Hill, *Proc. Florida State Hort. Soc., 1950,* 165.
(29) J. A. Berry, *Proc. Am. Soc. Hort. Sci., 33,* 224 (1936).
(30) R. H. Bedford, Canada Fisheries Research Board Prog. Repts., Pac. Biol. Sta. and Pac. Fish. Expt. Sta. No. 22, p. 14 (1934).
(31) G. Bennett *et al., Some Factors Affecting the Quality of Frozen Foods. II.,* Pennsylvania Agr. Exptl. Sta. Bull. 580, 1954.
(32) H. D. Michener, W. C. Dietrich, and P. A. Thompson, *Food Technol., 14,* 290 (1960).
(33) F. B. Engley, Jr., *Texas Rept. Biol. Med., 14,* 313 (1956).
(34) H. D. Michener and R. P. Elliott, *Advan. Food Res., 13,* 349 (1964).
(35) L. D. Witter, *J. Dairy Sci., 44,* 983 (1961).
(36) H. F. Smart, *Food Res., 2,* 515 (1937).
(37) C. F. Schmidt, R. V. Lechowich, and J. F. Folinazzo, *J. Food Sci., 26,* 626 (1961).
(38) K. Abrahamsson, B. Gullmar, and N. Molin, *Can. J. Microbiol., 12,* 385 (1966).
(39) H. Beerens, S. Sugama, and M. Tahon-Castel, *J. Appl. Bacteriol., 28,* 36 (1965).
(40) J. Forgacs, *Food Technol., 20,* 750 (1966).
(41) H. C. Mannheim, M. P. Steinberg, A. I. Nelson, and T. W. Kendall, *Food Technol., 11,* 384 (1957).
(42) T. Moran, *Proc. Roy. Soc. (London), Ser. B., 107,* 182 (1931).
(43) M. F. Gunderson, *Proc. Low Temperature Microbiology Symposium,* Campbell Soup Co., Camden, N. J., 1961, 299 (1962).
(44) J. C. Ayres, W. S. Ogilvy, and G. F. Stewart, *Food Technol., 4,* 199 (1950).
(45) R. P. Elliott and H. D. Michener, *Appl. Microbiol., 9,* 452 (1961).
(46) N. F. Dalton and D. H. Strong, *J. Food Sci., 32,* 116 (1967).
(47). R. E. Rogers and C. S. McCleskey, *Food Technol., 11,* 318 (1957).
(48) National Research Council, Food Protection Committee, *An Evaluation of Public Health Hazards from Microbiological Contamination of Foods,* (National

Research Council, Pub. 1195), 1964. N.A.S. Publishing Office, 2101 Constitution Ave., Washington, D. C., $2.

(49) American Public Health Assn., *Standard Methods for Examination of Dairy Products,* 11th ed., Am. Pub. Health Assn., New York, 1960.

(50) American Public Health Assn., *Recommended Procedures for the Bacteriological Examination of Sea Water and Shell Fish,* 3rd ed., Am. Pub. Health Assn., New York, 1962.

(51) American Public Health Assn., *Standard Methods for the Examination of Water and Wastewater,* 12th ed., Am. Pub. Health Assn., New York, 1965.

(52) American Public Health Assn., *Recommended Methods for the Microbiological Examination of Foods,* 2nd ed., Am. Pub. Health Assn., New York, 1966.

(53) Association of Food and Drug Officials of the U.S., Microbiological Examination of Precooked Frozen Foods, Assn. of Food and Drug Officials of the U.S., Quart. Bull. Supplement, 1966.

(54) Association of Official Analytical Chemists, *J. Assoc. Offic. Analytical Chemists, 49,* 246–250 (1966).

(55) U.S. Food and Drug Administration, *Bacteriological Analytical Manual* (loose leaf), Food and Drug Admin., U.S. Dept. Health, Education, and Welfare, Washington, D. C., 1966.

(56) W. C. Frazier, *Food Microbiology,* 2nd ed., McGraw-Hill, New York, 1967.

(57) W. C. Frazier, E. E. Marth, and R. H. Deibel, *Laboratory Manual for Microbiology,* 4th ed., Burgess Publishing, Minneapolis, Minnesota, 1968.

(58) K. H. Lewis and R. Angelotti, Ed., *Examination of Foods for Enteropathogenic and Indicator Organisms,* Washington, D. C., 1964 (U.S. Public Health Service Publication No. 1142).

(59) M. M. Galton, G. K. Morris, and W. T. Martin, *Salmonellae in Foods and Feeds, Review of Isolation Methods and Recommended Procedures,* U.S. National Communicable Disease Center, Atlanta, Georgia, 1968.

(60) H. B. Messinger, *J. Milk Food Technol., 26,* 328 (1963).

(61) P. A. Hartman and D. V. Huntsburger, *Appl. Microbiol., 8,* 382 (1960).

(62) H. H. Weiser, *Practical Food Microbiology and Technology,* Avi Publishing, New York, 1962.

(63) M. E. Parker and J. H. Litchfield, *Food Plant Sanitation,* Reinhold, New York, 1962.

(64) H. T. Meryman, Ed., *Cryobiology,* Academic Press, New York, 1966.

(65) R. P. Elliott and H. D. Michener, "Review of the Microbiology of Frozen Foods," in *Conference on Frozen Food Quality,* Albany, California, Nov. 4–5, 1960, U.S. Dept. Agr., ARS-74-21, p. 40.

(66) J. Farrell and A. H. Rose, *Advan. Appl. Microbiol., 7,* 335 (1965).

(67) W. Schmidt-Lorenz, "Behaviour of Microorganisms at Low Temperature," *Bull. Inst. Intern. Froid, 47,* 390, 1313 (1967).

(68) A. T. Miller, H. M. El-Bisi, and F. M. Sawyer, *Microbiological and Public Health Aspects of Prepared Frozen Foods,* Massachusetts Agr. Exp. Sta. Bull. 548, 1964.

(69) E. J. Bowmer, *J. Milk Food Technol., 28,* 74 (1965).

(70) National Conference of Salmonellosis, Atlanta, Georgia, 1964, *Proceedings,* National Communicable Disease Center, Atlanta, Georgia, 1965 (U.S. Public Health Service Publication No. 1262).

(71) P. R. Edwards and M. M. Galton, *Advan. Vet. Sci., 11,* 1 (1967).

(72) K. H. Lewis and K. Cassel, Jr., Eds., *Botulism, Proceedings of a Symposium,* Robert A. Taft Sanitary Engineering Center, Cincinnati, Ohio, 1964 (U.S. Public Health Service Publication No. 999-FP-1).

Chapter 4

Quality and Stability in Frozen Fruits and Juices

D. G. Guadagni

Vegetable Laboratory, Western Regional Laboratory, U.S. Department of Agriculture, Albany, California

I. QUALITY ATTRIBUTES IN FROZEN FRUITS

The classification of quality attributes in various foods and food products is discussed at length in volumes more or less exclusively devoted to this subject. It is not the intent of this brief discussion to examine quality attributes in detail or even to list or classify all of the attributes or properties which in some way or another contribute to that vague, intangible food characteristic called "quality." Furthermore, since the primary emphasis of this volume is on the stability of commercially available products, only brief mention will be made of the practical factors which affect quality of the originally frozen product.

A. Texture

In dealing with commercially frozen products, it is only natural to consider first the commercial freezing process itself. It is generally agreed by most workers that freezing is the best available method known at the present time for preserving the fresh color, taste, aroma, and nutritive value of many foods. For frozen piece fruits the most serious defect of the freezing process is its adverse effect on texture and structural appearance. This defect is least evident in properly harvested and ripened freestone peaches and most evident in certain varieties of strawberries and raspberries. The proposed mechanisms for the physical and chemical processes which irreversibly alter textural and structural characteristics of plant products are reviewed in a recent text on this subject (1). Whatever the reasons for these profound changes, most thawed frozen fruits leave much to be desired in textural quality when compared with fresh fruit. One of the most fruitful avenues for minimizing the textural–structural damage in frozen fruits has been found in the selection and breeding of varieties especially suited to freezing (2–6). Unfortunately, it is seldom possible to breed all of the commercially important quality characteristics into one variety. Economically, the most important characteristics are yield, disease resistance, ease of harvesting and handling the fresh fruit, size and appearance of the fresh fruit, etc. The varieties selected for commercial plantings are largely selected on the basis of these factors rather than on the texture, color, and flavor of the frozen product. Certainly the plant geneticists are trying to bring all of the factors which are important to frozen fruit quality into a single variety, but complete success has rarely been attained. Therefore, it is to be expected that most commercially available frozen fruits will, to varying degrees, suffer from lack of the most desirable texture after freezing and thawing.

1. Effect of Very Rapid Freezing

Another avenue for the improvement of texture in frozen fruits and vegetables has recently been explored, namely ultrarapid freezing (7,8). Very rapid freezing has been claimed to improve the texture of frozen peaches and strawberries, but no presently known processing procedures can produce a product whose texture is essentially equal to that of fresh fruit. In addition, previous work on freezing rates did not indicate any particular advantages derived from very rapid freezing (9). Consequently, until much more fundamental information is accumulated on the basic factors affecting texture and texture changes in frozen fruits, we will have to accept frozen fruits with less than ideal texture, even if we use the best varieties and processing techniques available at the present time.

B. Color or Appearance

1. Discoloration by Oxidation

Color or appearance of freshly frozen fruits is usually good, provided proper harvesting, handling, processing, and freezing techniques have been applied. During and shortly after thawing, however, light-colored fruits such as peaches, apples, and apricots are subject to rather serious color changes caused by enzymatic browning (10,11). These color changes could be minimized or completely eliminated by heat inactivation of the responsible enzymes, but generally the heat treatment necessary for complete inactivation of the enzymes changes the characteristic odor and taste from that of a fresh product to that of a cooked product. Obviously, this may not be too important for fruits intended for subsequent manufacture into cooked products such as pies, jams, and jellies, but for consumer-size dessert packs, a cooked odor and taste are unacceptable. The commonly used commercial method of minimizing these color changes during freezing and thawing of the raw fruit is by packing and freezing it in sugar sirups containing ascorbc acid (12–14). Using this technique, the primary factors involved in retaining the original color are the amount of ascorbic acid used and the degree of oxygen exclusion effected by the packing sirup, the package, and the packaging atmosphere. In nonhermetic containers such as the waxed composite cartons commonly used in the frozen fruit industry, the most important factor is a fruit-to-sirup ratio that will give complete sirup coverage of the slices and a container fill that leaves little or no headspace. Under these conditions, availability of oxygen to the fruit surfaces is minimized; oxygen diffusing through the package oxidizes the ascorbic acid in the sirup surrounding the fruit and is used up before it can oxidize the phenolic constituents in the fruit. In hermetic containers such as tin cans, these principles are still important, particularly when the products are packed under normal atmospheric pressure. When the sliced and siruped fruit is packed under a relatively high vacuum, the oxygen tension in the initial packaging atmosphere is greatly reduced and, consequently, susceptibility to oxidation in the package is also reduced. The foregoing considerations apply to the sealed packaged products, but once the package is opened and the product is directly exposed to air, the combination of atmospheric oxygen and active polyphenolase enzymes rapidly oxidize the ascorbic acid adhering to the surface of the slices and the normal flesh color turns to an unattractive brown (see Chapter 2). The amount of ascorbic acid added to the sirup is one of the factors which determines the time that the slices will retain their natural color after thawing. As with texture, the variety of fruit also influences both the rate and the extent of browning. Varieties lacking in polyphenolase enzyme or oxidizable phenolic constituents are less susceptible to browning than those which are high in either or both of these factors.

a. Peaches. In peaches, the Sunbeam variety has long been known for its resistance to color changes. The reason for its color stability has been traced to a very low concentration of the oxidizable phenolic compounds sometimes referred to as "catechol tannins" (15). The low browning tendency of certain varieties of apples has been related to their having little or no polyphenolase enzyme activity. Analysis of several peach varieties and breeders' selections has shown that the extent or severity of browning in frozen peaches is related to the concentration of the oxidizable phenolic compounds (16). Unfortunately, the varieties or strains which are best for natural color retention seem to lack some of the other important characteristics necessary for commercial acceptability. Other complications to the color problem in frozen peaches are the effects of growing area or climatic conditions, and cultural practices (17,18). For a given variety it has been generally observed that fruit grown in a relatively cool, cloudy area develops more oxidizable phenolic constituents than that grown in a hot, clear, dry area. Even in the same orchard, development of high tannin fruit has been associated with seasons which were predominantly cool and cloudy, in contrast to relatively warm sunny seasons (19). In certain varieties which are normally high in tannins, these climatic and cultural factors can also influence the taste of the frozen fruit as well as its color retention. The combination of high tannin and low pH values often causes a disagreeable astringent or puckery sensation which detracts from the acceptability of the frozen product. The relationship between astringency and tannin has also been observed in research on ripening bananas (20), and astringency problems are well known in ripening persimmons (21).

b. Cherries. Color and appearance problems in originally frozen RSP (red sour pitted) cherries are similar to those for the light-colored tree fruits in that they also contain polyphenol oxidases and oxidizable phenolic compounds in addition to the anthocyanin pigments which give this fruit its characteristic red color. The factors governing the oxidation of tannins in peaches, apricots, and apples also apply to cherries. In addition, the enzyme systems in cherries are also capable of oxidizing the anthocyanin pigments which are concentrated in the skins. Deviations from the bright red natural color of properly frozen RSP cherries can be the result of both anthocyanin pigment oxidation with resulting loss of natural redness, and the browning which occurs by oxidation of tannin-like phenolic compounds. Perhaps a more serious problem in frozen cherries is that dark specks or blotches appear on the surface of otherwise normal bright red fruit. These dark specks are a result of bruising of the fruit during harvesting, coupled with delay at relatively high temperatures prior to freezing (22–24). This is a good example of a quality defect in frozen fruits which has nothing to do with freezing or frozen storage, but is related to harvesting and handling practices prior to freezing. Improved methods of handling the fresh fruit from orchard to processing plant have greatly aided in minimizing this bruising problem in frozen RSP cherries.

C. Odor and Taste

Next to color and appearance, the odor and taste of frozen fruits, and particularly of juices, are probably the most important quality factors which determine the acceptability of the products. Taste and odor are of prime importance because consumers have become accustomed to expect the characteristic fresh odor and taste of raw fruits in the frozen products. While it captures these fresh quality characteristics, the freezing process does not normally add any new or undesirable tastes or odors. For example, the heat treatments required in canned and dried fruits and juices give the products tastes and odors which differ markedly from those of the fresh fruit. This does not imply that these canned and dried products are inferior, but only that their odors and tastes are decidedly different from those of fresh fruit. In frozen products, however, we are attempting to approach fresh odor and taste as nearly as possible, and therefore fresh fruit is the ideal standard for this quality attribute. Obviously, there can be a rather wide range of flavor quality even in freshly harvested fruits, and here again variety is one of the most important factors affecting fruit flavor. As mentioned previously, however, the decision to grow certain varieties in certain areas is based primarily on economic factors, and these do not always dictate production of varieties with the most desirable flavor. Therefore the original flavor of frozen fruits and juices can only be compared to that of their fresh counterparts of the same variety. With this limitation of within-variety comparison, the frozen product should certainly have the characteristic odor and taste of the freshly harvested fruit, except that it might have the modified taste imparted by sweetening agents such as sugar or sirups. Within this framework, maturity and preprocessing delay factors are probably among the most important ones affecting flavor in initially frozen fruits and juices. In the case of frozen peaches, it has long been believed that the best-flavored frozen product was obtained by using tree-ripened fruit (10). In commercial practice, however, harvesting at the tree-ripe stage causes severe losses from bruising and rotting, especially if the product must be stored a few days prior to freezing. It is also known that the texture of tree-ripened frozen peaches of certain varieties may be somewhat soft and mushy. Experience in the Pacific Northwest has shown that harvesting at the firm-ripe stage resulted in a frozen product of better texture and good natural flavor (25). This involves ripening of the fruit off the tree for from three to five days prior to processing for freezing. On the other hand, soft bush fruits such as strawberries must be ripened on the vine and frozen as rapidly as possible after harvesting in order to capture the desirable fresh odor and taste. Processing delays in excess of 24 hrs. after harvesting cause significant flavor losses in the initially frozen product.

1. *Effect of Sweeteners*

Sugar and sirup additions to frozen fruits have been discussed in terms of their ability to exclude oxygen from the fruit, and thus help preserve color and appearance. However, it is well known that most thawed frozen fruits are practically unpalatable without the addition of sugar, either prior to or subsequent to freezing and thawing. The thawed unsugared fruit has a pronounced sour, disagreeable taste which effectively masks fresh fruit flavor. There have been many statements in the literature that sugar or sirup addition is essential to preserve fruit flavor during freezing and storage. However, careful examination indicates that sugar in frozen fruits is primarily a sweetener and perhaps an enhancer of the natural fruit flavor. There is no evidence to indicate that the stability of sugared frozen fruits is superior to that of unsugared fruits provided all other factors are equalized. Studies on fresh and frozen strawberry puree have shown that sugar did not prevent development of off-flavor nor did it decrease the rate of off-flavor development during storage (26). While there is no question that sugar is necessary as a sweetener, there does not seem to be much difference as to the time at which the sugar is added. The kind of sugar used in frozen fruits traditionally has been sucrose (27,28). Experiments with other sweetners such as invert sugar, glucose, and various corn sirups have indicated that none of these can replace sucrose on a 100% basis. In general, about 25% of the sucrose could be replaced with these sugars without causing distinguishable flavor differences from the 100% sucrose packs (29–31).

D. Workmanship

In addition to the factors which are inherent in varietal characteristics, growing conditions, handling practices, processing conditions, and freezing rates, factors of workmanship also enter into determination of initial quality. These workmanship factors are primarily concerned with maintenance of the typical fresh appearance and character of the fruit and the sorting out of atypical, malformed, or blemished units and all foreign substances such as leaves, stems, stones, pits, etc. The value of workmanship in contributing to product quality has been discussed by Kramer and Twigg (32). Certain aspects of workmanship affect only initial quality of the frozen product, while others may also affect stability under adverse storage conditions. Misshapen pieces, presence of inert foreign materials, seed cells, seeds, pieces of skin, etc., may only affect initial quality by detracting from appearance and aesthetic value. On the other hand, a poor sorting job which allows an undue amount of moldy or contaminated fruit to enter the pack may also reduce product stability under unfavorable conditions.

II. STABILITY OF FROZEN FRUITS

A. Definition and Measurement

Before proceeding with the discussion of stability of frozen fruits and juices, it is essential that we clearly define what is meant by stability and describe exactly how it is measured. Some of the dictionary definitions of stability are "steadiness," "constancy," "fixedness," and other synonyms implying a condition of no change from a given original position or state of quality. Obviously the quality of frozen foods *does change* during storage, and the rate of change is dependent on temperature. Even though the changes which occur during brief periods of storage may not be measureable by any means at our disposal, we know that chemical reactions and changes of physical state are going on which will contribute to real quality changes, because, given enough time under the same conditions, a quality change does become evident and measurable. Although not entirely a satisfactory usage, we might arbitrarily define "the stable period" for frozen fruits or juices as that length of time at any given temperature during which no definite quality change has occurred. This immediately calls for another set of definitions telling what we intend to signify in this chapter by "quality" and "definite quality change." In the preceeding section we have briefly discussed some of the factors affecting initial frozen fruit quality, but in measuring the stable period (or in short, the "stability") we are limited to the original quality built into the initially frozen product. Obviously, unless this initially frozen product possesses a high degree of the desirable quality attributes, measurement of stability does not mean much. We are not particularly interested in how long it takes an already poor product to get a little worse, but we are greatly interested in how long it takes a good product to lose its original fresh taste, odor, or color, or to develop undesirable tastes, odors, or colors. Therefore it is extremely important that stability measurements should be made on products of good original quality, and it is equally important that control samples of the products be maintained as close to their original quality as possible for the duration of comparative stability measurements. For most frozen fruits, it has been found that no measurable changes occur for periods of 5 years or more when the products are held at −20°F. (−29°C.) or lower. Representative samples of each "lot" of the frozen fruit or juice to be tested will be held at this base temperature to serve as a control. Other representative samples from the same lot will be placed at higher temperatures such as 0, 10, or 20°F. (−18, −12, or −7°C.) for which a measurement of the stability of the particular product is desired. At intervals, the samples held at the higher temperatures are compared with the controls. In our work we have used a sensory difference test such as the triangle (33); the percentage of correct judgments is plotted against time in the form of a modified stimulus–response curve (Fig. 4.1). Time represents the stimulus, because the combination of time and temperature produces changes in the constituents responsible for differences in responses between control and experimental samples. In using the triangle test and trained sensory panels we have chosen the range of 70–80% correct judgments as representing the time required to produce a *definite change* in *quality* (34). For the purpose of our discussion of stability we may denote this time period at a given temperature as the stable period (or briefly, the "stability") at that temperature. (See Chapter 10 for a discussion of other measures of the stability of frozen foods.)

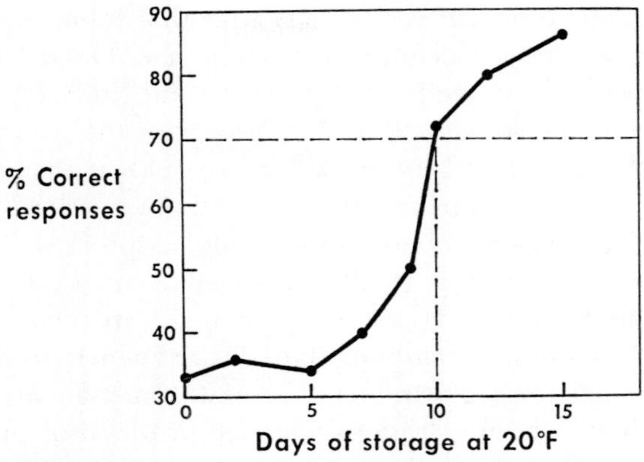

Fig. 4.1. Percentage of correct responses in triangle tests for flavor, as a function of storage time at 20°F. (—6.7°C.) for a commercial retail pack of frozen strawberries. Control sample held at —20°F. (—28.9°C.) (35).

1. "High-Quality Life"

The foregoing definition of period of stability makes it synonymous with the term "high quality life" as defined by Guadagni (35). It should be emphasized that statistically, this degree of discrimination between samples represents very highly significant differences. On the other hand, the change in quality represented by these differences leaves the changed product still well within the range of commercial acceptability provided the product was of good initial quality.

a. Commercial Acceptability. An estimate of commercial acceptability for retail packs of certain frozen fruits was obtained during studies in 1956 and 1957 from panels of quality control experts and other industry representatives (36). Since these people are directly involved in handling consumer complaints on the products they market, they should have a substantial basis for knowledge about the quality level which might cause a consumer complaint or rejection. They were asked to grade a series of commercially packed samples exposed to various patterns of time–temperature experience, and to indicate which samples were likely to cause consumer complaint. For the frozen strawberries, 70% of the quality control experts indicated that consumer complaints might occur after 6 weeks at 20°F., (−7°C.) while the trained laboratory panels had found a definite quality change after 2 weeks at that temperature. For this commodity, therefore, our laboratory "stable period" was extended at least threefold by the judgments of quality control experts. For frozen peaches, however, the quality control experts reported results which showed that color and appearance were the most important factors that could lead to consumer complaints (37). Their assessment of the degree of browning which might cause consumer complaint coincided very closely with that found by the trained panels as representing a definite quality change. On the other hand, raspberries were considered satisfactory by the industry groups for periods beyond 7 weeks at 20°F. (−7°C.), whereas the laboratory panels had found changes after between 2 and 3 weeks at that temperature (38). Therefore, except for peaches, it would appear that "commercial acceptability" extends well beyond the stable period as we have defined it. In actuality, however, it is idle to speculate on exactly how long a product can be stored at a certain temperature before it will be rejected by the consuming public. Many factors which are unmeasured and uncontrolled probably influence the consumer in his decision to buy or not buy. Not the least of these is price; also, many quality attributes are affected by factors other than storage temperature. Furthermore, there are no published data which would show that consumers' *buying habits* are not influenced by relatively small quality changes; the fact that a consumer does not go to the trouble of making a complaint is no assurance that he is satisfied with the product. He may simply reduce his consumption or stop using the item completely. Therefore, it we wish to make dependable measurements of product stability, we believe the preferred way is to use a criterion that is reproducibly measurable and that is affected by time and temperature alone.

Since a sensory difference test detects any perceivable sensory difference between samples held at different temperatures, a lack of stability or a change in quality may be due either to a single quality attribute such as odor, taste, texture, or color, or to a combination of any or all of these. The change may be measured in terms of a single quality attribute such as color or odor by requiring the panel to make a difference decision on visual color appearance alone or on smelling alone. However, when the product is placed in the mouth it is difficult or impossible to make an unbiased estimation of odor, taste, or texture independently of other attributes. The visual factor can be eliminated by conducting the test under appropriate colored lights or simply subdued lighting. For the most part stability will be based on the responses of trained panels to *flavor* differences when flavor is defined as the total sensory effect produced by eating the thawed fruit under controlled conditions. Experience with this procedure of measuring quality changes in frozen fruits has shown that it gives reproducible results (34). Therefore, it is possible to make valid comparisons of the stability of different lots of the same fruit, or of different fruits that have been stored under the same conditions. Unless the validity of the measuring technique is established, stability comparisons can be subject to large errors. The wide differences reported by different investigators for the stability of the same commodity can be ascribed partly to differences in the method of measuring this important characteristic.

B. Effect of Temperature

If we accept the definition of stability given in the previous section, we are now in a position to make some valid comparisons of the stability of different fruits and fruit products. While there is ample evidence that storage and handling of frozen foods in commercial channels seldom involves steady uniform temperatures, experimentally it is obviously essential to employ known steady temperatures in the study of frozen food stability. This is particularly important in studying the effect of temperature on stability or rate of quality change.

Stability, or "keeping quality," of frozen foods has been studied for many years (39). These studies have given rise to various general recommendations regarding storage of these products. The variance between some of the recommendations is easily understandable, considering the lack of proper investigational facilities and the diversity of measuring techniques. The data on average stability of frozen fruits and fruit products shown in Table 4.1 were all obtained in the same temperature-controlled facilities and by the same sensory measuring techniques. Therefore, we have reasonable assurance that these data represent valid differences in average stability within the given range of temperatures 0–20°F. (−18 to 7°C.). While this information is not directly usable in terms of commercial acceptability, it is necessary in determining the true effect of temperature on the rate of quality change. It also gives a good indication of the relative resistance of the various products to changes from their initial commercial quality when they are held at different temperatures.

C. Differences by Kind of Fruit and Type of Pack

It will be noted that regardless of the type of pack, the fruits listed in Table 4.1 were stable for at least one year at 0°F. (−18°C.), except for blueberry pie filling. Raspberries, whether packed in bulk with no sugar or in sirup-pack retail packages were stable for about 2 years at 0°F. Bulk pack strawberries, blackberries, and retail pack boysenberries were only slightly less stable at 0°F. (about 21 months). Pie fillings, except for blueberry, were stable for about one year at 0°F., but it is evident from the 0°F. data that no generalized statement with regard to stability applies to all types of packs or all products. At the higher temperatures, particularly 20°F. (−7°C.), raspberries and strawberries in the retail pack have a markedly lower stability than in the bulk pack. In raspberries at this temperature, the primary reason for the large

Table 4.1. Average Stability of Frozen Fruits and Fruit Products Held at Various Temperatures [a]

Product	Type of pack	Stablility, days		
		0°F.(– 18°C.)	10°F.(– 12°C.)	20°F.(– 7°C.)
Apples	Pie filling	360	250	60
Boysenberries	Pie filling	375	210	45
	Bulk, no sugar	405	125	45
	Retail, sirup	650	160	35
Blueberries	Pie filling	175	77	18
Cherries	Pie filling	490	260	60
Peaches	Pie filling	490	280	56
	Retail, sirup	360	45	6
Blackberries	Bulk, no sugar	630	280	50
Raspberries	Bulk, no sugar	720	315	70
	Retail, sirup	720	110	18
Strawberries	Bulk, dry sugar	630	90	18
	Retail	360	60	10

[a] Stability based on flavor changes, except for retail-pack peaches, for which it is based on color changes.

difference in stability is related to the transfer of acid, color, and other soluble constituents from the berries to the sirup (40–41). The transfer occurs by osmotic action of the partially thawed sirup. When a taster compares samples which had been held at –30°F. (–34°C.) with similar berries held at +20°F. (–7°C), it is readily apparent that the –30°F. samples have a more fresh, tart taste than those held at 20°F. At the higher temperature, a considerable proportion of the acid and other flavoring constituents will have been transferred from berries to sirup. This process occurs to varying degrees in any sirup-packed fruit product. The extent and rate of the transfer depend on storage temperature, concentration of sugar in the sirup, fruit-to-sirup ratio, and shape or piece size of the fruit. Conversely, as water and soluble constituents are removed from the fruit, the diluted sirup tends to be absorbed by the fruit, giving it a sweeter taste than corresponding samples held at –30°F. (–34°C.) will have. In the samples held at –30°F., essentially no transfer between fruit and sirup takes place except during the brief thawing period (40). In products packed without sugar, the flavor differences observed are due to chemical or enzymatic changes within the fruit itself rather than to transfer of constituents by osmotic action. In strawberries and raspberries the physical changes arising from solids transfer appear to be faster than the chemical or enzymatic changes at temperatures above 0°F. Other reasons for better stability in the bulk-pack products than in the retail packs are probably related to the following: The bulk packs for strawberries, raspberries, and blackberries were packed in sealed, enamel-lined 30-lb. tins, whereas retail packs were in 12-oz. and 1-lb. composite-type cartons. The tin containers are known to provide better protection from oxidative deterioration than the cartons. Furthermore, the much larger surface-to-volume ratio in the small retail containers would expose more of the product to the normal packaging atmosphere. Boysenberries, however, behave rather differently than raspberries, blackberries, or strawberries, particularly below 20°F. (−7°C.). At 20°F., the bulk pack still appears to be slightly more stable than the sirup pack, but at 10 and 0°F., (−12 and −18°C.), stability seems to be greater in the sirup pack. The bulk pack boysenberries were I.Q.F. (individually quick frozen) and packed in an ordinary, unlined cardboard box of 20-lb. capacity. This type of container provided practically no protection against oxygen, and hence oxidative reactions were only limited in rate by storage temperature rather than by a combination of low temperature and exclusion of oxygen. The retail pack of boysenberries, like that of raspberries, appears to be much more stable than the retail pack of strawberries. Comparing retail sirup packs in similar containers at 0°F. (−18°C.), raspberries and boysenberries are clearly more stable than strawberries or peaches (36–38). These differences are even more accentuated when the temperatures are raised to 10 and 20°F. (−12 and −7°C.). Peaches in composite cartons are particularly poor at the higher temperatures because of severe and rapid browning under these conditions. On the bulk packs held at 0°F., boysenberries are least stable, followed by strawberries and blackberries, and raspberries are the most stable (42–44). At the higher temperatures, the order is the same except for a switch between boysenberries and strawberries. Comparison of the fruits in the form of frozen pies shows that stability at 0°F. was best for peaches and cherries, followed by boysenberries, apples, and blueberries. Blueberries were markedly less stable than the other fruits. The stability patterns of frozen fruit pies held at temperatures above 0°F. were very similar to those obtained at 0°F. The striking difference found in the frozen fruit pies was the relatively low stability of blueberries compared to the other fruits (45).

D. Temperature Quotients

It should be emphasized that the data in Table 4.1 represent the average values from as many as 50 to 60 different "lots" of commercially packed products for any given fruit. They represent the means of stabilities which have been affected by many different variables, including varietal, climatic, packaging, processing, etc. The effects of some of these variables will be discussed later in connection with specific commodities. The average values of the properties of several different samples, however, are important in elucidating the true effect of temperature on rate of quality change. If the average stability, expressed in days, is plotted on a logarithmic scale against temperature for the various retail pack fruit products for which data are available, we obtain the kind of relationship shown in Fig. 4.2. These data indicate that within this range of temperatures there is an exponential relationship between stability and storage temperature, *i.e.*, the logarithm of stability in days varies linearly with temperature. The slopes of these average stability lines are measures of the "temperature quotients" expressing the stabilities of the various products. Temperature quotients for chemical reactions or physical processes are usually denoted Q_{10}

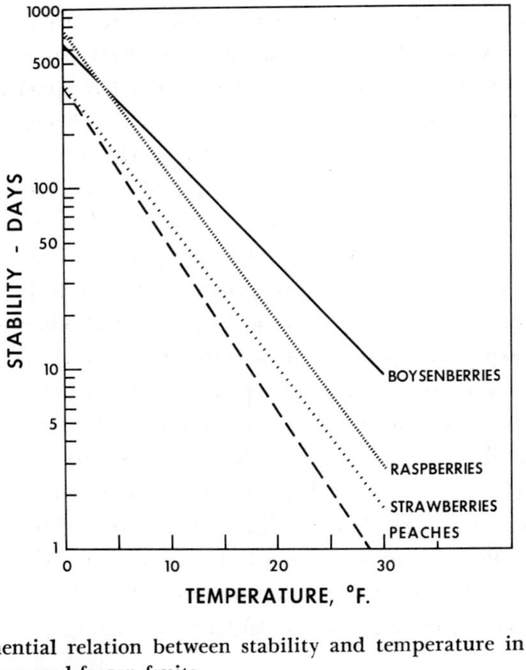

Fig. 4.2. Exponential relation between stability and temperature in retail packs of commercially prepared frozen fruits.

and represent the *ratio* of the rates at temperature intervals of 10°C. Since all of the data in the Time–Temperature Tolerance investigations were recorded with temperatures in °F., it is more convenient to express the temperature quotient as the *ratio* of stability at temperature t°F. to that at $(t+10)$°F., q_{10}. For example, the average stabilities of retail pack strawberries held at 0, 10, and 20°F. (−18, −12, −7°C.) are 360, 60, and 10 days, respectively; therefore, q_{10} for these strawberries over the temperature range 0–10°F. is 360/60 or 6, and between 10 and 20°F., it is 60/10 or also 6. This means that the rate of quality change in these frozen berries is increased by a factor of 6 for each 10°F. that the temperature is raised, or conversely, that stability is reduced by a factor of 6 for each 10°F. rise in temperature. For most frozen fruits and fruit products this approximately exponential relationship holds within the temperature range of 0–25 or 30°F. Therefore, if we know the stability of a frozen fruit at any temperature within this range, it is possible to predict its stability at any other temperature within the range from the value of q_{10} (46,47). Table 4.2 shows the q_{10} value for flavor and color changes in various frozen fruit products. These values give us a good indication of the effect of temperature on stability.

Table 4.2. Average Temperature Quotients (q_{10}) for Rates of Quality Change in Frozen Fruits

Product	Type of pack	Temperature range, °F	Temperature quotient, q_{10}, for: Flavor	Color
Apples	Pie filling	10–20	4.0	2.4
Boysenberries	Bulk, IQF	10–30	3.1	2.8
	Retail, sirup	10–30	5.1	3.6
	Pie filling	10–20	4.3	3.1
Blueberries	Pie filling	10–20	4.4	2.2
Cherries	Pie filling	10–20	4.7	4.4
Peaches	Pie filling	10–20	5.0	8.8
	Retail, sirup	10–25		7.8
Blackberries	Bulk, no sugar	0–30	3.1	2.8
Raspberries	Retail, sirup	10–30	3.6	6.6
	Bulk, no sugar	0–30	3.8	3.2
Strawberries	Retail, dry sugar	0–30	6.0	6.0
	Bulk, dry sugar	0–30	6.0	6.0

Relatively speaking, low values of the quotient mean that a change in temperature has a relatively small effect on stability, whereas a high q_{10} signifies that a change in temperature has a large effect on stability. For example, raspberries in the retail pack are very stable at 0°F. (720 days), but the q_{10} is high (about 6.4), so that the stability at 20°F. is only about 18 days.

The stability of a frozen product at any temperature can be calculated readily if investigation has shown that its stability varies exponentially with temperature, like those shown in Fig. 4.2, and the q_{10} for that product and its stability at some one temperature are given. In that case, the logarithmic plot is a straight line, q_{10} is proportional to the slope of the line, and one known point on the line (for example, the stability of the product at 0°F.) suffices to fix the stability line completely.

It may be noted that if temperatures are all expressed in °C., so that the quotient of stabilities at temperatures 10°C. apart is denoted by Q_{10}, $Q_{10}=q_{10}{}^{1.8}$ or $q_{10}=Q_{10}{}^{0.555}$.

It should be clear that q_{10} (or Q_{10}) alone is not a sufficient specification of the stability of a frozen product at various temperatures. The q_{10} for blueberry pie filling is only 2.2, while that for retail pack raspberries is 6.6, three times as great; however, while the stability of the raspberries is very high at 0°F. (720 days), this stability is so sensitive to a rise in temperature (corresponding to the high q_{10}) that at 20°F. the raspberries are no more stable than the blueberries (18 days).

This discussion is limited to the effect of steady temperatures; procedures for estimating the effect of a varying temperature are described in Chapter 10.

In general, the siruped or sugared retail pack fruits had the highest q_{10} values for flavor change. Except for strawberries, the same fruits packed in bulk without sugar had markedly lower q_{10} values for flavor changes. This difference between siruped and dry packs is at least partially explained by the solids transfer which occurs at the higher temperatures in sirup packs. As explained earlier, the transfer of acids and other flavoring constituents from fruit to sirup or sugar phase is rather rapid as the temperature is increased above 0°F. (−18°C.), and hence the q_{10} is high in comparison to nonsugared fruits. In general, the q_{10} values for flavor changes in fruit pies were somewhat higher than those for bulk pack fruits but lower than those for siruped retail packs. Therefore, it would appear that increases in temperature would affect flavor changes most in the retail packs, next in pies, and least in bulk packs.

In all of the fruits studied except peaches, the q_{10} for color changes was equal to or less than that for flavor, indicating that flavor was the quality attribute which was most affected by temperature changes. In peaches, however, color was by far the most important factor that was degraded. The q_{10} values for color-change in frozen peaches were higher than for any other fruit studied. Strawberries and raspberries also had relatively high q_{10} values for color change, but just as for flavor, transfer of color from fruit to sirup in the sugared and sirup packs was largely responsible for the high values.

It will be noted that if the q_{10} values were expressed as the traditional (Centigrade) Q_{10}, the values would be very much higher than the values quoted for ordinary chemical reactions, usually about 2–3. Undoubtedly there are several reasons for the large difference. First of all, we are not dealing with a single chemical reaction in a steady state, but with a process or series of processes for sugar- or sirup-packed fruits and concentrated juices. We are dealing at one extreme with a relatively solid product at −20°F., at the other with a relatively liquid product at 25°F., in which diffusional processes can go on with little hindrance. There is a continuously changing ratio of solid to liquid as the temperature is increased from 0 to 30°F. (−18 to −1°C.). This radical change of state over the temperature range of interest undoubtedly is a major factor in causing such abnormally high Q_{10} values.

E. Effect of Unsteady Temperature History

Up to this point we have considered stability at steady temperatures. Information presented in other chapters of this volume clearly shows that the commercial life of frozen foods is anything but a record of steady temperatures. Considerable discussion is found in the literature on the effect of fluctuating temperatures (48–50). Some investigators have reported reduced stability because of the fluctuation *per se,* while others have reported little or no significant effect due to fluctuation *per se*. Unfortunately it does not appear that any of these earlier studies adequately considered the temperature coefficients or quotients of the measured quality attributes over the range of fluctuating temperatures. Therefore it is impossible to determine whether observed differences between the effects of fluctuating and of steady temperatures were due to fluctuation *per se* or simply to differences between the *effective* steady temperature and the *mean* steady temperature of the fluctuating cycle. The effective temperature of a given fluctuating cycle depends on the amplitude of the cycle and the temperature coefficient or q_{10} for the particular quality attribute being measured (51). Thus cycles of high amplitude, coupled with high q_{10} values, could give the same quality effects as a steady effective temperature somewhat higher than the arithmetic mean temperature, and would not necessarily produce any effects due to fluctuation *per se*. When the *effective* temperature of a fluctuating cycle was considered, none of the "T-TT" (time–temperature tolerance) work on frozen fruits and fruit products showed any specific effects of fluctuation (52). The changes caused by any given fluctuating cycle were similar to the changes caused by the steady effective temperature of the cycle. When exposed to a fluctuating cycle, products with high q_{10} values changed at a rate equivalent to a steady temperature well above the mean temperature of the cycle. On this basis alone, appreciable temperature fluctuations are undesirable.

Chapter 10 discusses the requirements for calculating effective temperatures of any variable temperature system. In this discussion it is pointed out that the changes occurring in the food must be additive and commutative. Whether or not a particular product or series of products meets these conditions must be determined experimentally. Additivity and commutativity in frozen fruits were tested in a variety of ways, and within experimental error, quality changes were found to be additive and commutative. Thus frozen strawberries exposed to half of their stable period at 0°F. (−18°C.) (6 months), followed by half of their stable period at 10°F. (1 month), reached the end of their stable period after only 1 month at 10°F. (−12°C.) instead of the normal 2 months for frozen strawberries. Conversely, the same product held at 10°F. for 1 month, followed by 0°F. storage, reached the end of its stable period after only 6 months rather than the normal period of one year at this temperature. This simple example illustrates both additivity and commutativity. This relationship was demonstrated with other combinations of steady temperatures and other products. Therefore, the order of exposure of these products to different temperatures does not alter predictable stability based on steady temperature data. In other words, quality changes occurring within the temperature range of 0–25°F. (−18 to −4°C.) are similar, and preexposure to any temperature within this range does not cause any unexpected changes at any other temperature, either in kind or in magnitude. This is not only important in laying the basis for calculating the effects of variable temperatures, but is also important in the ordinary day-to-day management of stored products. It emphasizes the fact that the effects of temporary exposures to elevated temperatures are not erased by subsequent cooling to a low temperature. Perhaps of more importance, and less recognized, it also emphasizes that storage at 0°F. (−18°C.) has its limits. Strawberries stored at 0°F. for one year have already reached the end of their stable period. No matter how careful subsequent handling practices may be, the fact still remains that the product has reached the end of its stable period at a uniform steady temperature of 0°F. Any additional storage beyond this point merely adds additional changes on to what has already occurred. Storage of the same product at 0°F. for 9 months has, in effect, used up 3/4 of its stable period. Now, its stability at 10°F. (−12°C.) is only 15 days instead of 2 months, and only 2.5 instead of 10 days at 20°F. (−7°C.). These examples illustrate the importance of time as well as temperature in accumulating the quality changes which bring a product to the end of its stable period.

By definition, we have said that the stable period at any temperature is the *time* of storage required for 70–80% of the trained panel to distinguish between the stored and control samples. The time required to produce this degree of discrimination varies exponentially with temperature. The quality change represented by completion of the stable period at temperatures in the 0–30°F. (−18 to −1°C.) range, how ever, is approximately constant. This was confirmed by direct comparisons of products which had reached the end of the stable period at temperatures between 0 and 20°F. (−18 and −7°C.). No significant differences were found among the samples exposed to these various time–temperature combinations. The stable period for temperatures in this range represents a relatively constant quality change. A 10-day exposure at 20°F. (−7°C.) or a 60-day exposure at 10°F. (−12°C.) is no worse or better than a year's exposure at 0°F. (−18°C.) for frozen strawberries. Therefore, it is just as important to consider *time* as it is *temperature* in guarding the original quality of frozen fruits and fruit products. Brief periods at relatively high temperatures may be no worse than relatively long periods at temperatures in the 0°F. range.

F. Variability of the Product

The discussion so far has dealt with the effect of temperature on average stability of several different "lots" of each fruit or fruit product. Obviously, as with any biological material, the variation or deviations from the average values are considerable. The range of stability encountered in commercially packed frozen fruits is illustrated by the data in Fig. 4.3 for frozen strawberries. This graph was based on stability determinations for 55 different "lots" of berries held at a steady temperature of 20°F. (−7°C). Over 80% of the lots had stabilities in the range of 4–13 days at this temperature, with an average of 10 days for all "lots." For the smaller number of lots in which stability was measured at other temperatures as well as 20°F. the ratio of stabilities at other temperatures was somewhat similar to that at 20°F. The variability in strawberries depends on many factors, some of which

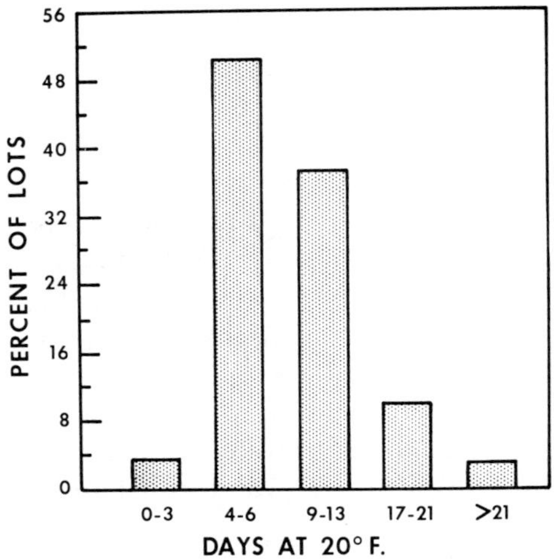

Fig. 4.3. Percentage of commercial lots of retail-size packages of frozen strawberries that became significantly different in flavor from their respective controls after various storage periods at 20°F. (—6.7°C.) (35).

will be discussed later. Variations in other frozen fruits such as raspberries, boysenberries, and blackberries, and the various fruit pies, were similar to those in strawberries. Some of the variation can be attributed to certain known factors, but many of the reasons for these variations are still unknown. In frozen retail-pack peaches we have an example of extreme variation in stabilty at temperatures above 0°F. (−18°C.). As mentioned earlier, the big stability problem with this fruit is the appearance of off-color. Even though 52 different commercial "lots" of frozen peaches were studied, there was greater within-lot than lot-to-lot variation in this particular fruit. The main reason, by far, for this degree of within-lot variability was variation in package net weight. Fig. 4.4 shows the distribution of package net weight. This, in turn, is directly related to degree of *container fill,* and inversely related to number of brown or discolored slices per package. This data was obtained on 12-oz. cartons, and shows the relatively wide range of weights encountered in standard commercial packages. While this variation in weight, or degree of container fill, does not affect browning to a great extent at 0°F. (−18°C.), the effect at 20°F. (−7°C.) is severe. Fig. 4.5 shows the effect of package net weight on the number of brown slices per package, for samples held 2 weeks at 20°F. There is a steep linear relationship between package net weight and brown slices per package. At the nominal net weight of 16 oz. each package contained five to six brown slices. The remarkable effect of package net weight on overall

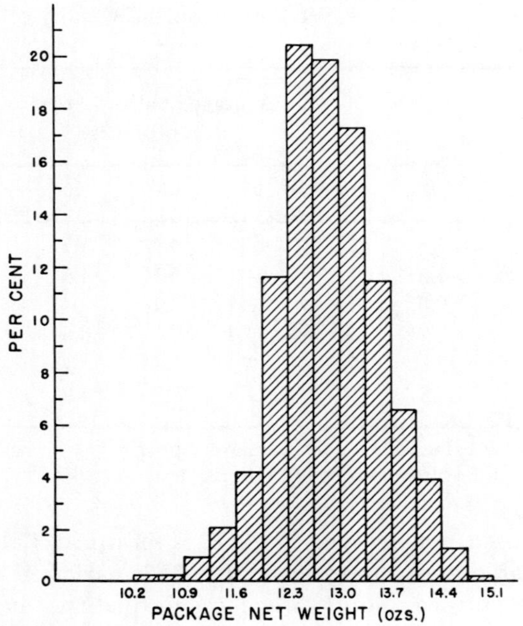

Fig. 4.4. Distribution of package net weight in a single "lot" of commercially packed frozen peaches (37).

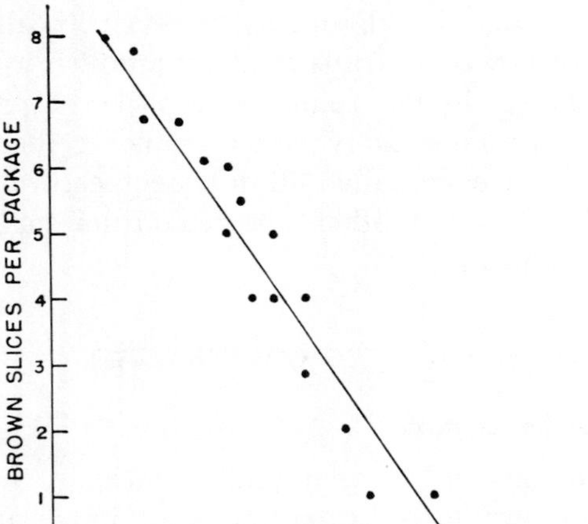

Fig. 4.5. Effect of net weight on number of brown slices per package of retail pack frozen peaches (37).

Table 4.3. Effect of Package Net Weight on the Quality of Frozen Peaches Stored at 20°F. (− 7°C.) (37)

Average net weight, oz.	Average quality scores,[a] days of storage						
	0	7	25	34	45	56	76
14.8–15.5	7.8	5.9 [b]	4.6	4.4	4.5	5.3	—
15.5–16.2	7.5	6.2 [b]	5.8	5.5	4.5	4.8	4.6
16.2–16.9	8.0	7.8	6.4 [b]	5.1	—	—	5.1
16.9–17.6	7.7	7.4	6.7 [b]	7.3	6.6	5.6	5.6
17.6–18.3	7.8	7.8	7.7	7.4	—	7.4	6.8 [b]
18.3–19.0	7.8	—	7.6	7.7	6.9	7.6	7.4

[a] 8.0 = good, 6.0 = fair, 4.0 = poor, 2.0 = very poor.
[b] Significantly different from control at the 5% level.

quality of frozen peaches held at 20°F. is shown in Table 4.3. It will be noted that initially (zero time at 20°F.), variation in net weight did not cause any significant differences in quality. During storage at 20°F., however, packages with low net weights changed significantly in quality after as little as one week at 20°F. On the other hand, packages which were overfilled by 2–3 ozs. did not develop significant quality differences for periods of 8–10 weeks at the same temperature. These data strongly suggest that if browning were not a problem, stability of frozen peaches at 20°F. would be relatively good, and stability variation would undoubtedly be much less than it is. While lot-to-lot variations for other fruits and fruit products were substantially less than that found in the commercial packs of peaches, stability variation in frozen commercially packed fruits is rather great. In some cases, such as this one of container fill in frozen peaches, the reasons can be identified, but in many others the reasons for large variations in stability remain obscure.

III. STABILITY OF FROZEN JUICE CONCENTRATES

A. Orange Juice Concentrate

Of all the commercially frozen juices which are marketed today, orange juice concentrate is by far the most important in terms of volume. It also appears to be the only juice in which extensive stability measurements have been made on commercially prepared products. Unfortunately, however, organized stability data on this product have not been published. Hence the relationship between temperature and stability of present-day commercial products is not too clear, except that 0°F. (−18°C.) still seems to be a generally desired and specified storage temperature. Generally it is considered that the flavor and color stability of frozen orange concentrate at 0°F. are longer than one year. In studies conducted at temperatures above 0°F., it has been concluded that flavor and color were not the limiting factors. Cloud loss (*i.e.*, rapid settling of the cloud in reconstituted juice) was considered to be the most important factor affecting acceptability and stability of the juice. Cloud loss (or even gelation) was more severe and occurred quicker in unheated than in pasteurized or heated juices. At storage temperatures above 0°F. cloud loss was the determining factor in stability of orange concentrates, even in the heated juices. Selection of cloud loss and or gelation of the juice as the limiting stability factor was made on a basis similar to that for the selection of color as the most important factor affecting stability of frozen peaches. It was and is still considered that the unattractive separation or clarification that occurs during cloud loss is more detrimental to consumer acceptability than slight flavor changes. Furthermore, in unheated juices, practically all investigators in this field agree that cloud loss or clarification precedes flavor loss at temperatures above 0°F. (53–58). The first published reports on storage stability of orange juice concentrate indicated that flavor stability was excellent at 0°F., and that flavor was better if the concentrate was packed and frozen under vacuum (59). Studies on clarification and gelation at temperatures ranging from 0 to 40°F. (−18 to +4°C.), reported that this defect occurred to varying degrees at temperatures above 0°F.; the rate of cloud loss increased with increasing temperature. Slight flavor changes followed the trend of clarification and gelation (60). The first published attempt to relate storage temperature and flavor stability quantitatively appeared in 1951 (61). The authors of this report confirmed previous data to the effect that cloud loss preceded flavor loss at temperatures above 0°F., and presented a table relating temperature to flavor loss. Unfortunately, the flavor evaluation part of the experiment was not carried beyond 204 days, and hence actual figures are given only for temperatures from 15 to 25°F. (−9 to −4°C.). At 10°F. (−12°C.) and lower, flavor stability was greater than 204 days. Flavor stability at 15, 20, and 25°F. (25°F. (−9, −7, and −4°C.) was reported as 175, 91, and 70 days, respectively. Storage of commercial samples for a period of 5 years at −4, 5, and 10°F. (−20, −15, and −12°C.) showed no cloud loss at −4°F., some loss at 5°F., and loss in ½ to 1½ years at 10°F. Flavor was regarded as normal after 5 years at −4°F., off at 5°F., and unpalatable at 10°F. (62).

1. Effects of Changed Processing

The most recent work on stability of commercially prepared orange juice concentrates gives data on more recent processing techniques. The most significant of these appear to be the practices of high-yield reaming and heat treating sufficiently to prevent cloud loss under normal handling conditions (63). In the 1957 season, when juices were made from oranges subject to the freeze, and prior to the high-yield, high heat-processing treatments, cloud stability was comparable to that of other years and flavor stability and cloud stability were found to be about equal. In 1960, however, with the new processing techniques in operation, cloud stability was much better than flavor stability. Presumably the extra heat treatment increased cloud stability, and the combination of heat and high-yield reaming reduced flavor stability. The flavor stability at 20, 15, 10, and 5°F. (−7, −9, −12, and −15°C.) varied from 26 to 52, 26 to greater than 52, 39 to 52, and 39 to greater than 52 weeks, respectively. Therefore it appears that this product is extremely stable even at relatively high storage temperatures; perhaps the initial flavor quality of the juice may be more of a problem than its stability. Where different combinations of temperatures were used in studying stability of orange juice concentrate, the changes under different temperature conditions were considered to be additive and cumulative (57, 63).

Recently the development of new methods of essence recovery have led to the production of aroma concentrates which possess much of the characteristic fresh orange aroma (64). Some commercial producers of orange concentrate have used some form of aroma or essence solution in their concentrates in recent years. Studies on the flavor stability of orange concentrates with added essence have been very limited. The meager information available indicates that orange concentrates with added aroma solution have a significantly better aroma immediately after freezing than similar concentrates with added cutback juice (65). On storage at 0°F. (−18°C.) however, the superiority in aroma of the samples with added essence disappears. After six months at 0°F. no significant differences were observed between samples prepared with cutback juice and with aroma solution.

B. Apple Juice Concentrate

A host of other concentrated juices and combinations of juices are found in modern-day frozen food cabinets. These range from the rather old and well-known grape juice concentrate to various exotic mixtures of tropical fruit juices, punches, and even synthetic mixtures.

However, the literature reveals little or no information on stability or storage characteristics of these products. Presumably, the reason for lack of published research activity on these products is their limited volume, and perhaps there have been no serious problems in their storage and distribution.

Over the years a small amount of work has been reported on frozen apple juice concentrate (66–68). These studies were entirely devoted to products of laboratory or pilot plant production, and indicated that four fold apple juice concentrate could be stored for at least a year at 0°F. (−18°C.) without significant loss of color or flavor. Recently, frozen apple juice concentrate with added essence has been marketed commercially. One of the early questions with regard to this product was whether it stored better with or without the added essence. Studies on the commercial samples indicated that there were no advantages in storing the concentrate and essence separately (69). Fourfold concentrate with added essence was stable for over 2 years at 0°F. (−18°C.), 1 year at 10°F. (−12°C.) and 2 to 4 months at 20 to 30°F. (−7 to −1°C.). Concentrates without added essence had about the same stability at 0°F., but were definitely less stable at 20 and 30°F. When essence was added to the latter concentrates just prior to tasting, apparent stability was very similar to that of the concentrates stored with essence. This indicates that addition of essence to the stored, stripped concentrate effectively masked the flavor changes which occurred during storage at the higher temperatures.

Storage of the 100-fold apple essence also presents problems which apparently have not been recognized. Studies on the undiluted essence at temperatures ranging from −30°F. (−34°C.) to 72°F. (22°C.) indicated a great degree of variability among samples stored at the higher temperatures. The only sure way to maintain original odor strength and quality was to store at −30°F. While temperatures above −30°F. appeared to exert a certain amount of influence on stability, other unknown factors definitely affect the stability of apple essence.

IV. FACTORS AFFECTING STABILITY

Many of the factors affecting stability of frozen fruits during storage at temperatures of 0°F. (−18°C.) and above are the same as those which affect initial quality. These have already been discussed in general terms in the earlier sections of this chapter. However, for certain fruits, some factors have such a profound effect on stability that further examination is merited.

A. Exclusion of Oxygen

In a previous section we have seen that among commercial retail packs of frozen peaches, container net weight was the most important measured factor related to incidence of browning. This was due to the fact that packages filled to zero headspace did a much better job of excluding oxygen from the fruit surfaces than normally filled packages. In the few instances where commercial retail samples were packed in cans, browning was much less than in composite cartons. Even at relatively high temperatures, the rate of browing in cans was minimal compared to that in composite cartons (Fig. 4.6). Similar differences between cans and composite cartons were found in retail packs of red sour pitted cherries (70). The color improvement obtained by a hermetic container immediately suggested that at least partial vacuum packing would further improve color retention at unfavorable temperatures (71). Fig. 4.7 shows the effect of packing at different degrees of vacuum on rate of color change in frozen peaches stored at 20°F. (−7°C.). The Hunter "b" value is a measure of yellowness; it is evident that 15–25 in. of vacuum was sufficient to maintain natural yellow color up to 6 weeks. As pointed out earlier, if color changes are minimized in frozen peaches, stability at temperatures above 0°F. (−18°C.) could be greatly improved. In addition to controlling brown-

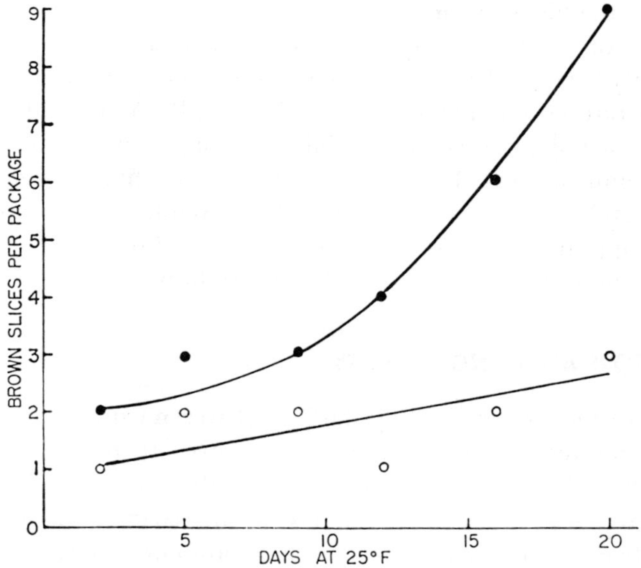

Fig. 4.6. Effect of container on incidence of browning in retail packs of frozen sliced peaches (71). (●) Metal end composite; (○) enamelled can.

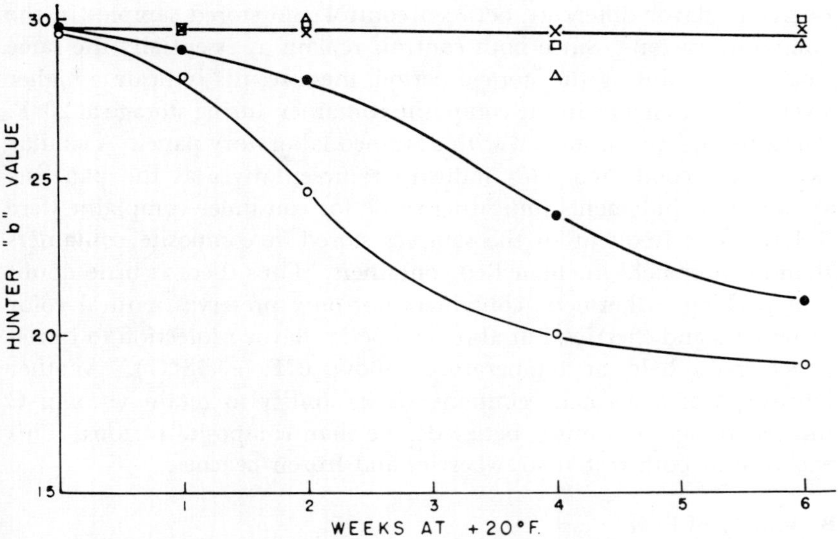

Fig. 4.7. Effect of vacuum packing on rate of color change in frozen peaches (71). (○) no vacuum; (●) 8 inches vacuum; (△) 15 inches vacuum; (□) 20 inches vacuum; (×) 25 inches vacuum.

ing in frozen peaches and RSP cherries, packing in hermetic containers was beneficial to flavor retention in frozen strawberries. Fig. 4.8 shows the effect of container on flavor retention during storage at 20°F. (−7°C.). The higher percentage of correct flavor judgments for the composite carton as compared to enamelled cans indicates a greater

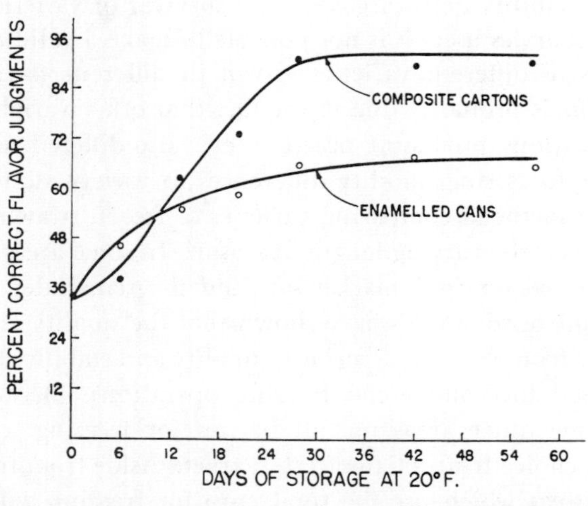

Fig. 4.8. Effect of container on percentage of correct flavor judgments during storage at 20°F. (—6.7°C.) of retail packs of frozen strawberries (36).

degree of flavor difference between control and stored samples in the composite cartons. Since both controls remain at essentially the same quality level during the storage period, these results indicate a higher level of flavor change in the composite container during storage at 20°F. These results were obtained with a trained laboratory panel. A similar experiment conducted with industry representatives as the panelists showed that judgments indicating cause for consumer complaint were 3–4 times as frequent in the samples stored in composite containers than in those held in enamelled containers. Thus there is little doubt that packing in hermetic containers not only preserves natural color in peaches and cherries, but also gives better flavor protection to frozen strawberries held at temperatures above 0°F. (−18°C.). Another advantage of a hermetic container is its ability to retain vitamin C during storage to a much better degree than composite cartons. This was true in both frozen strawberries and frozen peaches.

B. Variety of Fruit

While climatic changes during different seasons are known to affect initial quality of certain fruits, results of the "time–temperature tolerance" studies on frozen fruits did not indicate any significant seasonal effects on stability of frozen peaches, strawberries, or other berries. Insofar as it was possible to compare different varieties grown in the same area in the same season, however, some varieties were found to be much more stable than others. Unfortunately, one of the most stable of the strawberry varieties is no longer available on the commercial market. This is another example of the fact that factors other than quality and stability help to govern the survival of varieties. For the commercial samples used, it is not possible to make a valid comparison of stabilities of different varieties grown in different sections of the country. This is primarily due to the fact that other variables such as growing conditions, processing practices, etc., also differed, and hence it is impossible to ascribe stability differences to variety alone. Another complicating factor in comparing varieties of frozen strawberries from different areas is the varying length of season. In some areas, the choice fruit is harvested for fresh-market sale and the remainder is used for freezing. Controlled studies have shown that the quality and stability of this choice fruit were far superior to quality and stability of the field-run fruit used for commercial freezing operations. Because of late ripening, some other areas use all berries for freezing, rather than selecting the choice fruit for the fresh market. Aside from true varietal differences, areas which use the total crop for freezing will generally have a better-quality frozen product.

In frozen RSP cherries one of the most readily recognizable changes in fruit held at temperatures above 0°F. (−18°C.) is an increase in firmness. While the increase in firmness or toughness encountered in samples held at temperatures of 10–25°F. (−12 to −4°C.) was not sufficient to be objectionable, it was readily detectable by the trained laboratory panels. As shown in Fig. 4.9, the increase in firmness closely paralleled the change in flavor. However, the predominant change was in texture, and this was related to a decrease in the methoxyl content of the pectin in the stored frozen cheries (72). As pointed out earlier, once the product is placed in the mouth it is virtually impossible to assess different quality attributes such as texture and flavor independently. Therefore, the main reason for the close relation between texture and flavor is probably due to the bias effect of the texture changes rather than to any real causative relationship between texture and flavor.

C. Effect of Packaging

In bulk packs and frozen fruit pies made therefrom, the largest differences in stability of any one fruit were ascribed to packaging and processing (44). The biggest differences in stability of boysenberry pies were traced to packaging. Frozen boysenberry pies made from berries packed in unlined cartons were only about half as stable as pies made from berries packed in enamelled tins. This was true for storage temperatures of 0–10°F. (−18 to −12°C.), and affected both color and flavor, where each of these quality attributes was judged independently.

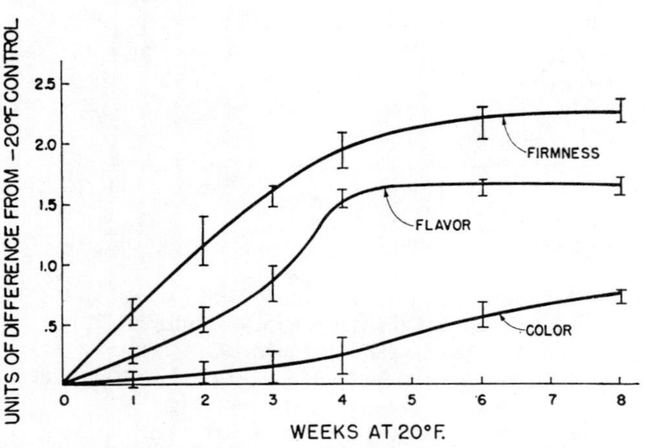

Fig. 4.9. Effect of storage at 20°F. on color, firmness, and flavor changes, of retail packs of frozen RSP cherries. Brackets represent total range of differences obtained with 7 "lots" of cherries. Units represent differences in average rating between control and stored samples (70).

As in retail packs, protectiveness of the container cannot be overemphasized as an important factor in the stability of such frozen fruits as peaches, cherries, and boysenberries.

D. Effect of Processing Method

In frozen apple pies, differences in processing of the bulk pack fruit from which the pies were made caused startling differences in stability at all storage temperatures in the 0–30°F. (−18 to −1°C.) range. The differences were primarily in flavor stability, and varied by more than two-fold. Factors other than processing of the fruit, such as variety, container, freezing, etc., were similar in both cases, so that the observed stability differences were largely, if not entirely, due to processing. In one case the apples were water blanched, and in the other they were subjected to a sulfite-vacuum impregnation treatment prior to freezing. The vacuum treatment not only gave fruit of better stability, but also of better initial quality. These are some of the factors which are rather easily identified as affecting quality and stability of these products. Unfortunately there are many other factors, still unknown, which account for much of the variation in quality and stability of frozen fruits and juices. Isolation and identification of these factors requires a great deal more research, using the latest techniques and most advanced equipment available to the investigator.

REFERENCES

(1) H. T. Meryman, *Cryobiology,* Academic Press, London, 1966.
(2) J. S. Caldwell, J. M. Lutz, and H. H. Moon, *Food Ind., 4,* 402 (1932).
(3) J. S. Caldwell, J. M. Lutz, H. H. Moon, and C. W. Culpepper, U.S. Dept. Agric., Tech. Bull. 731 (1940).
(4) D. G. Sorber, in *The Freezing Preservation of Foods,* 3rd ed. Vol. I. Avi Publishing, Westport, Connecticut, 1957, p. 466.
(5) E. R. Wolford, J. A. Sacklin ,and C. D. Schwartze, *Food Technol., 15,* 152 (1961).
(6) H. E. Thomas and E. V. Goldsmith, Calif. Agric. Exp. Sta. Bull., 690 (1945).
(7) Anon., *Canner/Packer, 131,* No. 13, 31 (1962).
(8) R. C. Webster, E. J. Benson, and W. H. Lucas, *Quick Frozen Foods, 25,* No. 5, 35 (1962).
(9) F. A. Lee, W. A. Gortner, and J. W. Whitcomb, *Food Technol., 3,* 164 (1949).
(10) M. A. Joslyn and L. A. Hohl, Calif. Agric. Exp. Sta. Bull., 703 (1948).
(11) J. D. Ponting and M. A. Joslyn, *Arch. Biochem., 19,* 47 (1948).
(12) J. C. Bauernfeind and C. F. Siemers, *Quick Frozen Foods, 7,* No. 12, 46 (1945).
(13) D. K. Tressler and C. Dubois, *Food Indus., 16,* 701 (1944).
(14) J. G. Woodroof, E. Shelor, S. R. Cecil, and I. Atkinson, Ga. Agric. Expt. Sta. Bull., 251 (1947).
(15) Z. I. Kertesz, N. Y. Agric. Sta., Geneva, Tech. Bull., 219 (1933).
(16) D. G. Guadagni, D. G. Sorber, and J. S. Wilbur, *Food Technol., 3,* 359 (1949).
(17) M. A. Blake and O. W. Davidson, *Proc. Am. Soc. Hort. Sci., 39,* 201 (1941).

(18) M. A. Blake, *Proc. Am. Soc. Hort. Sci.*, *40,* 153 (1942).
(19) D. G. Guadagni and C. C. Nimmo, *Food Technol.*, *7,* 59 (1953).
(20) H. R. Barnell and E. Barnell, *Am. Botany,* N. S., IX, 77 (1945).
(21) M. A. Joslyn and J. Goldstein, *J. Agr. Food Chem.*, *12,* 511 (1964).
(22) R. T. Whittenberger and C. H. Hills, *Food Technol.*, *7,* 29 (1953).
(23) C. H. Hills, R. T. Whittenberger, W. F. Robertson, and W. H. Case, *Food Technol.*, *7,* 32 (1953).
(24) R. L. LaBelle and J. C. Moyer, *Food Technol.*, *14,* 347 (1960).
(25) A. M. Neubert, J. H. Schultz, C. L. Bedford, and G. H. Carter, Wash. Agr. Expt. Sta. Circ. 63 (1948).
(26) D. G. Guadagni, *Food Res.*, *19,* 396 (1954).
(27) E. H. Wiegand, Oreg. Agric. Exp. Sta. Bull., No. 278 (1931).
(28) H. C. Diehl, J. R. Magness, C. R. Gross, and V. B. Bonney, U.S. Dept. Agric. Tech. Bull., 148 (1930).
(29) M. A. Joslyn, *Food Technol.*, *3,* 8 (1949).
(30) M. A. Joslyn and L. A. Hohl, Bull. Calif. Agric. Exp. Sta., 703 (1948).
(31) J. F. Caul and L. B. Sjoström, *Quick Frozen Foods, 13,* No. 10, 59 (1951).
(32) A. Kramer and B. A. Twigg, *Fundamentals of Quality Control for the Food Industry,* Avi Publishing, Westport, Connecticut, 1962.
(33) D. R. Peryam and V. W. Swartz, *Food Technol.*, *4,* 390 (1950).
(34) D. G. Guadagni, *Food Technol.*, *11,* 471 (1957).
(35) D. G. Guadagni, "Cold Storage Life of Frozen Fruits and Vegetables as a Function of Time and Temperature," in *Low Temperature Biology of Foodstuffs,* Pergamon Press, Oxford, England, in press.
(36) D. G. Guadagni, C. C. Nimmo, and E. F. Jansen, *Food Technol.*, *11,* 389 (1957).
(37) D. G. Guadagni, C. C. Nimmo, and E. F. Jansen, Food Technol., *11,* 33 (1957).
(38) D. G. Guadagni, C. C. Nimmo, and E. F. Jansen, *Food Technol.*, *11,* 633 (1957).
(39) D. K. Tressler and C. F. Evers, *The Freezing Preservation of Foods,* 3rd ed., Avi Publishing, Westport, Connecticut, 1957.
(40) D. G. Guadagni and C. C. Nimmo, *Food Technol.*, *11,* 604 (1957).
(41) D. G. Guadagni, S. H. Kelley, and L. L. Ingraham, *Food Res.*, *25,* 464 (1960).
(42) D. G. Guadagni, N. J. Downes, D. W. Sanshuck, and S. Okano, *Food Technol.*, *15,* 207 (1961).
(43) D. G. Guadagni, J. Harris, and K. M. Eremia, *Food Technol.*, *17,* No. 2, 103 (1963).
(44) D. G. Guadagni, K. M. Eremia, S. H. Kelley, and J. Harris, *Food Technol.*, *14,* 148 (1960).
(45) D. G. Guadagni, J. Harris, and S. Okano, *Food Technol.*, *17,* No. 7, 114 (1968).
(46) J. Kuprianoff, *Kältetechnik.*, *8,* 102 (1956).
(47) R. Plank, *Intern. Congr. Refr. Proc., 10th (Copenhagen), 3,* 45 (1959).
(48) J. G. Woodroof and E. Shelor, *Food Freezing, 2,* No. 4, 206 (1947).
(49) A. Hustrulid and J. D. Winter, *Agr. Eng.*, *24,* 416 (1943).
(50) W. A. Gortner, F. Fenton, E. Volz, and E. Gleim, *Ind. Eng. Chem.*, *40,* 1423 (1948).
(51) S. Schwimmer, L. L. Ingraham, and H. M. Hughes, *Ind. Eng. Chem.*, *47,* 1149 (1955).
(52) D. G. Guadagni and C. C. Nimmo, *Food Technol.*, *12,* 306 (1958).
(53) O. W. Bissett, M. K. Veldhuis, and N. B. Rushing, *Food Technol.*, *7,* 258 (1953).
(54) R. B. Guyer, W. M. Miller, O. W. Bissett, and M. K. Veldhuis, *Food Technol.*, *10,* 10 (1956).

(55) G. J. Keller, R. G. Rice, R. J. McColloch, and E. A. Beavens, *Food Technol., 8,* 195 (1954).
(56) R. J. McColloch and R. G. Rice, *Food Technol., 9,* 70 (1955).
(57) R. J. McColloch, R. G. Rice, M. B. Bandurski, and B. Gentili, *Food Technol., 11,* 444 (1957).
(58) M. A. Joslyn and W. Pilnick, in *The Orange, Its Biochemistry and Physiology,* W. B. Sinclair, Ed., University of California Press, Berkeley, 1961, p. 373.
(59) R. H. Cotton, W. R. Roy, C. H. Brokaw, O. R. McDuff, and A. L. Schroeder, *Proc. Florida State Hort. Soc., 60,* 39 (1947).
(60) E. L. Moore, R. L. Huggart, and E. C. Hill, *Proc. Florida State Hort. Soc., 63,* 165 (1950).
(61) C. W. Dubois and T. J. Kew, *Refrig. Eng., 59,* 772 (1951).
(62) T. J. Kew, *Proc. Florida State Hort. Soc., 70,* 182 (1957).
(63) T. J. Kew and M. K. Veldhuis, *Food Technol., 16,* No. 10, 119 (1962).
(64) J. L. Bomben, J. A. Kitson, and A. I. Morgan, Jr., *Food Technol., 20,* No. 9, 125 (1966).
(65) H. C. Mannheim, J. L. Bomben, D. G. Guadagni, and A. I. Morgan, Jr., *Food Technol, 31,* No. 3A, 151 (1967).
(66) R. K. Eskew, G. W. M. Philips, R. P. Homiller, C. S. Redfield, and R. A. Davis, *Ind. Eng. Chem., 43,* 2399 (1951).
(67) C. C. Nimmo, L. H. Walker, and V. S. Seamans, *Food Technol., 6,* 31 (1952).
(68) L. H. Walker, D. C. Patterson, and V. S. Seamans, *Food Technol., 8,* 108 (1954).
(69) D. G. Guadagni and J. Harris, *Food Technol., 21,* No. 3A, 136A (1967).
(70) D. G. Guadagni, C. C. Nimmo, and E. F. Jansen, *Food Technol., 12,* 36 (1958).
(71) D. G. Guadagni and C. C. Nimmo, *Food Technol., 11,* 43 (1957).
(72) M. Gee and R. M. McCready, *Food Res., 22,* 300 (1957).

Chapter 5

Quality and Stability of Frozen Vegetables

Robert L. Olson

Western Utilization Research and Development Division, Agricultural Research Service, U.S. Department of Agriculture, Albany, California

William C. Dietrich

Vegetable Laboratory, Western Regional Research Laboratory, U.S. Department of Agriculture, Albany, California

I. INTRODUCTION

Freezing is our closest approach to preserving fresh quality in foods. If the best technologies are applied, vegetables, after blanching, freezing, and cooking are not much different from those only cooked. At low enough temperatures we can protect and maintain the initial quality of frozen vegetables for very long periods of time. In five years at −20°F. (−31°C.), green beans, peas, cauliflower, and spinach have shown no measurable or detectable loss of color or flavor and no measurable change in chemical constituents or physical attributes.

Frozen vegetables are basically stable enough to maintain high quality from one harvest season to the next when held at 0°F. (−18°C.). However, rigorous evaluations with trained test panels can detect minor color and flavor deterioration for many types of these frozen vegetables within a year. The testing methods to do this involve direct comparison of a stored sample with a protected sample from the same lot—protected by storage at or below −20°F. (−31°C.). It is not likely that the first perceptible change in color or flavor quality could be detected even by experts in the absence of the direct comparison.

Foods frozen for commercial use are perishable and their deterioration at temperatures usually encountered in the commercial cold chain of storage and distribution can frequently be detected. Controlled experiments that simulated some measured commercial temperature experiences have caused readily measured quality changes in frozen vegetables.

In this chapter we will discuss quality factors of frozen vegetables, the subjective evaluation of flavor and color in relation to quality stability, objective measurements related to flavor and color quality, and the significance of quality instability in the commercial handling of frozen vegetables. We will depend to a considerable extent on the data published in a series on time–temperature tolerance (T-TT) of frozen foods based on invetigations at the Western Regional Research Laboratory of the U.S. Department of Agriculture (1–4). (These references are to publications in *Food Technology* of the T-TT series on green beans, peas, spinach, and cauliflower. Some information stemming from the research leading to these publications, but not previously reported, is included. These articles will not be cited again but information obtained from other sources will be indicated.)

Most vegetables must be blanched before freezing to prevent development of bad flavor in subsequent storage. Good commercial practice has included blanching of vegetables when needed. This technology has been reviewed recently (5–7). The T-TT investigations on frozen vegetables covered the instability of commercially preserved products which were adequately blanched. The instability of unblanched and underblanched vegetables was not included, nor is it in this chapter.

Research over the past 40 years, along with commercial practice, has shown that blanching to inactivate catalase and peroxidase stabilizes texture, ascorbic acid, and flavor during freezing storage of peas, corn, lima beans, green beans, carrots, cauliflower, Brussels sprouts, spinach, and some other vegetables. Mechanical arrangements to obtain uniform heating of all pieces and control of piece size is important to shorten the heating time for adequate blanch. Longer blanching will lessen chlorophyll retention (green color) and may give a softer texture than desired. Quality control with respect to time–temperature effects on raw material from field to processing and during the blanching operation must be effective at all times to insure a high-quality frozen product. The complexities of enzyme inactivation by blanching and correlation of stability with chemical constituents of blanched products are not yet completely understood and remain a fruitful area for research (8).

In the T-TT investigations, frozen food samples were procured from industry so that findings would have increased significance to commercial operations. Applicability of results was broadened by selecting samples over several harvest years, from different processors and areas of production. These samples also represented the principal varieties then being grown commercially. Furthermore, stability tests were made over a wide temperature range so that effects of storage temperature could be, for the first time, measured and evaluated rather precisely.

The T-TT investigations of vegetables included for *green beans,* 19 lots obtained at 12 packing plants during 1952 through 1956, including four varieties from six major growing areas with considerable variation in blanching (both water and steam) and conventional-cut and French-cut styles; for *peas,* 14 lots, eight packing plants, 1952 through 1957, four varieties and four growing areas; for *spinach,* six lots, two packing plants, 1955 through 1956, two varieties, leaf and chopped styles; and for *cauliflower,* 10 lots, four packing plants, 1955 through 1957, two varieties, two growing areas. Experimental storage studies were conducted with steady, fluctuating, and intermittent temperature conditions in the range of −40 to 40°F. (−40 to 5°C.). Times of exposure at different temperatures were selected so as to locate the time of occurrence of a perceptible difference in color or flavor from that of the −20°F. reference sample, and also longer times, such as to cause somewhat extreme levels of deterioration.

There was much variation in stability between lots of a particular frozen vegetable. Much of the variability in green bean color stability was found to be related to conditions of blanching, and chopped spinach was found to be less color stable than leaf spinach. For the other vegetables the source of variability was not discovered and data developed to describe product stability were of a general nature.

No significant differences in stability were found to be related to variety, harvest year, growing area, or processing conditions (except for green bean blanching conditions and spinach chopping). Interactions of all these factors were not rigorously investigated, however.

Other published investigations of stability of frozen vegetables are valuable as a corroboration of the T-TT data and provide information on vegetables not included in the T-TT study. Studies usually were conducted on experimental packs of vegetables representing a single variety and a single condition of culture with storage at only one or two temperatures. In view of large variability between individual commercial packs, it is not to be expected that stability data from single experimental packs of frozen vegetables can be interpreted as being specifically descriptive of the commodity.

The most authoritative source of information on frozen-vegetable stability exists in the more than 30 years of commercial experience and includes information on relationships among varieties and season with processing procedures. Unfortunately, such information is not adequately documented and can only be recovered as a collection of general statements from industry spokesmen. While this body of knowledge cannot be tapped adequately to include in this chapter, it is hoped that information printed here will serve as a foil to sharpen the knowledge of those experienced in the field and also provide a basic source of information for those who are entering or have recently entered this field without the benefit of long experience.

II. QUALITY EVALUATION AND MEASUREMENT

The quality factors of frozen vegetables to be discussed in this chapter are those that can be observed and measured. They include such organoleptic attributes as color, appearance, and flavor, and chemical and physical factors such as ascorbic acid content, ratio of chlorophyll to pheophytin in green vegetables, soluble brownish pigments in cauliflower, and color reflectance. This definition of quality is more limited than that discussed by Couden (Chapter 11). We cannot very easily measure those subjective value aspects of quality that would tell us how a given degree of quality difference would influence the marketing decisions of buyers as to price and quantity of purchases.

A. Subjective Attributes

By test panel procedures we can measure differences in subjective quality attributes, such as flavor and color, and describe the instability of frozen vegetables under given storage conditions by using such measurements. A most useful organoleptic evaluation is the determination of the first perceptible change caused by storage; this can be done by comparing a treated sample (that is, a sample subjected to a defined storage condition) with a protected control sample from the same lot. (Procedures are described by Guadagni in Chapter 4.)

In attempts to obtain an absolute indication of storage deterioration, industry-trained panels judged appearance of samples in various degrees of change and indicated, from a remembered standard, which ones were *(a)* of high quality, *(b)* acceptable but not of high quality, and *(c)* likely to cause consumer complaint. Rough estimates thus obtained were used in some of the T-TT studies to serve as points of correlation with objective measurements and test panel evaluations of deterioration.

Changes in textural qualities of frozen vegetables would also be evaluated by determining the first perceptible difference from a control lot. Preliminary investigations with conventionally frozen peas, green beans, and corn have revealed no significant texture changes when frozen products are subjected to severe time–temperature abuse, including thawing and refreezing several times. Some types of frozen vegetables prepared by extremely rapid freezing techniques have improved textural quality (9, 10) that may be subject to texture deterioration with poor storage and handling. However, definitive research in this area is only now being initiated and no published results are available.

B. Objective Measurements

Chemical and physical changes that occur as frozen vegetables deteriorate can be measured and tested for their relationship to changes in quality attributes that are judged subjectively and they are meaningful only when so correlated.

Perhaps the most useful objective measurement is percentage chlorophyll conversion to pheophytin in green vegetables. In processed green vegetables, as this chemical conversion increases, it is accompanied by a change in color from a bright green to a dull olive green. In the extreme, these vegetables change to a brownish green. Generally all of the green pigment in living plants is chlorophyll. Only when a tissue is disrupted or no longer respiring does the conversion to pheophytin occur. Therefore, if reasonable care is taken in handling vegetables after harvest, all the green pigment is still chlorophyll as the processing commences. With blanching of vegetables, life and respiration stop and the conversion of chlorophyll to pheophytin begins. Thus, the absence of converted chlorophyll at the start of processing provides a bench mark that is useful in comparing different lots, treatments, or samples. Conversion is promoted by processing as well as storage deterioration and it is not possible to tell in an unknown sample whether the color loss occurred before freezing or in frozen storage. However, if vegetables are handled properly and processed with some degree of uniformity, the color loss during processing can be standardized and all additional change attributed to storage instability.

Campbell (11) demonstrated that color changes occurring in frozen peas were due to chlorophyll conversion. In 1940 Mackinney and Weast (12) proposed the use of chlorophyll conversion to measure adverse color changes in frozen vegetables and they described a method for determining the ratio of chlorophyll to pheophytin. In 1957 Dietrich *et al.* (13) described the application of chlorophyll conversion measurement to the T-TT study. Later Dietrich (14) published details of the method which was a modification of the Dutton *et al.* (15) extraction and the Mackinney and Weast (12) spectrophotometer measurement.

In the Mackinney and Weast (12) method the concentration of green pigment need not be determined; only the ratio of chlorophyll to pheophytin must be and is determined spectrophotometrically. This allows some freedom in handling samples. Nevertheless, the method most suitable for frozen vegetable studies (14) is a tedious analysis and more suitable to a research laboratory than to the quality control laboratory of a frozen food processor.

Ascorbic acid is easily measured in the laboratory and reduction in ascorbic acid content has long been used in food processing research as a general indicator of oxidative food deterioration. In addition, ascorbic acid oxidation in stored frozen vegetables is important because it represents a loss of nutritional value. In the presence of oxygen, ascorbic acid is oxidized to dehydroascorbic acid (DHA) which is further oxidized to diketogulonic acid (DKA). DHA has residual antiscorbutic action so oxidation of ascorbic acid, measured chemically, does not represent a complete loss of the antiscorbutic value. Since the oxidative reactions are temperature sensitive they may be used to measure quality change related to time–temperature experience in handling of frozen vegetables. The original ascorbic acid content of vegetables is quite variable between lots so there is no original bench mark for comparison as there is with chlorophyll conversion. Ascorbic acid analysis of a control sample of the test lot is required for comparison. In strawberries, the original total of ascorbic acid, DHA, and DKA content is relatively constant (16) and this bench mark can be used in evaluating processing and storage deterioration. Dietrich *et al.* (13) found that this method is not applicable to frozen vegetables because the total concentration of ascorbic acid and its oxidation products decrease during storage of

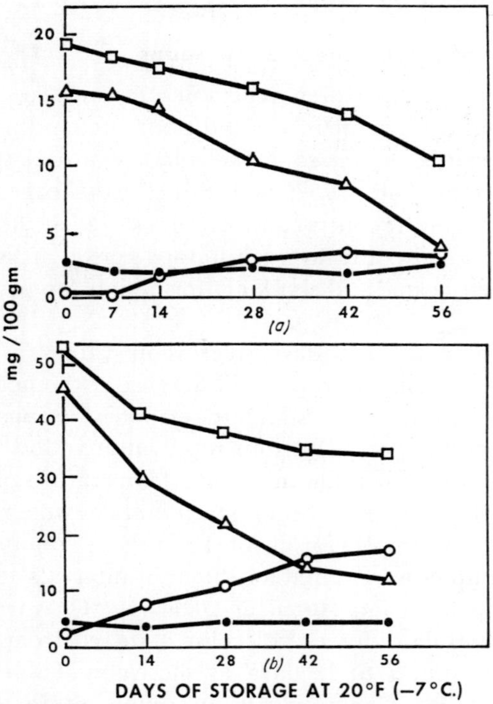

Fig. 5.1. Effect of storage at 20°F. (—7°C.) on (△) AsA, (●) DHA, (○) DKA, and (□) the sum of these components in (*a*) Laxton peas and (*b*) March cauliflower (1955) (13).

frozen peas and cauliflower (Fig. 5.1). Initial levels of ascorbic acid of green beans were too low to provide a reliable indication of change.

The color of frozen cauliflower changes with storage and the rate of color development is temperature dependent. The water:acetone-soluble brown pigments of cauliflower can be measured by a photocolorimeter or spectrophotometer as an indication of storage deterioration. Extractable color varied from lot to lot so comparative values based on a protected reference sample or an initial value obtained at time of processing would be necessary to ascertain storage deterioration.

The same generality holds for reflectance color measurements. The Hunter Color and Color Difference Meter and the Gardner Color and Color Difference Meter (HCM) have been used extensively in studying deterioration of frozen food color. Data vary considerably from lot to lot and no absolute correlations with color quality have been developed. However, the progressive changes in reflectance color parameters when compared with protected controls provide a useful description of quality change for given lots of frozen vegetables.

III. TEST RESULTS FOR SOME FROZEN VEGETABLES

A. T-TT Study of Frozen Green Snap Beans

Of the most popular frozen vegetables, green beans have relatively low stability under commercial conditions of handling. The rate of deterioration can be expressed as the number of days storage at a particular temperature before 75% of a trained panel concurs that a difference exists between the control and samples stored at the higher temperature. Table 5.1 shows this average storage period at five temperatures for frozen green beans, and similar information for peas, spinach, and cauliflower.

Data in Table 5.1 are days (regression values as logarithms of observed data) to a first perceptible difference at the noted storage temperatures. For green beans standard errors of estimate represent a scatter of individual points for four representative lots and are such as would occur with a variation in storage temperature of about plus or minus 4°F. (2°C.). For green beans, seven samples were ranked at a time for color and four samples for flavor. Each series contained a protected control and samples were stored for different intervals at a temperature. Fourteen green bean lots stored at 20°F. (−7°C.) varied in the time before a perceptible difference of color occurred from 3.5 to 14 days, and for flavor from 7 to 42 days, as shown in Table 5.2. Summary information on first perceptible flavor difference for green beans at other temperatures and for peas, cauliflower, and spinach is shown in Table 5.3.

For 10 lots of green beans, other than the four lots summarized in Table 5.1, storage at 20°F. (−7°C.) caused the first perceptible color and flavor differences to show up at averages of 6 and 27 days, respectively. The rapid rate of deterioration emphasizes the desirability of

Table 5.1. Days in Storage at Various Temperatures Required to Bring About a Perceptible Change in Quality [a]

	Beans		Peas		Cauliflower		Spinach	
°F.	Color	Flavor	Color	Flavor	Color	Flavor	Color	Flavor
0	101	296	202	305	58	291	350	150
10	28	94	48	90	18	61	70	60
15	15	53	23	49	10	28	35	30
20	8	30	11	27	6	13	20	20
25	4	17	5	14	3	6	7	8

[a] Source: green beans (1), peas (2), cauliflower (3), and spinach (4).

Table 5.2. Average Color and Flavor Ranks for Beans Stored at 20°F. (– 7°C.)

	Time, days							
Lot no.	0 [a]	3½	7	14	21	28	42	56
				Color				
3	1.1	—	2.5 * [b]	2.4	4.3	4.7	6.0	7.0
4	1.3	—	1.9 *	3.2	4.1	4.6	5.9	7.0
5	1.6	—	1.7	2.9 *	3.9	4.9	6.0	7.0
6	1.0	—	2.0 *	3.0	4.2	4.8	6.0	7.0
7	1.3	—	1.9 *	2.9	3.9	5.0	6.1	6.9
8	1.2	—	1.9 *	2.8	—	4.1	5.7	5.2
9	1.3	—	1.9 *	2.8	4.4	4.6	6.2	6.8
10	1.4	1.7	2.8 *	4.0	5.1	6.1	6.9	—
11	1.2	2.2 *	2.7	3.9	5.1	5.9	7.0	—
12	1.1	1.9 *	3.0	4.0	5.0	6.0	7.0	—
13	1.5	2.1	2.4 *	4.1	4.9	6.0	7.0	—
14	1.4	2.0	2.7 *	4.0	5.0	6.0	7.0	—
15	1.6	1.6	2.8 *	4.0	5.2	5.8	7.0	—
16	1.2	2.1 *	2.9	3.8	5.0	6.0	7.0	—
				Flavor				
3	1.6	—	—	—	—	2.7 *	2.8	2.9
4	1.6	—	—	—	—	2.2	3.0 *	3.2
5	1.4	—	—	2.1 *	—	3.0	3.5	—
6	1.9	—	—	2.0	—	2.5	3.6 *	—
7	1.5	—	2.1 *	2.9	—	3.5	—	—
8	1.6	—	2.2	2.8 *	—	3.3	—	—
9	1.5	—	2.2	2.8 *	—	3.4	—	—
10	2.1	—	—	2.3	—	2.7	2.9 *	—
11	2.0	—	—	2.2	—	2.8	3.0 *	—
12	1.8	—	—	2.0	—	2.9 *	3.3	—
13	1.7	—	—	2.7 *	—	2.8	2.9	—
14	2.0	—	—	2.1	—	2.8	3.1 *	—
15	1.6	—	—	2.0	—	2.9 *	3.6	—
16	2.0	—	—	2.4	—	2.4	3.2 *	—

[a] 0 days assigned to control sample stored at —20°F.
[b] Asterisks indicate first significant difference. Source: (1).

maintaining a low temperature in the storage and handling of green beans.

Extensive data on the conversion of chlorophyll to pheophytin indicate that the measurement of this change is a very useful way to provide objective measurement of stability of frozen green snap beans. When

Table 5.3. First Perceptible Flavor Difference. Arithmetic Mean and Coefficient of Variability

Vegetable	Storage temp., °F.[a]	Mean number of days to a flavor change	Number of lots	Coefficient of variability, %
Green beans	0	285	7	47
	10	150	4	40
	20	28	14	47
	30	7	4	51
Peas	0	265	5	35
	10	120	3	50
	20	30	13	36
	30	4	3	37
Spinach	0	271	4	36
	10	128	4	66
	20	35	4	74
	30	10	2	—
Cauliflower	0	353	5	4
	10	61	5	43
	20	17	7	38
	30	2	3	50

[a] 0°F., —18°C.; 10°F., —12°C.; 20°F., —7°C.; and 30°F., —1°C.

the chlorophyll conversion data from the same four lots of frozen beans used for the data in Table 5.1 were analyzed, the variability was equivalent to the average effects of a plus or minus 2°F. (1°C.) difference in storage temperature. The rates of chlorophyll deterioration for six lots of stored frozen green beans are plotted against temperature in the range of 0–40°F. (−18 to 4°C.) in Fig. 5.2 (S=change per day in logarithm of chlorophyll percentage retention).

Differences in stability of chlorophyll in frozen beans were substantially related to the degree to which unaltered chlorophyll had been preserved at the time of freezing and this in turn was related to the blanching procedure. All lots tested had been adequately blanched as measured by peroxidase inactivation. Those blanched at a relatively low temperature (below 200°F. (93°C.)) and for long periods of time (more than about 1.5 min.) suffered a greater conversion of chlorophyll to pheophytin prior to freezing than those that had been blanched at higher temperatures for shorter periods. Chlorophyll deteriorated more rapidly in subsequent storage of beans in which more chlorophyll had

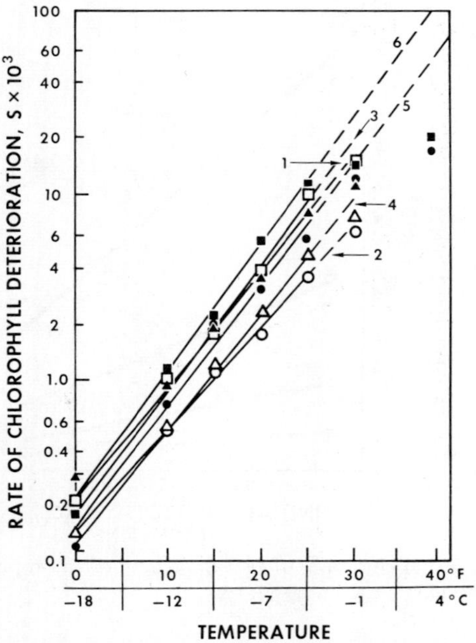

Fig. 5.2. Effect of storage temperature on the rate of chlorophyll deterioration. Points are observed values *S;* lines are drawn from regression values *S′* and plotted as a semilogarithmic graph (1). Lot 1 (▲); 2 (○); 3 (□); 4 (△); 5 (●); 6 (■).

been converted to pheophytin during processing than in those that had preserved more chlorophyll initially.

No substantial variation among the 19 lots of commercially packed green beans tested was directly attributed to variety, growing area, harvest year, or type of cut (regular or French style). There appeared to be no difference between water and steam blanching, although the most stable lots of the study, which had the highest retention of chlorophyll through the processing operation, happened to be water-blanched at near boiling temperature. By chance there were no steam-blanched samples among the 19 lots in which blanching conditions were comparable to the high-temperature, short-time water blanched samples.

A detailed comparison of six lots of frozen beans produced the data shown in Fig. 5.3. A consistent relationship is seen in the chlorophyll deterioration rate at four different temperatures. Green beans with an initial high retention of chlorophyll had a relatively low rate of deterioration during subsequent storage in the frozen state. At the higher storage temperatures the difference in deterioration rates among these lots became very substantial. Nineteen storage studies using four of these six lots and ten other lots, but not using any of the experimental

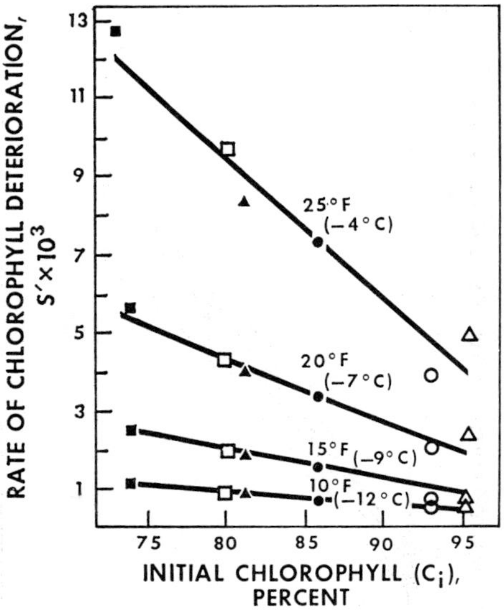

Fig. 5.3. Effect of initial chlorophyll level C_i on rate of chlorophyll deterioration (1). Lot 1 (▲); 2 (○); 3 (□); 4 (△); 5 (●); 6 (■).

data involved in establishing the relationship illustrated in Fig. 5.3, provided a good verification of the effect of initial chlorophyll retention on subsequent rate of chlorophyll deterioration in storage. Differences between observed chlorophyll retention and that calculated on the basis of the relationship shown in Fig. 5.3 averaged only 1.6%.

Based on the data of Fig. 5.3, an empirical equation was developed to predict the mean rate of chlorophyll conversion to pheophytin for frozen beans at temperatures between 0 and 25°F. (−18 and −4°C.) if the initial chlorophyll retention at the time of freezing is given:

$$S_c = (119.0 - 0.55\ T - C_i)\ e^{(0.173\ T - 12.21)}$$

where S_c = the change per day in the logarithm of the chlorophyll percentage retention, by this empirical relation, T = storage temperature °F., and C_i = initial percentage chlorophyll retention.

Laboratory tests (17) indicated that chlorophyll loss during blanching is linear with respect to time at a given temperature and that blanching temperature and degree of blanch affect chlorophyll retention in subsequent storage within certain limits of enzyme inactivation. In short, the key to better initial color and color stability in frozen green beans was found to be short-time, adequate blanching to peroxidase inactivation in the temperature range 200–212°F. (93–100°C.). Table 5.4 indicates that water blanching at 200°F. (93°C.) can inactivate peroxidase while retaining over 90% of the chlorophyll; at 212°F. (100°C.), nearly

Table 5.4. Chlorophyll Retention, %, in Water-Blanched Green Beans

Blanching temperature		Blanching time, min.[a]				
°F.	°C.	1	2	3	4	5
190	88	*	*	88.1	83.4	81.3
200	93	*	91.5	87.0	—	78.7
212	100	94.4	89.5	84.6	—	75.5

[a] Asterisk denotes inadequate blanch to inactivate peroxidase. Source: (17)

95%. Stability tests of these sample packs of frozen beans confirmed the finding that percentage chlorophyll loss during blanching was related to rate of subsequent loss in storage.

Subjective evaluations of color differences between individual samples (same lot, same storage temperature, different intervals of storage) were compared with differences in chlorophyll retention and other objective measurements. When about 4% difference in chlorophyll retention existed between green bean samples, the change in color could be detected by 85% or more of the panel. When the difference was 3.5% or less, concurrence of difference judgments ranged from 50 to 84%. The differences in objective measurements that correlate with the perceptible color difference for green beans, peas, spinach, and cauliflower are shown in Table 5.5.

In storage at a given temperature, a color change was detected before a flavor change. The chlorophyll difference that corresponded with the first perceptible flavor difference was about four times as large as that which produced the first perceptible color change.

Individuals with industry experience and responsibility for frozen food quality were assigned as members of a panel to judge the appearance of various samples from six of the frozen bean lots. The results were not highly consistent. Three of the six lots, representing three varieties, two growing areas, two packers, and three harvest years, had initial chlorophyll retention in the range of 80–85%. Independent judgments for these three lots indicated that quality was high initially but a majority of panel members would judge the product as no longer of the highest quality when storage conditions had resulted in a conversion of about 25% of the chlorophyll to pheophytin so that only 75% of the chlorophyll was retained. For all six lots, frozen green beans retaining less than 65% of the chlorophyll were judged by some industry people as likely to cause consumer complaint; when only 50–60% of the chlorophyll was retained a majority of the industry people

Table 5.5. Initial Values and Differences in Objective Measurements that Correlate with the First Perceptible Color Difference [a]

	Chlorophyll retention, %		HCM "a" value, units		Ascorbic acid, mg./100 g.	
	Initial	Difference	Initial	Difference	Initial	Difference
Green Beans 19 lots	83	3.5	14.2	0.5	8	0.7
Peas 14 lots	96	1.5	19.2	0.4	20	1.5
Spinach 6 lots	92.5	2.2	— [b]	— [b]	40	5.6
			HCM hue angle,°			
Cauliflower 10 lots	— [c]	— [c]	107	2.8	49	10.0

[a] Source: green beans (1), peas (2), spinach (3), and cauliflower (4).

[b] Correlation of first perceptible difference with change in HCM parameters was good only within lots.

[c] Not applicable for cauliflower.

believed that complaints would be likely to follow. Therefore, a conservative estimate of consumer complaint level would be drawn at about 50% chlorophyll retention. On the other hand, it is likely that few industry people would judge frozen beans to rate as high quality unless they retained more than 75% of the chlorophyll. Based on these data, we can say that substantially more chlorophyll conversion than is necessary just to make a perceptible color difference is required for the industry panel to judge that high initial quality of frozen green beans is lost.

Frozen green beans are generally low in reduced ascorbic acid content. Nineteen commercial lots ranged from 13.4 down to 3.8 mg./100 g. Ascorbic acid loss was generally linear with time, and the rate of deterioration increased exponentially with storage temperature in the temperature range between 0 and 25°F. (−18 and −4°C.). Analyses for reduced ascorbic acid were not sufficiently reproducible from package to package to provide as accurate a description of quality change during storage as the measurement of chlorophyll conversion.

Initial ascorbic acid content of peas was higher than that of green beans. Ascorbic acid loss in stored frozen peas better illustrates this point. Fig. 5.4 is the calculated regression of days (on a logarithmic scale) of storage on temperature corresponding to a 1.5 mg./100 g. change in ascorbic acid concentration.

Regression lines of HCM "a" values (on a logarithmic scale) on storage time were linear at a given temperature as a measurement of color deterioration of both beans and peas.

Differences in ascorbic acid content between samples and of HCM "a" values between samples showed the relationship of the objective measurement to the subjective evaluation (Table 5.5). However, differences were small relative to experimental error and to sample variability. These values were not found as useful as chlorophyll retention as objective measurements of quality change in stored green vegetables.

When frozen beans were subjected to cyclic fluctuations of storage temperatures, subjective evaluations of color and flavor deterioration indicated that rates of change were similar to those of samples stored at constant temperature at the arithmetic mean of the two temperature extremes. Objective measurements of chlorophyll deterioration indicated that deterioration rates were slightly higher than they would be at mean temperature and agreed well with theoretical considerations expressed by Schwimmer *et al.* (18). Table 5.6 compares theoretical with observed values for the effective constant storage temperatures for cyclic fluctuations. Observed data were the basis for interpolated values using the relationship shown in Fig. 5.3 for chlorophyll deterioration rates. Similarly, chlorophyll deterioration rates for frozen green beans at the extremes of cyclic temperature fluctuations were used in calculating the theoretical effective constant temperatures.

Table 5.6. Effective Constant Storage Temperatures, °F. Having Rates of Chlorophyll Deterioration Equivalent to Observed Values Under Cyclic Fluctuations

	Fluctuating temperatures, °F.									
	−5 to 5		−10 to 10		0–20		15–25		10–30	
Lot No.	A[a]	B[a]	A	B	A	B	A	B	A	B
1	—	—	5.8	4.4	—	—	21.7	22.5	21.2	22.7
2	—	—	4.9	2.9	—	—	21.3	23.0	20.3	22.2
3	1.2	—	5.4	4.8	15.7	15.5	22.5	22.4	22.9	23.9
4	1.5	—	6.2	5.1	14.8	13.8	23.0	20.1	22.4	20.1
5	0.1	1.7	—	—	14.4	14.8	—	—	—	—
6	2.5	2.5	—	—	12.5	12.2	—	—	—	—
Mean	1.3	2.1	5.6	4.3	14.4	14.1	22.1	22.0	21.7	22.0

[a] A: Interpolated value by use of Fig. 5.3; B: calculated value by the method of Schwimmer *et al.* (18). Source: (1).

When storage temperature was varied in patterns, deterioration was found to be a summation of time–temperature experiences by all of the criteria of quality used in the T-TT study. The order of presentation of such experiences was of negligible importance.

B. T-TT Study of Peas

Average storage times for six lots of frozen peas at which the first perceptible color and flavor differences were found are given in Table 5.1. The storage temperature range for these samples was 0–25°F. (−18 to −4°C.). At the storage temperature near the bottom of this range, which is appropriate for commercial frozen foods, about 50–60% more time was required to produce a perceptible difference in flavor than in color. At higher temperatures it took nearly twice as long to detect flavor difference as color difference. At −10°F. (−25°C.) no detectable flavor or color differences were observed after 2 years of storage.

Two objective methods were used to measure the change in color that coincides with the first perceptible difference in visual appearance. First, an average value for 14 lots was calculated for the change in objective values between a control and a stored sample at several temperatures for each lot in which a minimum perceptible visual color difference occurred. See Table 5.5 for average values. Second, an indirect relationship of visual panel evaluation with objective measurements was derived for several temperatures after the equivalent of 202 days' storage at 0°F. (−18°C.) for the six lots from which the subjective evaluation data in Table 5.1 were derived. By this second method a difference of 1.8% chlorophyll conversion, 0.4 HCM "a" value and 1.5 mg./100 g. ascorbic acid content occurred corresponding with the first perceptible difference in color. The close check of values by these two methods lends some dignity to them despite the inadequacy of the sample studied relative to the total commercial pack of frozen peas and the inherent variation in composition of different samples.

The amount of variation among the 14 lots of frozen peas was substantial. For example, the range of chlorophyll deterioration rates at 20°F. (−7°C.) was almost as large as the difference in rates for a temperature range of plus or minus 5°F. (3°C.) for a single typical lot. Eleven lots of frozen peas stored at 20°F. (−7°C.) ranged in chlorophyll deterioration rate from 0.032×10^{-3} to 1.034×10^{-3} (change per day of log chlorophyll percentage) with an average value of 0.464×10^{-3}. No specific causes were found for the variation in stability between lots. Unlike the investigation of green beans, no significant relationship of initial chlorophyll retention to rate of subsequent conversion of chlorophyll during storage was found. In 14 commercial lots of frozen peas, blanching conditions were all at relatively high temperatures for short time periods, ranging from 195 to 210°F. (91 to 99°C.) for 1.0–1.5 min. Initial chlorophyll retention ranged from 92.5 to 97.4%. Thus, the lots were relatively uniform in chlorophyll retention and were initially all of high quality.

Industry panels evaluated six lots of frozen peas stored at 20°F. (−7°C.) and indicated that most panelists would consider peas to be no longer of high quality when chlorophyll retention had decreased to about 85%. For half the lots, the loss of high quality was judged to have been reached at even higher levels of chlorophyll retention. The 85% chlorophyll retention represents a 7–10% decrease from the initial chlorophyll retention of high-quality frozen peas. Thus, 4–6 times the chlorophyll deterioration that would represent the first perceptible difference in color would be required to reach the industry panel judgment of lost high quality.

The 14 lots of frozen peas in the T-TT study had an average initial ascorbic acid content of 20 mg./100 g. At 0°F. (−18°C.) almost four years of storage were required to reduce ascorbic acid by 50%; at 10°F. (−12°C.), less than a year; and at 20°F., less than two months. (Comparable data for spinach: three years, one year, and four months; and for cauliflower: two years, six months, and less than two months.)

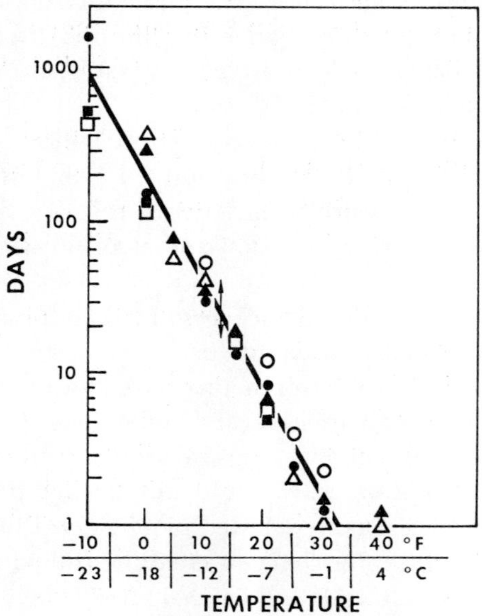

Fig. 5.4. Influence of temperature on time required to produce a 1.5 mg./100 g. change in ascorbic acid concentration in frozen peas (2). Lot 1 (○); 2 (●); 3 (△); 4 (▲); 5 (□); 6 (■).

Frozen peas in transparent polyethylene bags lost green color and chlorophyll more rapidly at 20°F. (−7°C.) than at 0°F. (−18°C.) and much more rapidly when the product was illuminated than in the dark (19).

Changing the order of occurrence of two different steady temperatures did not significantly affect the total deterioration of frozen pea samples. Fluctuation of temperature *per se* had little or no effect on deterioration of peas held under regular 24-hour sine wave cycles with temperature intervals of 10 and 20°F. (6 and 11°C.) magnitude in the range of −10 to 20°F. (−25 to −7°C.).

C. T-TT Study of Spinach

Different lots of spinach varied in stability, but no significant stability differences in frozen spinach were found specifically due to harvest year, variety, growing area, or blanching conditions. However, chlorophyll deterioration for chopped spinach at any given temperature was about twice that found for leaf spinach. Since the comparison was made with leaf and chopped spinach commercially packed from the same raw material, it is believed that greater physical manipulation of the chopped spinach created conditions whereby chlorophyll deteriorated at a faster rate in subsequent storage. Spinach converts chlorophyll to pheophytin about twice as fast as peas and a quarter as fast as beans in frozen storage. The smallest change in chlorophyll conversion detectable by trained panels is least in peas, greatest in beans, and intermediate in spinach. See Table 5.5.

A 50–75% loss in ascorbic acid occurred in spinach in one year when it was stored at 10°F. (−12°C.) but only 10% or less was lost at 0°F. (−18°C.). The temperature coefficients in the range of 0–25°F. (−18 to −4°C.) of chlorophyll conversion and ascorbic acid loss were about the same.

Extreme variability between packages of frozen spinach was observed, and only broad trends were valid for comparisons between lots and between spinach and other frozen products. Variable amounts of ice in packages, ratio of leaf to stem, and differences in concentration of ascorbic acid between leaf and stem are known factors contributing to variability. Deterioration rates could not be determined with good precision. Two of the lots were tested more rigorously than the others and data on flavor and color change were quite comparable. These data are plotted in Fig. 5.5 and summarized in Table 5.1. Unlike other vegetables in the T-TT study, spinach flavor change was perceptible sooner than color change.

The rate of chlorophyll conversion increased exponentially with

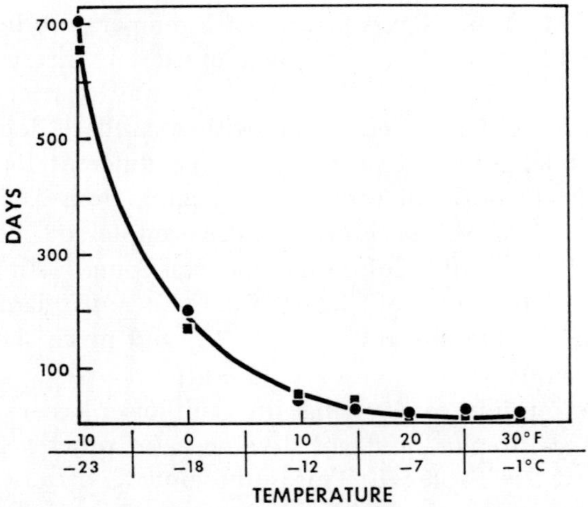

Fig. 5.5. Effect of temperature on stability of spinach flavor. Number of days to 75% concurrence of judgment there was a difference in flavor from that of samples protected by storage at —20°F. (—29°C.) (3). Lot 5 Viroflay (■); 6 Savoy (●).

temperature; a 10-fold increase in rate occurred with an increase of about 15°F. (8°C.) in storage temperature, *i. e.,* q_{10} was about 4.6. Rate of ascorbic acid loss also increased exponentially with temperature at about the same rate. Measurement of HCM color attributes did not change at a constant rate during periods tested. Within lots there was a high degree of correlation between percentage conversion of chlorophyll and logarithmic values of HCM "a," saturation index, and hue angle, but when data from separate lots were combined the correlation was lost.

Industry panel judgments of spinach appearance indicated that judgments were correlated with chlorophyll conversion and, less well, with HCM values for "a," hue angle, and saturation index. Considerable color quality deterioration ("consumer complaint" level) would be expected after eight months' storage at 10°F. (−12°C.) and after two months at 20°F. (−7°C.). On the basis of first perceptible difference testing, it is thought that somewhat shorter storage periods than these would result in customer dissatisfaction with the flavor of the spinach.

D. T-TT Study of Cauliflower

There are several appearance defects (discolorations) which may appear in cauliflower, including sunburn and bruising. The development of a tan color during storage in the frozen state appears to be a distinct type of discoloration and is specifically related to the conditions of storage. Like other measurements of instability of frozen vegetables, the rate of discoloration is exponential with temperature in the range of 0–25°F. (−18 to −4°C.). The discoloration is measurable in a water–acetone extract; in the T-TT studies it was expressed in terms of optical density of the filtered extract with a standard depth of solution (2 cm.) and light source (420 mμ). Nine different lots of cauliflower had initial optical density values ranging from 0.065 to 0.144 $E^{130\%}_{2cm., 420m\mu}$. Direct comparisons between samples of the same lot indicated that a perceptible color difference was found with changes in optical density in the range of 0.025–0.050. This much damage would occur in about two months at 0°F. (−18°C.) and much shorter times at higher temperatures.

The first perceptible color change in cauliflower occurs at shorter time intervals than color or flavor difference for other vegetables of the T-TT study. See Table 5.1. This is undoubtedly because it is much easier to judge between shades of a light color than of a dark color. Industry panel judgments considered samples to change from "high quality" to "acceptable but not high quality" at difference levels of objective measurements that would occur only after approximately two years of storage at 0°F. (−18°C.).

E. The Pennsylvania State Study of Frozen Food Stability

For frozen vegetables other than the four discussed above, published studies of stability have been more limited in scope. An example is the Pennsylvania State Agricultural Experiment Station's (PS) (20,21) investigation of factors affecting quality of frozen peas, lima beans, asparagus, green beans, broccoli, wax beans, spinach, corn, and cauliflower, as well as some fruit and meat. Experimental packs, mostly of foods grown under experimental conditions on the University farm, were used. Many, but not all, of the vegetables were of commercial varieties. Each product was tested for only one harvest year, except that peas and lima beans were studied for two successive years. Samples were frozen at two freezing rates and evaluated after storage at three temperatures, 10, 0, and −20°F. (−12, −18, and −31°C.). These tests provided a simulation of commercial processing. A comparison between the PS the T-TT findings is possible where the same vegetables were used and provides a basis for broader interpretation than possible without this comparison. Ascorbic acid loss and subjective evaluation of color and flavor are points in common.

In the PS study of frozen peas (20) two harvest years (1948 and 1949) and four blanching times at 200°F. (93°C.) were included. The Freezonian variety (not included in the T-TT study, which included only Thomas Laxton and Dark Seeded Perfection varieties) was used and was commercially grown, vined, and washed. In the PS 1948 pack the ascorbic acid loss was 21 and 10% in one year of storage at 10°F. (−12°C.) for 3– and 6–min. blanching; this loss was extremely low compared to that found in either the T-TT study or the PS 1949 pack of peas. In the latter, 53 and 57% losses were measured for 2– and 4–min. blanched frozen peas under these storage conditions; this represents close agreement with the average 50% loss in 10 months found in the T-TT studies. The low values for ascorbic acid loss in the PS 1948 samples cannot be readily explained. They weaken the significance of other findings of that year's experimentation.

The subjective evaluation in the PS study (20) was made by palatability scoring over a 1–5 range (the T–TT study used difference testing.) Scores were summarized but no data on statistical significance were given. The authors state that eight months' storage at 10°F. (−12°C.) caused a deterioration relative to other samples and that deterioration was recorded, but at a slower rate, for storage at 0°F. (−18°C.). In the T–TT studies, the first perceptible color difference in frozen peas stored at 10°F. (−12°C.) was in 1.5 months; flavor difference in three months. We conclude that the PS study used appraisal methods that would not detect differences as small as were observed in the T-TT study.

In the PS study (21) ascorbic acid loss in stored frozen green beans was about 55% in four months at 10°F. (−12°C.) and 25% in 12 months at 0°F. (−18°C.). These are comparable to the T-TT study in which 50% of the ascorbic acid was lost in four months at 10°F. (−12°C.) or 16 months at 0°F. (−18°C.).

Flavor and color change of green beans, spinach, and cauliflower (21) were detected in four months at 10°F. (−12°C.). Cauliflower and spinach suffered greater reduction in scores than the other vegetables tested. In the T-TT study the first perceptible differences for these three vegetables were observed earlier than four months. In all vegetables tested, the scoring method used at PS indicated longer storage time required for a significant score change than was required for the first perceptible difference in the T-TT study.

The PS study (20) also showed apparent changes in tenderness of asparagus, broccoli, cauliflower, and corn after 12 months' storage at 10°F. (−12°C.). No equivalent change was observed for green beans, spinach, wax beans (21) or green peas (20), although scores for tenderness of green and wax beans were lowered slightly by 12 months' storage at 10°F. (−12°C.). Preliminary evaluations in the T-TT study did not uncover significant changes in tenderness of stored peas or beans.

The PS study (20,21) summarized stability of frozen vegetables by indicating that four months' storage at 10°F. (−12°C.) would cause a 50% loss of ascorbic acid and lowering of palatability scores. Of the nine vegetables studied, none were less stable than cauliflower and spinach. These vegetables were the least stable of the four studied in the T-TT series. The PS results indicated neither blanching conditions nor freezing rates had significant effects on stability of frozen vegetables. The T-TT studies, on the other hand, found significant effects of method of blanching on the stability of frozen green beans.

F. Studies of Other Vegetables

Frozen corn-on-th-cob and Brussels sprouts cannot be easily classified as to stability. As in most other vegetables, their stability is closely related to blanching method and adequacy of blanch. The large pieces and variable heat penetration of these two vegetables make it difficult to inactivate destructive enzymes without overheating the product. When properly blanched, these vegetables appear to be at least as stable as the others discussed above. Dietrich and Neumann (22) minimized destructive aspects of blanching for Brussels sprouts by preheating them to about 130°F. (54°C.) before steam or water blanching at near boiling temperature. This reduced the required time of exposure to the high temperatures by 15–25% and reduced conversion of chlorophyll during blanching. Water blanching yielded products that were more stable against ascorbic acid loss and chlorophyll conversion than those blanched in steam under comparable time and temperature conditions.

Woodroof and Shelor (23) indicated a continual loss of natural appearance, color, aroma, and flavor in asparagus, green snap beans, corn, and some fruits when held in storage for one year or more. They found that the deterioration rate was directly related to storage temperature. All the commodities except green snap beans were found to lose firmness and to increase in defrosting leakage with storage. They found that 0°F. (−18°C.) is an acceptable storage temperature for one year and believed that temperatures above that are unsatisfactory.

Frozen French-fried potatoes are considered by industry spokesmen to be very stable. A recent review on frozen potatoes indicates that potatoes are relatively more stable than most other frozen vegetables when held in storage at 0°F. (−18°C.) (24).

IV. SIGNIFICANCE OF QUALITY INSTABILITY

The T-TT study indicates that the first perceptible loss of initial quality of many frozen vegetables occurs in less than a year's storage at 0°F. (−18°C.). The exponential temperature effect on deterioration rates means that a few degrees difference in temperature for long-term storage can cause a great difference in quality retention. On the other hand, since time is also an important factor, it follows that short exposures to temperatures above 0°F. (−18°C.) are no more damaging than long exposures at lower temperatures. The T-TT studies have provided rate information about product deterioration that should be useful in determining the time and temperature conditions necessary to protect the initial quality of frozen vegetables.

If first perceptible difference in quality of green beans is noted after 100 days' storage at 0°F. (−18°C.) and industry judgment that the initial high quality is two or three times as long, it may be of value to use a lower temperature than 0°F. (−18°C.) for long-term storage of green beans between harvest seasons. Storage at lower temperature would reduce color loss in the part of the pack that would remain unsold at the end of the year. On the other hand, if the temperature of green beans were to rise to 10°F. (−12°C.) during delivery and be brought back to 0°F. (−18°C.) within a day, only about a 28th of the change that would be perceptible would occur, corresponding to about four days longer storage at 0°F. (−18°C.). This small loss, in itself, might not warrant replacement of delivery trucks with new, better equipped trucks. However, every bit of deterioration is additive and a rise in temperature during delivery, failure to reduce temperature rapidly after delivery, and carelessness in holding low temperature during marketing could add up to a significantly deteriorated product at time of purchase. It is conservative to assume that frozen foods in the hands of consumers are less well cared for than in the marketing cold chain. Any deterioration that precedes the sale of frozen foods reduces the reserve against consumer mishandling. Assurance of high-quality frozen foods reaching the dinner plate depends upon maintaining to the maximum the initial quality in all commercial handling.

Some differences exist between the stabilities of various frozen vegetables. For long-term storage of cauliflower and green beans, and perhaps spinach, a temperature lower than 0°F. (−18°C.) is probably warranted. However, it would be inefficient to distribute and market frozen vegetables at different specified temperatures. The commercially accepted 0°F. (−18°C.) seems appropriate for orderly delivery and market distribution as long as quick delivery and rapid turnover in retail markets is maintained.

Experimental judgments of frozen food stability obtained from industry panels were useful in showing that the first perceptible difference measured by trained panels would probably not represent a commercially detectable loss of initial high quality.

Variability between commercial packs of frozen vegetables is great. If causes and effects can be found, there will be much hope that initial quality of frozen vegetables can be better preserved under existing conditions of the market cold chain. Knowledge obtained in the T-TT study of the effects of blanching conditions on frozen green bean stability undoubtedly has led to improved commercial practice and a more stable product. If we knew the cause of the variability in stability of pea and spinach lots that was exhibited in the T-TT study, perhaps equivalent improvement in stability of these products could be developed. Hopefully, the mere reporting of the wide variability that exists should lead to improvement in these products as observant processors direct their attention to the problem.

The publication indicating changes in firmness of frozen foods during storage (20,21,23) emphasize the need for further evaluation of frozen food texture stability. Improved and vastly different methods for extremely rapid freezing (such as liquid nitrogen, fluidized bed, and liquid immersion freezing) now are becoming commercially feasible. It is important to consider how these methods can improve initial textural quality of frozen vegetables and how stable such quality improvements can be in the subsequent storage and handling of products.

REFERENCES

(1) W. C. Dietrich, M-D. Nutting, R. L. Olson, F. E. Lindquist, M. M. Boggs, G. S. Bohart, H. J. Neumann, and H. J. Morris, *Food Technol., 13,* 136 (1959).
(2) M. M. Boggs, W. C. Dietrich, M-D. Nutting, R. L. Olson, F. E. Lindquist, G. S. Bohart, H. J. Neumann, and H. J. Morris, *Food Technol., 14,* 181 (1960).
(3) W. C. Dietrich, M. M. Boggs, M-D. Nutting, and N. E. Weinstein, *Food Technol., 14,* 522 (1960).
(4) W. C. Dietrich, M-D. F. Nutting, M. M. Boggs, and N. E. Weinstein, *Food Technol., 16,* No. 10, 123 (1962).
(5) M. A. Joslyn, *Econ. Botany, 15,* 347 (1961).
(6) M. A. Joslyn, "The Freezing of Fruits and Vegetables," in *Cryobiology,* H. T. Meryman, Ed., Academic Press, London, 1966.
(7) R. L. Olson and W. C. Dietrich, "Vegetables: Characteristics and the Stability of the Frozen Product," in *The Freezing Preservation of Foods,* 4th ed., Vol. 2, D. K. Tressler, W. B. Van Arsdel, and M. J. Copley, Eds., Avi Publishing, Westport, Connecticut, 1968, ch. 4.
(8) S. Schwimmer, *Canner/Packer, 133,* No. 7, 48A–48B (1964).
(9) E. R. Wolford and M. S. Brown, *Food Technol., 19,* No. 7, 109 (1965).
(10) M. S. Brown, *J. Sci. Food Agr., 18,* 77 (1967).
(11) H. Campbell, *Food Res., 2,* 55 (1937).
(12) G. Mackinney and C. A. Weast, *Ind. Eng. Chem., 32,* 392 (1940).
(13) W. C. Dietrich, F. E. Lindquist, J. C. Miers, G. S. Bohart, H. J. Neumann, and W. F. Talburt, *Food Technol., 11,* 109 (1957).
(14) W. C. Dietrich, *Food Technol., 12,* 428 (1958).
(15) H. J. Dutton, G. F. Bailey, and E. Kohake, *Ind. Eng. Chem., 35,* 1173 (1943).

(16) D. G. Guadagni and S. H. Kelly, *Food Technol., 12,* 645 (1958).
(17) W. C. Dietrich, R. L. Olson, M-D. Nutting, H. J. Neumann, and M. M. Boggs, *Food Technol., 13,* 258 (1959).
(18) S. Schwimmer, L. L. Ingraham, and H. M. Hughes, *Ind. Eng. Chem., 47,* 1149 (1955).
(19) A. D. Shepherd, *Food Technol., 13,* 539 (1959).
(20) N. B. Guerrant, E. R. Hartzler, J. E. Nicholas, J. S. Perry, J. G. Garey, J. F. Murphy, M. L. Dodds, and G. Bennett, Pa. Agr. Exp. Sta Bull., *565* (1953).
(21) G. Bennett, J. F. Cone, M. L. Dodds, J. C. Garey, N. B. Guerrant, J. G. Heck, J. F. Murphy, J. E. Nicholas, J. S. Perry, R. T. Pierce, and M. D. Shaw, Pa. Agr. Exp. Sta. Bull., *580* (1954).
(22) W. C. Dietrich and H. J. Neumann, *Food Technol., 19,* No. 7, 150 (1965).
(23) J. G. Woodroof and E. Shelor, *Refrig. Eng., 56,* 514, 550, 552 (1948).
(24) I. C. Feustel and R. W. Kueneman, "Frozen French Fries and Other Frozen Potato Products," in *Potato Processing,* 2nd ed., W. F. Talburt and O. Smith, Eds., Avi Publishing, Westport, Connecticut, 1967, p. 340.

Chapter 6

STABILITY OF FROZEN POULTRY MEAT AND EGGS

L. E. Dawson

Food Science Department, Michigan State University, East Lansing, Michigan

I. INTRODUCTION

Poultry meat products have traditionally been sold primarily as whole birds, either fresh ice packed or fresh frozen. In recent years, however, much interest has been exhibited in producing and merchandising an increasing volume of cooked meat and combination poultry food products. Fig. 6.1 shows this rapid increase in volume of products certified for canning or further processing by the USDA in official plants. Since all products entering into interstate commerce must be produced in official plants, these data represent a majority of poultry food products so produced. During the ten-year period ending in 1966, turkey meat and young chicken meat in such products increased by over 400%, and total poultry meat used for these purposes increased 330%.

The production of frozen eggs, on the other hand, remained relatively constant, as shown in Fig. 6.2. This undoubtedly reflects a lack of product development for frozen eggs or an increase in the substitution of egg solids for frozen eggs. One major deterrent to greater utilization of frozen and dried eggs has been the persistent *Salmonella* problem. Only recently have developments in pasteurizing procedures been effective for egg albumen (Cunningham and Lineweaver (1)) so that greater assurance of quality can now be given.

The significant increase in utilization of frozen cooked poultry has

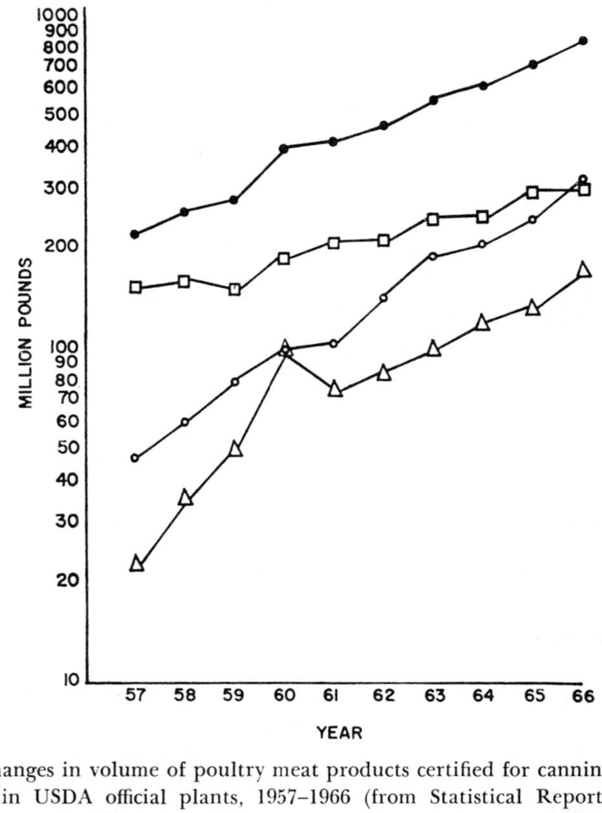

Fig. 6.1. Changes in volume of poultry meat products certified for canning or further processing in USDA official plants, 1957–1966 (from Statistical Reporting Service, USDA). (●) Total; (□) mature chickens; (○) turkeys; (△) young chickens.

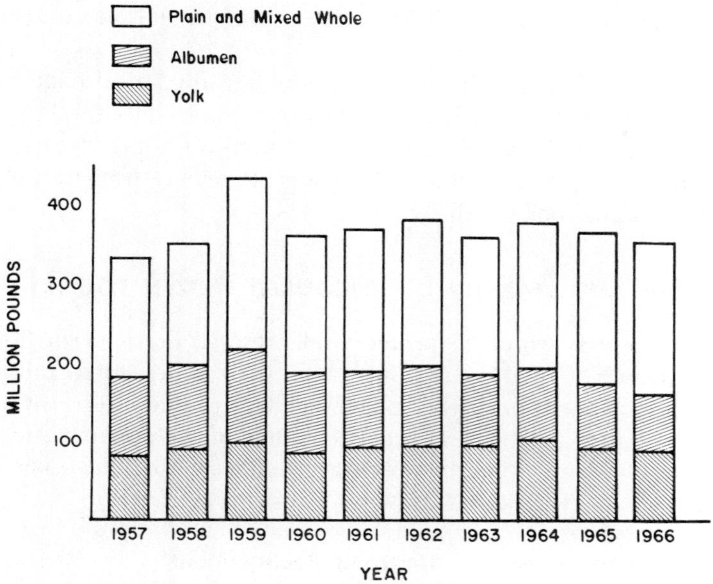

Fig. 6.2. Changes in volume of frozen egg products produced in the U.S., 1957–1966 (from Statistical Reporting Service, USDA).

been possible only because of the technological developments and improvements in initial quality of such meat and the increased stability of the resulting food products. Initial successes have stimulated increased research in product development, so that today numerous firms throughout the U.S. are producing and merchandising an increasing number of products.

Ready-to-cook poultry is now the predominant form in our markets. Although a major portion of ready-to-cook poultry is still sold as whole birds, a significant and increasing percentage is being merchandised as cutup pieces. In 1966, the volume of cutup poultry processed in federally inspected plants was 1.1 billion, 7.5 million, and 121.3 million pounds for young chickens, mature chickens, and turkeys, respectively.

Although young chickens (fryers) are now produced on a fairly uniform year-around basis, turkeys and mature hens are still processed seasonally. Fluctuations in production volumes, seasonally or otherwise, efficiencies in marketing procedures, increased offerings to consumers, and increased quality and stability of products have resulted in an increase in the production of frozen ready-to-cook whole or cutup products. Major contributions have also resulted from more effective packaging materials and methods, and closer attention to freezing methods and storage conditions.

Microbial problems, although affected by freezing and storage conditions, will not be stressed here (see Chapter 3). In this respect, scrupulous care should be exercised in the sanitation practices during processing and preparation, and during thawing, preparation for cooking, and postcook handling.

II. QUALITY AND STABILITY OF UNCOOKED FROZEN POULTRY

Poultry meat is frozen to protect and preserve its inherent "fresh" flavor and other quality attributes. As with other food products, special precautions must be taken to freeze and store properly packaged poultry products. However, poultry meats present a number of unique problems, some of which are not recognized in other food products.

A. Consumer Preference for Unfrozen Poultry Meat

Consumers appear to prefer fresh fryers over frozen fryers, although they frequently buy in quantity lots for later home freezing. Over three-fourths of the consumers in a three-state survey preferred fresh fryers, with better flavor given as the leading reason (Brant *et al.* (2)). Fresh fryers were also preferred to cutup quick-frozen fryers by a majority of 1500 tasters (3 of 5 preferred the fresh) after the frozen birds had been held at 0°F. (−18°C.) for 9 months; loss of chicken *se* or to slow freezing rates. Snyder *et al.* (4) found no flavor difference (Mountney *et al.* (3)).

Flavor or loss of flavor has frequently been attributed to freezing *per se* or to slow freezing rates. Snyder *et al.* (4) found no flavor difference between hens frozen fast or slow and stored 13 months, and Stewart *et al.* (5) found no palatability differences in broilers frozen at different rates.

B. Color—Preference and Influences

Light-colored poultry meat is preferred by a majority of consumers. This light color is favorably associated with degree of finish, and is influenced by scalding temperatures and rate of freezing. The appearance of frozen poultry may also change sufficiently during storage to affect consumer preference. MacNeil (6) reported that brine-frozen (−43°F., −42°C.) turkey products decreased in preference scores (appearance only) through six months of storage, whereas those frozen in air blast at −40°F. (−40°C.) were preferred to those frozen at 0°F.,

(−18°C.), −20°F. (−29°C.), and in brine. The products frozen in brine became abnormally white and chalky in appearance after extended storage at 0°F. (−18°C.).

When the epidermal layer of skin (cuticle) is removed during feather removal, the carcasses appear darkened after freezing, apparently because of increased transparency of the frozen skin (Klose and Pool (7)). Slowly frozen carcasses exhibit a darkened appearance and rapidly frozen ones a light appearance. The discoloration involved in slow freezing has been reported to take place in the flesh rather than the skin. Baker (8, 9) froze flesh and skin at −20°F. (−29°C.) and both were white, while after freezing at 20°F. (−7°C.) the flesh was dark while the skin was light. Appreciable and easily detectable optical changes may, however, take place in both the skin and in the surface layer of the flesh (7).

While temperature is the main factor influencing rate of freezing and color of resulting product (8), the movement of the surrounding atmosphere is also very important. Lightness and uniformity of appearance in air-blast frozen poultry increase markedly when the air velocity at −20°F. (−29°C.) increases from 0 to 700 ft./min. (van den Berg and Lentz (10); Lentz and van den Berg (11)).

Packaged, ready-to-cook, 140°F. (60°C.) scalded, Broad Breasted Bronze tom turkeys of two ages (13–14 weeks and 6 months) were frozen in a blast tunnel under conditions varying with respect to prechilling, blast-air temperature and velocity, and presence of insulative boxing. For equivalent degrees of finish (amount of fat in skin layer) faster rates of freezing result in a lighter frozen appearance (7). Optimum appearance is most critically affected by the degree of finish and by freezing in a box versus the open-air blast. Moderate effects result from a variation in air temperature and air velocity. Close attention to these factors is essential for production of optimum frozen appearance in poultry processed by current practices. No appreciable difference in frozen appearance has been observed between ice-chilled turkeys and those frozen immediately from the warm eviscerated state.

C. Bone Darkening Defects

Bone discoloration in frozen poultry, "bone darkening," is a condition induced by freezing and thawing chicken fryers. It seldom occurs in other classes of chickens (especially old chickens) and has not been reported in turkeys, ducks, or geese. After freezing and thawing fryers, a bloody discoloration occurs around the bones in the vicinity of the knee joint, leg, thigh, back and second joint of the wing. (Brant and Stewart (12)). Cooking causes the color to change from blood red to deep grey, brown, or in severe cases, black. Red hemoglobin is oxidized to brown methemoglobin (Koonz and Ramsbottom (13)). Although this change does not affect aroma, flavor, or texture of the cooked meat (Ellis and Woodruff (14)), it becomes a serious problem due to the undesirable appearance.

A number of methods have been investigated for minimizing or eliminating this dark discoloration. Among the most promising prefreezing treatments is microwave heat for 2–3 min. (Essary (15); Ellis and Woodruff (14)). The latter authors also suggested that precooking before freezing would minimize bone darkening. Cooking by steam heat (10 lb. for 5–10 min.), steam at 210°F. for 15–20 min., radiant heat at 350°F. (177°C.) for 25–30 min., or deep fat frying at 350°F. (177°C.) for 15–20 min. before freezing would minimize this undesirable discoloration.

D. Fat Stability

It is well known that turkey fat is much less stable than chicken fat. This difference in stability assumes economic importance to the extent that it is reflected in the seriously shorter frozen storage life of turkeys, and in the relative unacceptability in foods of rendered turkey fat as compared with chicken fat.

1. Diet Influences Fat Composition

Of the three recognized chemical factors governing the resistance of natural fats to rancidification, *i.e.*, content of fatty acids, their differing degrees of unsaturation, and the content of natural antioxidants, fatty acids have received more attention than antioxidants in most of the previous studies on poultry fat. Studies by Hilditch *et al.* (16), Cruickshank (17), Kummerow *et al.* (18), Chu and Kummerow (19), and Klose *et al.* (20,21) have shown that the fatty acid composition of the depot fat in poultry, and its corresponding stability are directly related to the amount and fatty acid composition of the ingested fat. Nutter *et al.* (22) compared certain chemical and physical constants of depot fats from chickens and turkeys fed unspecified, and presumably different, commercial diets, and found no differences that were consistent enough or large enough to explain the great difference in stability between commercial chicken fat and turkey fat. They suggested the possibility that the inferior stability of turkey fat is due to a lower content of natural antioxidants, such as tocopherols. Criddle and Morgan (23,24) fed tocopherols to turkeys at dietary levels of 0.01, 0.1, 1, or 2% for 2.5–35 days before slaughter and found corresponding measurable increases in tocopherol content of the tissues and storage life of the frozen carcasses.

When chickens and turkeys were held on a natural dietary level of tocopherol a greater efficiency of deposition of tocopherol was found in the carcass fat in the chicken than in the turkey, and greater fat stability. A dietary supplement of 0.1% tocopherol for periods of five and ten weeks increased the tocopherol content and stability of carcass fat in both species, but the superiority of chicken over turkey was maintained or enhanced. The stabilizing effect of tocopherol supplements was confirmed by chemical and organoleptic tests on carcasses stored at +10°F. (−12°C.) for nine months. It is concluded that tocopherols play a major role in establishing the stability of poultry fat (Mecchi *et al.* (25)).

Chickens and Beltsville White turkeys grown on the same diet until they were 10, 17, and 24 weeks of age were analyzed for fatty acid composition, tocopherol content, and stability of their crude fat (Mecchi *et al.* (26)). Analyses of the depot fats confirmed that tocopherol deposition and stability were greater in the chicken fats than in the turkey fats. One- and two-week supplementation of the diet with tocopherol increased content and stability to a marked extent in chickens but only slightly in turkeys. Fatty acid composition of the depot fats was essentially the same for chickens and turkeys. Crude fat extracts of the muscle tissue showed good agreement in fatty acid composition between chickens and turkeys, but very little evidence of the consistent relation between tocopherol content and stability that was found for the depot fat.

Since the fatty acid composition of the carcass fat is a reflection of the fatty acid composition of the dietary fat, it is evident that the type of dietary fat can affect the storage life and eating quality of the carcass (Klose *et al.* (27)). Differences in stability were obtained at dietary fat levels in the range of those usually found in commercial rations, but the effects were not marked unless fats were present that contained appreciable amounts of highly unsaturated fatty acids.

There appear to be other factors in addition to contents of crude tocopherol and the commonly occurring fatty acids that contribute to differences in stability of chicken and turkey fat. Typically fishy flavors and odors, for instance, in roasted turkey meat, which can be caused by feeding fish (oil) products, can also be produced in the absence of fish products by a highly unsaturated vegetable oil, linseed oil. The fishy flavor is present in the roasted, freshly slaughtered turkey and apparently increases very little if any in intensity during storage (20).

2. *Antioxidants Affect Stability*

The extent of protection provided by antioxidants applied as solutions or suspensions in various carriers to the natural and cut surfaces of eviscerated turkeys in frozen storage has been reported by Klose *et al.* (27,28). Peroxide development in the fat was well correlated with subjective scores for rancid off-flavor in the cooked turkey. Butylated hydroxyanisole, propyl gallate, and combinations of these antioxidants markedly reduced fat deterioration in frozen storage. Of the solvents or carriers used, a 3% aqueous gelatin solution, in which the antioxidants were suspended at about 0.15% total concentration, was found to be most effective. The 3% gelatin coating alone provided some protection, but not nearly so much as the gelatin coating containing suspended antioxidants. Prooxidant effects produced by some samples of solvents suggest that care should be exercised in the selection of carriers for the antioxidants. The results indicate that coatings containing antioxidants offer considerable promise for increasing the frozen storage life of turkeys, especially when large surfaces are exposed, as in eviscerated and cutup poultry.

Use of proper antioxidants as a spray, dip, or injection will decrease the development of undesirable flavors or odors related to rancidity. Nordihydroguaiaretic acid (NDGA) as a spray on turkeys will inhibit oxidation, and the rate depends on treatment and storage time (Pool *et al.* (29)). NDGA in propylene glycol reduced peroxide formation and off odors, but did not affect the flavor of cooked meat.

3. *Quality Related to Time–Temperature Variations*

Poultry products are subjected to a variety of environmental conditions during processing, freezing, storage, and distribution. Knowledge of the effects of each environmental variable on quality changes is essential for the production of optimum quality products and for maintaining that quality throughout marketing channels.

Uncooked chicken roasters, ducklings, and turkeys can be successfully stored for at least a year without serious objectionable changes. When these products were packaged in conventional plastic bags, or in a hot melt resin for 12 months, the peroxide number of the fats was found to increase; however, aroma, flavor, and acceptability of the cooked meat did not decline during a normal storage time (Dawson *et al.* (30)).

It has been assumed that poultry products with high fat contents (ducks and geese) would undergo excessive deterioration during frozen storage, especially rapid development of rancidity. Well packaged, ready-to-cook ducks and geese, however, when stored at 0°F. (−18°C.) or below, will maintain their original high quality for 8 to 12 months (Klose *et al.* (31)).

During frozen storage studies of cutup fryers and cutup fowl, slight off odor in thawed raw meat is the first quality defect to be detectable, followed by off-flavors in cooked products (Klose *et al.* (32)). Chickens packaged under different conditions had a high-quality storage life of at least 6 months at 0°F. (−18°C.). However, a large difference in stability between two commercial packs of fryers was a reflection of the large difference in package permeability to oxygen. Chickens in the more permeable package had less than one-third of the storage life of chickens in the less permeable package. Stewing hens held in a package of permeability intermediate between those of the two fryer packages had a corresponding intermediate storage life. Peroxide development in fat of stewing hens was correlated with the level of organoleptic deterioration (Tables 6.1 and 6.2).

A storage test of cut-up fryers in tin cans with controlled atmospheres was used to establish independently the deteriorative effects of oxygen (air) availability and moisture loss. Moisture loss in itself was found to have little effect on organoleptic deterioration or chemical changes involving oxygen consumption. Access to oxygen of the air, on the other hand, contributed as much to rate of deterioration as a substantial rise in temperature. Appreciable deterioration and the presence

Table 6.1. Comparative Deterioration in Poorly Packaged Fryers Held at Various Temperatures (32)

	Months to develop significantly[a] more deterioration at higher (or fluctuating) temperature					
Quality evaluated by panel[b]	+20 vs. −30°F.	+10 vs. −30°F.	0 vs. −30°F.	−10 vs. −30°F.	0 to +20 fluctuation vs. +10°F.	−10 to +10 fluctuation vs. 0°F.
Odor of raw meat	<1	3	6	9	3	9
Flavor of cooked meat	3	6	>12	27	>12	18
Flavor of cooked liver	6	>12	>18	>27	>12	>18

[a] Data represent times in months at which the difference between higher (or fluctuating) temperature and control was significant at a probability level of about 1%.

[b] Eight trained panel members ranked four replications of each comparison at each storage period.

Table 6.2. Comparative Deterioration in Packaged Cutup Stewing Hens Held at Various Temperatures (32)

	Months to develop significantly[a] more deterioration at higher (or fluctuating) temperature					
Quality evaluated by panel[b]	+20 vs. −10°F.	+10 vs. −10°F.	0 vs. −30°F.	−10 vs. −30°F.	0 to +20 fluctuation vs. +10°F.	−10 to +10 fluctuation vs. 0°F.
Odor of raw meat	3	6	9	18	9	>18
Flavor of cooked meat	4	7	>12	>18	12	>18
Flavor of cooked skin	4	7	12	>18	9	>18
Odor of cooked broth	5	6	12	>18	>12	>18

[a] Data represent times in months at which the difference between the two indicated temperature conditions was significant at a probability level of about 1%.

[b] Six to 7 trained panel members ranked 4 replications of each comparison at each storage period.

of carbon dioxide in nitrogen packs indicated that anaerobic reactions are also involved.

Tests at +20, +10, 0, −10, and −30°F. (−7, −12, −18, −23, and −34°C.) provided evidence to support a recommendation for storage at 0°F (−18°C.) or below. Storage life at +20°F. (−7°C.) ranged from less than one month to six months for the various packaging conditions, at +10°F. (−12°C.) from 3 to 10 months, while at 0°F. (−18°C.), storage life was at least six months in the poorest package and about two years in the best package. Within reasonable storage periods, a series of temperature fluctuations had practically the same (only slightly greater) deteriorative effect on quality as a steady temperature equal to the arithmetic mean of all of the fluctuations.

Quality changes in ready-to-cook turkeys over a storage period up to 18 months were reported by Klose *et al.* (33). Experimental temperature conditions included constant temperatures of −30, −10, 0, and +10°F. (−34, −23, −18, and −12°C.) and a temperature that fluctuated from −10 to +10°F. (−23 to −12°C.) in 24 hr. Evaluation of the roasted carcasses by a trained taste panel (Table 6.3) revealed small but consistent increases in rancidity with increase in the constant storage temperature. Results of the organoleptic tests were for the most part consistent with the peroxide development. Values for the fluctuating (−10 to +10°F.) condition in most cases fell near the value for a constant 0°F. (−18°C.) condition. From a practical standpoint, however, it appears that the effect of the fluctuating temperature condition on rancidification does not differ in any important way from the effect of a constant temperature equal to the mean of the fluctuations.

The fluctuation of −10 to +10°F. (−23 to −12°C.) every 24 hr is, of course, much larger and more frequent than would be expected in normal commercial operations. This fact gives greater weight to the conclusion that a fluctuating temperature in most respects is no more harmful than a constant temperature equal to the mean of the fluctuations. It must be emphasized, however, that this conclusion for the present should be considered valid only for the temperature range studied, and it ignores undoubted effects of temperature fluctuations

Table 6.3. Influence of Temperature Condition on Rancidity Development in Roasted Turkeys Stored for 12 Months (33)

Packaging	Storage temperature, °F.	Average rank [a] (1=least rancid; 4=most rancid) Light meat	Dark meat	Skin
Wax paper	−10	2.1	2.3	1.8
	0	2.4 *	2.4	2.6 * **
	+10	3.0	2.8	3.3
	−10 to +10	2.4	2.4	2.2 **
Polyethylene	−10	2.4	2.4	1.9
	0	2.2	2.5	2.6
	+10	2.9	2.7	2.8
	−10 to +10	2.4	2.4	2.7
Polyethylene	−30	2.0	2.1	1.7
	−10	2.1 * *	2.4	2.1 **
	+10	3.1	2.9	3.7 **
	−10 to +10	2.8*	2.5	2.5 **

[a] Values represent means of ranks assigned to roasted birds by six or seven judges. There were six replications of each tasting for the wax paper groups and four replications for all other groups. All statistically significant differences are indicated by an asterisk (5% level) or double asterisk (1% level of probability).

and gradients on frost formation within the package. In somewhat higher temperature ranges, where partial thawing and microbial growth become important factors, additional work is needed to establish the relative importance of temperature fluctuations.

The results emphasize that statements on storage life, to have real value, must accurately define the intimate product environment as governed by packaging as well as by temperature conditions.

III. STABILITY OF FROZEN COOKED POULTRY MEAT

A. Frozen Fried Chicken

The quality of fried chicken is optimum immediately after cooking. With an increase in demand for more convenience foods, however, the practice of precooking and freezing fryers for distribution has increased. Changes in quality have been attributed to freezing and holding environment, to initial source of the fryers, and to precooking treatments. Cutup and fried chickens from different commercial plants and those laboratory processed behave similarly during frozen storage (Hanson *et al.* (34)). Off-flavors develop about equally in commercial lots held under the same conditions, and laboratory-prepared samples develop similar off-flavors at a corresponding time. Processing equipment, coatings, packaging materials, and presence of an antibiotic appear to be less important than storage conditions in stability of frozen fried chicken.

During storage, fried chickens first lose "freshly cooked" flavor, then develop a "warmed over" flavor, and eventually rancid flavor. Fig. 6.3 shows that off flavor was noted in precooked fryers held at 20°F. (−7°C.) in two weeks, staleness in less than two months, and rancidity in two months. Products stored at 0°F. (−18°C.) did not develop detectable off flavors until two months of storage, staleness in four months, and rancidity after eight months. These products were stored in "air," and compared with fryers held at −20°F. (−29°C.) in nitrogen.

Flavor changes develop at a faster rate in meat than in skin (34). Although flavor changes in skin could be masked by the coating or seasoning ingredients, the greater increases in peroxide values from meat than from skin suggested an actual greater rate of flavor change in the meat itself.

Flavor of precooked fryers is much more stable in nitrogen atmospheres than in air. Table 6.4 shows the dramatic effect of inert atmosphere for the prevention of off flavor and rancidity. Even at 20°F. (−7°C.) precooked fryers can be held for 12 months without adverse

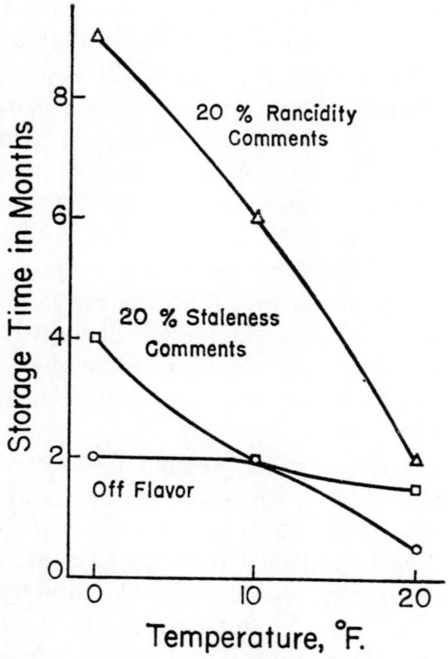

Fig. 6.3. Relative times to indicated flavor change in fried chicken at 0, 10, and 20°F. in retail packages, in nitrogen and air atmospheres (from Hanson *et al.* (34)).

Table 6.4. Influence of Inert Atmosphere in Retarding Flavor Change in Frozen Fried Chicken Meat (34)

Storage temperature		Time (months) for development of flavor change			
		Off-flavor [a]		Rancidity [b]	
°F.	°C.	N_2	Air	N_2	Air
20	− 7	12	0.5	>12	2
10	−12	>12	2	>12	6
0	−18	>12	2	>12	9

[a] Off flavor is indicated at time when flavor differences between experimental samples and controls (— 30°, N_2) were distinguished at statistically significant levels.

[b] Rancidity is indicated at times when it was recognized and noted in 20–25% of the judgments.

flavor changes when stored in nitrogen. Oxidation of cooked chicken during frozen storage may also be partially inhibited by a precooking treatment with polyphosphate (6½% in chill water) (Thomson (35)).

Fried chicken that has been frozen and reheated can be distinguished from freshly fried chicken by slight differences in texture that seem to be recognized as tenderness rather than juiciness changes. After 4–6 months storage at 0°F. (−18°C.), the meat develops a stale or "warmed over" flavor (Hanson (36)).

Skin, because of its high fat content, has frequently been incriminated as a source of off-flavor and rancidity development. Katz *et al.* (37) showed that phospholipids from muscle contained more long-chain fatty acids than phospholipids from skin and depot fat, and the meat contained sufficient polyunsaturated fatty acids to be of major concern in maintaining stability of poultry meat.

Storage flavor changes are also related to precooking treatments. Both loss of desirable chicken flavor and development of off-flavors are directly related to the degree of cooking. Carlin *et al.* (38) deep-fat fried broiler pieces and then packaged them in Cryovac, polyethylene, or aluminum prior to storage up to 15 weeks at 0°F. (−18°C). The chicken deep-fat fried for 10 min. declined in desirable flavor and increased in off-flavors more than the control and those cooked 3 min. All cooked pieces declined in quality more than uncooked similarly stored pieces (Fig. 6.4). Packaging treatment did not influence flavor change or development of off-flavors.

Fried chicken can be processed, frozen, and stored for 6 months or longer, provided that cooking time is limited, the product is packaged in inert gas or in the presence of minimum of oxygen, and held at approximately 0°F. (−18°C.).

B. Other Chicken Products

Bake-and-serve chicken loaves frozen and stored at −22°F. (−30°C.) remain acceptable for at least 4 months, and frozen chicken sticks for 1 month. Rancidity was not detected at these times and temperatures; however, thiobarbituric acid (TBA) values were relatively high.

C. Precooked Goose Products

Geese are known to contain a large amount of fat. This fat, rendered during cooking, is oftentimes objectionable, but this objection could be overcome by utilizing geese in precooked frozen convenience foods, provided the flavor of goose meat remains stable. Stability of frozen cooked goose and gravy, stored in air or nitrogen at 20, 10, and 0°F. (−7, −12, and −18°C.) is comparable to that of turkey prepared in

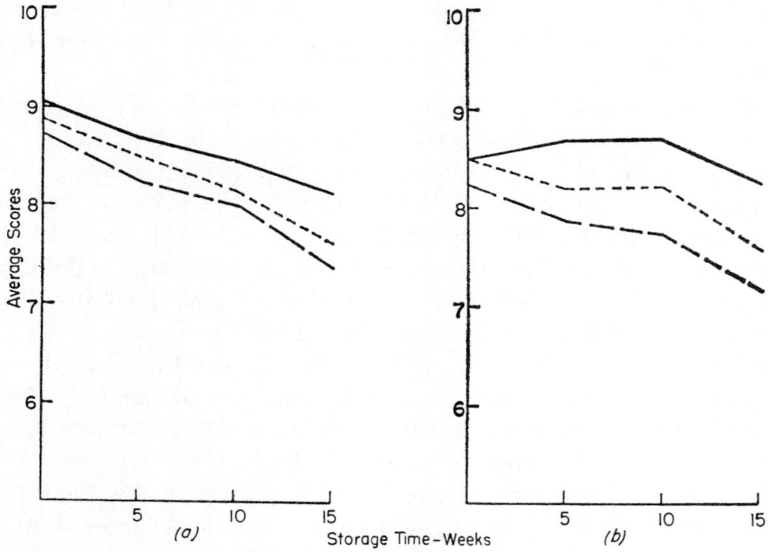

Fig. 6.4. Changes in (*a*) meat flavor and (*b*) in skin flavor of ((– – –) 3 min.; (- - -) 10 min.) precooked and (—) untreated frozen broilers stored at 0°F. (Each experimental point represents an average of 5 judges' scores for 9 replicates.) (From Carlin *et al.* (38)).

the same manner (Bean and Hanson (39)). Off-flavors have been detected in samples in air atmosphere in 2 months at 20°F. and 6–9 months at 10 and 0°F. Nitrogen-packaged products remain acceptable for 6 months at 20°F. and 9–12 months at 10 and 0°F.

Goose meat, prepared in the same way as fried chicken, is less stable than goose and gravy. Use of a 2% oxygen pack, comparable to commercial nitrogen pack, prevented off-flavors in meat, but skin was rancid after 6 months of storage. A nitrogen pack prevented off-flavors in both meat and skin for one year at 0°F. (–18°C.). Cooked goose livers were stable for over 6 months at 0°F. in air.

D. Precooked Turkey, Turkey Pies, and Dinners

Little research has been reported relative to the stability (other than microbial) of precooked whole turkeys or turkey roasts. It is essential, until additional research is conducted, to relate stability data from other poultry products to these turkey products for process or storage decision-making purposes.

Precooked and reheated turkey roasts (4.5 lb.) were found to lose flavor and develop off flavors when stored 7 months at 0°F. (–18°C.) (Martinsen and Carlin (40)). TBA values did not increase significantly, but panel members detected quality changes. Cooking rate, altered by use of different oven temperatures, was not related to these flavor changes.

Hanson and Fletcher (41) evaluated turkey dinners and turkey pies, both laboratory and commercial preparations, for flavor and other changes after storage at temperatures ranging from −30 to 20°F. (−34 to −7°C.). Stale and rancid flavors developed in 1½ to 3 months in samples held at 20°F., and significant loss of ascorbic acid in peas in the turkey dinners occurred in 3 months at 20°F. Mild stale flavors and occasional rancidity flavors developed in these products after 6 and 12 months at 10°F. (−12°C.) and below. Thus these products are 3 to 6 times as stable at 10 as at 20°F. Under the conditions used, antioxidant addition did not cause sufficient improvement in flavor stability of the meat to warrant its inclusion in dinners stored for one year at 0°F. (−18°C.) or 6 months at 10°F. (−12°C.).

Temperatures fluctuating from 0 to 20°F. (−18 to −7°C.) and from −10 to 10°F. (−23 to −12°C.) did not cause flavor changes differing significantly from those in products held at the mean temperatures of 10 and 0°F.

Peroxide values generally increased with increasing storage time and temperature, but did not show sufficient correlation with rancid flavor development to be useful for prediction purposes.

Although pies and dinners held for one year at 10°F. (−12°C.) or lower showed flavor changes only part of the time, for optimum retention of the desirable qualities, exposure to 20°F. (−7°C.) should be avoided and storage at 10°F. or lower should be limited to 6 months.

Frozen pies should be kept frozen until heated prior to serving. Defrosting to any degree results in adverse effect on quality (Fanelli and Gunderson (42)).

E. Creamed Chicken and Turkey

Sufficient rancidity may develop during the cooking (simmering) of turkey to be readily detected after it has been creamed. The degree of rancidity development differs among lots of turkeys. The presence of an antioxidant (Tenox II) during cooking of the turkey to be used in precooked products markedly retards the development of rancid flavors and peroxide during both cooking and frozen storage. Addition of the antioxidant during preparation of the cream sauce rather than during cooking of the turkey less markedly retards development and rancid flavors during storage. Thus the greatest overall benefit is obtained by adding the antioxidant at the beginning of the cooking process.

Creamed turkey prepared from turkeys cooked with the antioxidant

Tenox II showed little or no rancid off-flavor and low peroxide values immediately after freezing, or after storage periods from 4 to 12 months. Off-flavor scores of creamed turkey cooked with antioxidant and stored at 10 or 0°F. (−12 or −18°C.) were very low through 12 months of storage. When temperature was 0°F. for 6 months, then 10°F. for 6 months, off-flavor scores increased but never approached the degree of off-flavor in products cooked without the antioxidant (Lineweaver *et al.* (43)).

A number of antioxidants are effective in lessening the development of off-flavors and rancidity in turkey meat. Butylated hydroxyanisole (BHA), propyl gallate, and citric acid may be effective either alone or in combination for retarding peroxide development. Due to bird-to-bird and sample-to-sample variations, the relative degree of effectiveness of those antioxidants needs further study.

The flavor of frozen creamed chicken is influenced by the class and breed of chickens used. Meat from roasters, used in creamed products, is less flavorful than meat from older birds (Hanson *et al.* (44)). The method of cooking turkey meat for creamed products (roasting, simmering, or pressure cooking) does not influence flavor or off-flavor of the creamed product; however, fat from roasted turkeys when used in white sauce results in a more rancid product after 3 to 6 months of storage at 0°F. (−18°C.).

When turkey and chicken fats are compared for use in creamed products, the turkey fat is found to have more off-flavor when it is stored at 10°F. (−12°C.) for 4 to 12 months. Chicken fat was comparable in stability to hydrogenated vegetable oil, but turkey fat developed more off-flavor (44).

Packaging treatment also affects stability of creamed chicken or turkey. When the product is stored in a sealed can at temperatures from +10 to −30°F (−12 to −34°C.), no differences were found due to temperature. Products wrapped in MSAT cellophane, placed in a carton, and overwrapped developed off-flavors faster at the higher temperatures.

These results indicate that turkey cannot be substituted for chicken without adding an antioxidant; in general, turkey is not recommended for creamed products without the addition of an antioxidant, especially if it is to be stored for 4 months or longer.

F. Sauces, Gravies, and Thickened Desserts

A number of cooked foods containing egg or starchy thickening agents exhibit undesirable texture changes after freezing and frozen storage. Liquid separation occurs on thawing, and the sauces, gravies, and thickened desserts have a curdled appearance; the cooked egg white shrinks, toughens, and separates into layers. These texture defects occur in different degrees in thawed foods and become greater with frozen storage (Hanson (44a)).

1. Sauces and Gravies

Sauces and gravies are important in this connection because they increase the flavor stability of frozen cooked foods by displacing air in the package, thus retarding oxidative flavor changes. Hanson *et al.* (45) showed that turkey and chicken gravies, prepared with poultry fat and with broth substituted for part of the milk, have shown stabilities similar to that of white sauce. After eight months storage at 0°F. (−18°C.), gravies thickened with waxy rice flour showed little or no separation after thawing. Those thickened with wheat flour or waxy corn starch showed the curdled appearance and liquid separation found in white sauces thickened with those products.

The type of thickening agent was found to be a principal factor in the development of a curdled appearance and liquid separation in sauces and gravies subjected to freezing and frozen storage. The use of amylopectin starches and flours minimizes these defects, but since amylopectin starches in contact with some waxy cereal flours have a long paste character, only certain flours appear suitable for use in sauces and gravies which are to be frozen. Waxy rice flours appear to be superior to waxy corn and waxy sorghum flour, but if appearance and liquid separation are important only in heated sauces and gravies, then waxy white corn and rice flours are essentially interchangeable.

Separation and curdled appearance of thawed sauces can be prevented for 11 months at 0°F. (−18°C.) with waxy rice flour, and this defect can be reduced in products stored up to 5 months by using a mixture of wheat and waxy rice flours, especially if the sauce is homogenized.

2. Thickened Precooked Frozen Foods

Undesirable texture changes are encountered in the freezing and storage of cooked foods thickened with eggs or the usual starchy thickening agents. The texture of puddings and custard-type desserts changes from a smooth to a grainy or lumpy consistency when these foods are thawed immediately after freezing. When the products are subjected to frozen storage, a curdled appearance and visible liquid separation develop; and the original smoothness is not completely regained even in the foods that can be reheated for serving (Hanson *et al.* (46)).

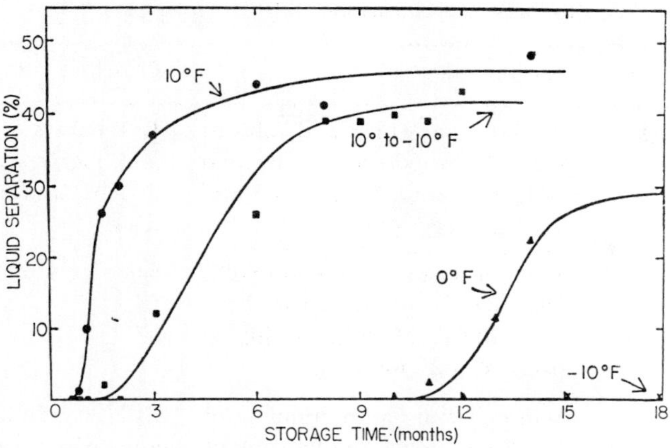

Fig. 6.5. Effect of constant and fluctuating temperatures on storage stability of frozen white sauce thickened with waxy rice flour (from Hanson *et al.* (46)).

Previous studies have established the beneficial effects of waxy rice flour in improving stability of these products. The storage time without texture changes is increased, and products that have separated regain their original smoothness on reheating. By using waxy-rice flour, changes can be prevented at 0°F. (−18°C.) for a year in sauces and gravies and for several months in thickened desserts. However, at 10°F. (−12°C.) texture changes generally occur within a few weeks.

Liquid separation in white sauce, as affected by storage conditions, is shown in Fig. 6.5. However, a low and uniform temperature is essential for good stability in frozen white sauce, even if it has been thickened with waxy rice flour.

IV. EGGS AND EGG PRODUCTS

A. Uncooked Egg Products

Frozen egg products are available as frozen whole eggs, frozen egg whites, or frozen egg yolks. Variations in these basic products have resulted from differences in user specifications, and additional products such as frozen fortified whole eggs, quick-whipping frozen whites, frozen sugared yolks, and frozen salted yolks are now available. Specifications for these products are shown in Table 6.5.

Cooked egg whites, however, become toughened and rubbery, and develop an unpleasant flavor during storage. The freezing of egg white has little visible effect on its properties, apart from shrinking the thick white to some extent and thus increasing the volume of the thin white in a mixture of the two (Brooks and Taylor (47)). Henry and Barbour

Table 6.5. Egg Products: Specifications and Principal Uses (57)

Egg product	Specifications	Principal uses
Frozen whole eggs	A mixture of whites and yolks in natural proportions, with no additives; contains a minimum of 25.5% egg solids.	By bakers in cakes, milk pies, cookies, sweet doughs, and other pastries.
Frozen fortified whole eggs	Whole eggs to which extra yolks and sugar, salt, or syrup have been added. Made according to packers' own formulas.	Same as whole eggs.
Standard frozen whites	Whites with a minimum of 11.5% egg solids, and fat content not over 0.03%.	By bakers in angel food and white cakes; meringues and icings; by candy manufacturers.
Quick-whipping frozen whites	Whites that are specially processed prior to freezing to produce quicker whipping than regular whites.	By bakers in angel food cakes.
Frozen plain yolks	Yolks with a minimum of 45% egg solids and no additives. "Dark yolk" must show No. 4 or No. 5 NEPA color.	In egg noodles which by law must contain 5½% egg solids; also in baby foods.
Frozen sugared yolks	Yolks with a minimum of 43% egg solids and containing 10% sugar.	By bakers in cakes, milk pies, sweet goods, and other pastries; by manufacturers of French ice cream; also baby foods.
Frozen salted yolks	Yolks with a minimum of 43% egg solids and containing 10% salt.	In mayonnaise.

(48) and Bailey (49) found that egg white could be stored in the frozen state for considerable periods of time (up to 177 days) with little change in the beating properties. More recently, Clinger *et al.* (50) have determined the volume of angel cakes prepared from egg white after freezing and storing it at various temperatures. Fresh egg white produced slightly larger cakes than frozen egg white, but there was no appreciable difference between the cakes prepared from egg white which had been frozen and stored at temperatures ranging from 10 to

−50°F. (−12 to −45°C.) and which had been kept for periods ranging from 1 to 54 weeks.

Pearce and Lavers (52) stored frozen white and frozen yolk for three months at −10°F. (−23°C.). After thawing, the products were mixed with an equivalent amount of fresh yolk or fresh white, respectively, and the volumes of foams (after the addition of sugar) and of sponge cakes prepared from the mixtures were measured. The freezing and storage of egg white had little or no effect on foam volume, although a slight increase was observed after the second month; cake volumes increased slowly throughout the storage period. The freezing of yolk barely lowered the foam volume, and there was a further decrease of about 8% during two months of storage followed by an increase during the last month; cake volumes increased slowly throughout the period of yolk storage.

When egg yolks are frozen at temperatures below 22°F (−5.5°C) an irreversible change occurs in physical characteristics; the yolks become gummy and partially coagulated, with considerable change in viscosity. Addition of substances to lower the freezing point of the yolk mixture, such as salt, sugar, or glycerin, will prevent this gelation or thickening property after freezing and thawing. As shown in Table 6.5, the common commercial treatment is to add 10% sugar or salt to the yolk before freezing.

Frozen whole eggs, frozen yolks, and frozen whites are equal to fresh eggs for use in leavening and thickening qualities, according to tests made on angel food cakes, sponge cakes, butter cakes, and baked custards. Frozen eggs stored 6 months are as desirable as those stored 6 weeks (Erikson and Boyden (53)).

The type of container used to store the frozen eggs has little effect on qualities of the products made from the eggs, except for flavor. Some indication has been found that eggs stored in cardboard cartons or in laquered tin containers have slightly less desirable flavor than when stored in glass.

Four per cent glycerin, 4% corn syrup, 4% sodium chloride, 2% sodium chloride, or 1% sodium chloride, added to egg yolks to lower their freezing point, were only partially effective in preventing gelation after thawing.

Very little gumminess was found in yolks frozen and stored 6 weeks. All the yolks stored 6 months were slightly gummy. Yolks to which nothing had been added showed definite coagulation. Four per cent glycerin was slightly better than the other anti-gelation agents in preventing gumminess. All the yolks could be beaten to a creamy consistency with the electric mixer, but small particles of coagulated protein persisted in spite of lengthy beating. These particles appeared in the cooked products, but did not seem to affect in any consistent manner the qualities for which the products were scored. If appearance had been rated, some of the judges might have found the particles objectionable, especially in the custards.

While flavor is probably the most important criterion by which consumers judge the acceptability of eggs, it cannot be concluded from the tests in this experiment that flavor of eggs, as they are used in baked products, is impaired enough by freezing to be detected easily by the average consumer. Differences between products made with frozen eggs and fresh eggs were so slight that for all practical purposes, it may be said flavor is not affected by freezing.

A number of reports indicate the good functional stability of egg after freezing and thawing. Frozen eggs and yolks (except untreated yolks) retain to a high degree the functional properties necessary for satisfactory performance in plain cakes and custards, as compared with fresh yolks (Jordan *et al.* (54)).

Equally good angel food cakes can be made from fresh and frozen whites, although thin whites have been found to produce better angel cakes than thick whites. Some bakers prefer aged whites (frozen 6 months) over freshly frozen or "green" whites. They claim that better angel cakes result. Miller and Vail (59) found no difference, however, in cake volume or score of cakes baked with fresh, recently frozen, or egg whites frozen 6 months. No functional differences were found between whites frozen at −10 and −50°F. (−23 and −45°C.); no difference due to temperature was found after times ranging from 1 to 54 weeks.

Pasteurization of egg white prior to freezing (1) does not impair functional properties of the product (Clinger *et al.* (50)). Cake volume resulting from use of frozen egg whites was nearly as good as the volume produced with fresh egg whites, and acceptability scores were as good or better.

Frozen eggs stored at 0°F. (−18°C.) for 2 months were used in a standard custard recipe, evaluated for quality, and compared with custards made from freshly spray-dried eggs and freeze-dried eggs. A large consumer panel was unable to distinguish any difference between custards made from fresh eggs and those made from frozen eggs (Kelly (55)). The quality of crusts, color, aroma, and flavor of custards from frozen eggs is not different from those made from fresh eggs (Miller and Winter (56)). Egg nog from frozen eggs is comparable to that made from fresh eggs (Fleet *et al.* (57)). Pasteurization of whole egg or yolk, followed by freezing, does not impair the emulsifying properties and stability of emulsions for mayonnaise. As per cent defrosted yolk in mayonnaise was decreased from 13.5 to 5.5%, stiffness and stability of mayonnaise declined, according to Miller and Winter (58).

B. Hard-Cooked Eggs

When cooked eggs are frozen, the problem area is a reverse of that encountered in the freezing of uncooked eggs. The white, rather than the yolk, changes in texture and presents a serious problem. Diced hard-cooked eggs are frequently used in products such as chicken à la king, but they cannot be successfully used if the product is to be frozen (Davis *et al.* (60)). The white becomes rubbery, granular, and watery, and separates into small clumps or layers. The quality of the cooked egg white varies with factors that affect ice crystal size. Large ice crystals leave large holes in the cooked white, and the water frozen out of the cooked white is not taken up again upon thawing.

C. Egg-Containing Desserts

Egg white or egg yolk in the formula causes no problem in the freezing of baked angel cakes, sponge cakes, and layer cakes. Lemon pie filling thickened with eggs, flour, and cornstarch can be frozen and stored for longer than 6 months at 0°F. (−18°C.) or 4 months at 10°F. (−12°C.) (Hanson (61)). Custard and pudding-type desserts, and cake and cream puff fillings in which eggs are used as thickening agents, show serious texture defects when frozen. Both soft and baked custards exhibit curdling and liquid separation when they are thawed the day after being frozen (Hanson *et al.* (62)). A similar defect occurs in cake fillings containing eggs. Desserts suitable for freezing, after replacing part of the eggs with waxy rice flour, have been prepared. The curdling and separation can be prevented for storage (0°F) of 2–4 months for custard-type desserts by using waxy rice flours. The methods developed for producing stable desserts of the soft-custard types are adaptable to formulation of cake and cream puff fillings stable to freezing and storage.

Frozen baked custard has not been satisfactorily produced. Other precooked products containing eggs that do not freeze satisfactorily are the soft meringues used on pies and cake frostings of the seven-minute type. The meringues shrink and toughen, the frosting becomes frothy and spongy (Fenton and Darfler (63)).

Damage to cooked egg white caused by freezing was also reported by Davis *et al.* (60) to be due to the mechanical effects of ice crystals formed. Damage was reduced by supercooling before freezing, or by the addition or finely divided calcium carbonate; a mixture of egg yolk and egg white for use in frozen foods could be prepared with satisfaction.

Hard-cooked eggs cannot be used in food mixtures such as chicken à la king, that are to be frozen, because the white becomes rubbery, granular, and watery and separates into small clumps or layers (Woodruff (64)). This writer also reported that cooked yolk is practically unchanged after freezing.

Other egg-containing cooked products which have been found acceptable after freezing and short storage periods are instant French toast, chiffon pies, and frozen western egg (Baker *et al.* (65)).

REFERENCES

(1) F. E. Cunningham and H. Lineweaver, *Food Technol., 19,* No. 9, 136 (1965).
(2) A. W. Brant, R. H. Forsythe, and M. H. Swanson, *Food Technol., 19,* No. 4, 229 (1965).
(3) G. J. Mountney, R. E. Brown and W. C. Hurley, *Poultry Sci. 39,* 287 (1960).
(4) E. S. Snyder, J. H. L. Truscott, and C. R. Waxgood, *Poultry Sci., 30,* 383 (1953).
(5) G. F. Stewart, H. L. Hanson, B. Lowe, and J. J. Austin, *Food Res., 10,* 16 (1945).
(6) J. H. MacNeil, M. S. thesis, Michigan State University, East Lansing, Michigan, (1958).
(7) A. A. Klose and M. F. Pool, *Food Technol., 10,* 34 (1956).
(8) R. C. Baker, *Poultry Sci., 33,* 1040 (1954).
(9) R. C. Baker, *Poultry Processing Marketing, 61,* No. 3, 14 (1955).
(10) L. van den Berg and C. P. Lentz, *Food Technol., 12,* 183 (1958).
(11) C. P. Lentz and L. van den Berg, *Food Technol., 11,* 247 (1957).
(12) A. W. Brant, and G. J. Stewart, *Food Technol., 4,* 168 (1950).
(13) C. H. Koonz and J. M. Ramsbottom, *Food Res., 12,* 393 (1947).
(14) C. Ellis and J. G. Woodroof, *Food Technol., 13,* 533 (1959).
(15) E. O. Essary, *Poultry Sci., 38,* 527 (1959).
(16) T. P. Hildrith, E. C. Jones, and A. J. Rheed, *Biochem. J., 28,* 786 (1934).
(17) E. M. Cruickshank, *Biochem. J., 28,* 965 (1934).
(18) F. A. Kummerow, W. Wingerd, G. Jacobson, and D. Muller, *Poultry Sci., 29,* 768 (1950).
(19) T. K. Chu and F. A. Kummerow, *Poultry Sci., 29,* 846 (1950).
(20) A. A. Klose, E. P. Mecchi, H. L. Hanson, and H. Lineweaver, *J. Am. Oil Chemists Soc., 28,* 162 (1951).
(21) A. A. Klose, E. P. Mecchi, G. A. Behman, H. Lineweaver, F. A. Kratzer, and D. Williams, *Poultry Sci., 31,* 354 (1952).
(22) M. K. Nutter, E. E. Lockhart, and R. S. Harris, *Oil Soap (Egypt), 20,* 231 (1943).
(23) J. E. Criddle and A. F. Morgan, *Fed. Proc., 6,* 247 (1947).
(24) J. E. Criddle and A. F. Morgan, *Proc. Soc. Exptl. Biol. Med., 78,* 41 (1951).
(25) E. P. Mecchi, M. F. Pool, G. A. Behman, M. Hamachi, and A. A. Klose, *Poultry Sci., 35,* 1238 (1956).
(26) E. P. Mecchi, M. F. Pool, M. Nonaka, A. A. Klose, S. J. Marsden, and R. J. Lillie, *Poultry Sci., 35,* 1246 (1956).
(27) A. A. Klose, H. L. Hanson, E. P. Mecchi, J. H. Anderson, I. V. Streeter, and H. Lineweaver, *Poultry Sci., 32,* 82 (1953).
(28) A. A. Klose, E. P. Mecchi, and H. L. Hanson, *Food Technol., 6,* 308 (1952).

(29) M. F. Pool, H. L. Hanson, and A. A. Klose, *Poultry Sci., 29,* 347 (1950).
(30) L. E. Dawson, J. A. Davidson, E. H. Farmer, R. J. Evans, E. Jones and W. Westveer, *unpublished mimeo,* Michigan State University, East Lansing, Michigan, 1958.
(31) A. A. Klose, A. A. Campbell, and H. L. Hanson, *Poultry Sci., 39,* 1136 (1960).
(32) A. A. Klose, M. F. Pool, A. A. Campbell, and H. L. Hanson, *Food Technol., 13,* 477 (1959).
(33) A. A. Klose, M. F. Pool, and H. Lineweaver, *Food Technol., 9,* 372 (1955).
(34) H. L. Hanson, L. R. Fletcher, and H. Lineweaver, *Food Technol., 13,* 221 (1959).
(35) J. E. Thomson, *Food Technol., 18,* No. 11, 147 (1964).
(36) H. L. Hanson, *Proc. Inst. Am. Poultry Ind.,* Fact Finding Conference (1954).
(37) M. A. Katz, L. R. Dugan, Jr., and L. E. Dawson, *J. Food Sci., 31,* 717 (1966).
(38) A. F. Carlin, R. M. V. Pangborn, O. J. Cotter, and P. G. Homeyer, *Food Technol.,* 13, 557 (1959).
(39) M. L. Bean and H. L. Hanson, *Poultry Sci., 41,* 243 (1962).
(40) C. S. Martinsen and P. F. Carlin, *Proc. Inst. Food Technol.,* Abstract (1967).
(41) H. L. Hanson and L. R. Fletcher, *Food Technol., 12,* 40 (1958).
(42) M. J. Fanelli and M. F. Gunderson, *Food Technol., 15,* 419 (1961).
(43) H. Lineweaver, J. D. Anderson, and H. L. Hanson, *Food Technol., 6,* 1 (1952).
(44) H. L. Hanson, H. M. Winegarden, M. B. Horton, and H. Lineweaver, *Food Technol., 4,* 430 (1950).
(44a) H. L. Hanson, Proceedings of Symposium, *Precooked Frozen Foods,* QM. Food & Container Institute, 18 (1955).
(45) H. L. Hanson, A. A. Campbell, and H. Lineweaver, *Food Technol., 5,* 432 (1951).
(46) H. L. Hanson, L. R. Fletcher, and A. A. Campbell, *Food Technol., 11,* 339 (1957).
(47) J. Brooks and D. J. Taylor, *Food Investigation Report 60.* Her Majesty's Stationery Office, London 55 (1955).
(48) W. C. Henry and A. D. Barbour, *Ind. Eng. Chem. (Industr.), 25,* 1054 (1933).
(49) M. I. Bailey, *Ind. Eng. Chem. (Industr.), 27,* 973 (1935).
(50) C. Clinger, A. Young, I. Prudent, and A. R. Winter, *Food Technol., 5,* 167 (1951).
(51) J. W. Koudele and E. C. Heinsohn, *Kansas Agr. Expt. Sta. Bull. 423,* 46 (1960).
(52) J. A. Pearce and C. G. Lavers, *Can. J. Res., 27,* 253 (1949).
(53) S. E. Erickson and R. E. Boyden, *Kentucky Agr. Expt. Sta. Bull., 635,* 13 (1955).
(54) R. Jordan, B. H. Luginbill, L. E. Dawson, and C. J. Echterling, *Food Res., 17,* 1 (1952).
(55) J. A. Kelly, M. S. thesis, Michigan State University, East Lansing, Michigan, 1960.
(56) C. Miller and A. R. Winter, *Poultry Sci., 29,* 88 (1950).
(57) O. Feet, T. I. Hedrick, L. E. Dawson, and H. E. Larzelere, *Mich. Agr. Expt. Sta. Bull., 46,* 80 (1963).
(58) C. Miller and A. R. Winter, *Food Res., 16,* 43 (1951).
(59) E. Miller and G. Vail, *Cereal Chem., 20,* 528 (1943).
(60) J. G. Davis, H. L. Hansen, and H. Lineweaver, *Food Res., 17,* 393 (1952).
(61) H. L. Hanson, *Poultry Processing Marketing, 60* (6) (1954).
(62) H. L. Hanson, K. D. Nichita, and H. Lineweaver, *Food Technol., 7,* 462 (1953).
(63) F. Fenton and J. Darfler, *Cornell Univ. Bull. 692* (1946).
(64) J. G. Woodroof, *Quick Frozen Foods, 8,* No. 9, 90 (1946).
(65) R. C. Baker, L. B. Darrah, and J. M. Darfler, *Poultry Sci., 45,* 1017 (1966).

Chapter 7

QUALITY AND STABILITY OF FROZEN BAKERY PRODUCTS

James W. Pence

Cereals Laboratory, Western Regional Research Laboratory, U.S. Department of Agriculture, Albany, California

I. INTRODUCTION

Commercial freezing of bakery products on a significant scale began in the late 1940's and early 1950's, but substantial sales of frozen products direct to consumers required another decade to develop (1,2). Early use of freezing was chiefly to achieve economies in processing or distribution. Products to a large extent were defrosted before sale. The success of several pioneering companies in preserving high-quality frozen products through extended marketing channels greatly helped acceptance of frozen bakery products. By 1966 sales of frozen bakery products and unfrozen fruit pies had reached a value of $400 million (3).

In general, bakery products freeze well under conditions suitable for other frozen foods, but important differences exist in their tolerance to variations in freezing, frozen storage, and defrosting conditions. For example, retention of the crumb resiliency characteristic of freshly baked lean products, such as bread and rolls, is greatly affected by rates of freezing and defrosting. Cakes and other rich products are quite tolerant to variations in these conditions. Some properties of cakes are even improved by freezing. Successful freezing of bakery products therefore requires careful and informed attention to all aspects of the operations. The following review of available information on quality and stability of major classes of bakery products is designed to assist both the student and the practitioner.

II. BREAD AND ROLLS

The noncommercial freezing of bread and rolls may have a longer history than that of any other baked product. The practice is alleged to have been common for many years in rural households of northern Europe and North America (4). The long persistence of the practice suggests that results were generally good. Research findings and recent commercial experience also indicate that excellent results can be obtained, if recommended procedures are carefully followed. The texture characteristics of freshly baked breads, for example, are more sensitive to variations in freezing practices than are those of any other baked product.

A. Types of Quality Change and Mechanisms

Freshly baked bread and rolls have several distinguishing features that characterize the product and that change rather rapidly in the first few hours after baking. Prominent among these is a soft, resilient crumb texture.

Bread doughs going into the oven have a moisture content of about 44%. As baking and crust browning proceed, the outer surface of the loaf becomes quite dry and crisp. Regions just beneath the crust also dry to a considerable extent, but the moisture gradient at the completion of baking is surprisingly steep. Up to within an inch or so of the crust, the inner crumb still has the moisture content of the original dough (5). Heat penetration during baking is slow enough that the temperature at the center core of the loaf reaches the boiling point of water only for a very few minutes before the baking is complete (6). Thus, there is little driving force to move moisture outward from the center of the loaf.

The rigidity that bread crumb attains during baking arises largely from the uptake of water by starch granules in the dough. These become swollen and flexible enough to bend around the periphery of the gas cells, but the granules seldom rupture or lose their visible organization. The protein forming the continuum in the dough is resistant enough to heat denaturation so that the structure of the crumb is still quite soft and permanently deformable at the completion of baking. As the crumb cools to room temperature, however, the crumb attains enough resiliency to spring back into its original shape from a moderate deforming stress.

As fresh bread ages, the crumb gradually becomes firmer, less resilient, and harsh to the touch. Within a day or two the crumb loses its capacity to spring back from a deformation. The structure becomes embrittled enough to rupture and crumble easily. The principal organoleptic defect is that the crumb now feels dry in the mouth, even though there may have been no significant loss of moisture by the crumb. This physical staling change, however, is aggravated by loss of moisture (7,8).

Accompanying these firming changes of the inner crumb is a gradual decrease in the moisture gradient near the crust. Moisture moves outward to soften the crisp, dry crust and change its fresh character. A fairly steep moisture gradient persists for several days, however, and may contribute to a visual defect in the crumb during frozen storage that is commonly referred to as white rings. This will be discussed further below.

The rate at which bread crumb firms is temperature dependent and appears to be governed chiefly by changes taking place within the starch granules or in the solubilized starch that may have become leached from the granule during mixing, fermentation, and baking. The phenomenon of bread staling often is attributed to retrogradation of the starch, a phenomenon normally associated only with amylose, the straight chain molecules of starch. This seems quite reasonable for any amylose starch that is displaced from the granule structure in the crumb. Some students of the phenomena feel, however, that amylose remaining in the distorted and swollen granules has not lost its original intermolecular associations enough to be considered dispersed or retrogradable. They feel, rather, that the straight chain portions of the branched amylopectin, less well organized to begin with perhaps, become reassociated during crumb aging to harden the structure (9). Solubility and X-ray diffraction studies on bread crumb are, in fact, consistent with increasingly crystallinelike associations of starch in bread crumb as it ages, whether or not amylose or amylopectin are primarily involved (10,11).

Flavor changes that occur as freshly baked bread ages are perhaps more subtle than the physical changes and are frequently omitted when the term staling is applied. The chemistry of bread flavor is ill understood, but several mechanisms for flavor deterioration can be hypothesized. These would include loss of components due to volatility, oxidation of reactive components, and slow continuation of aggregative processes such as the Maillard browning reaction. All these processes are slowed by lower temperatures, so that flavor preservation in bread would be expected to be helped by freezing.

B. Effects of Freezing and Defrosting on Quality Factors

Many studies have shown that temperature affects both the firmness of bread crumb and the rate at which firmness increases with storage time. Perhaps the earliest systematic studies were conducted by Katz (12), who showed that the rate of firming was negligible at 140°F. (60°C), or above, and maximum at 36–37°F. (1–2°C.). The rate decreased at either higher or lower temperatures. Fig. 7.1, derived from later work (13), illustrates these relationships.

Best retention of initial softness and resiliency of bread crumb obviously requires that freezing be accomplished as rapidly as possible.

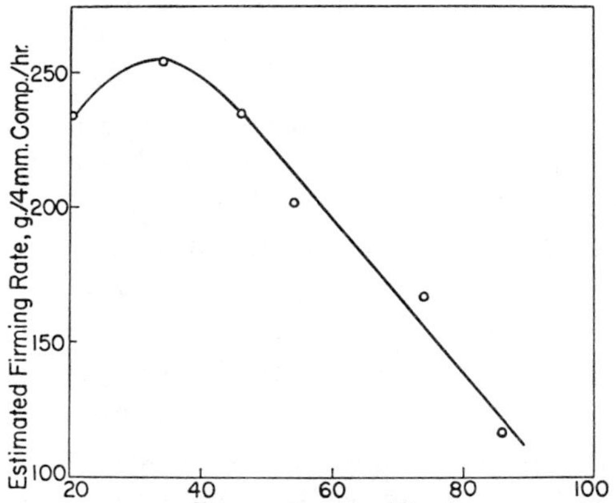

Fig. 7.1. Effect of temperature on the firming rate of commercial white bread (13). Reproduced by permission from *Cereal Chem., 32,* 519 (1955), copyright American Association of Cereal Chemists.

Similarly, bread should be frozen as soon after baking as possible. A very large part of the total firming that bread will undergo during normal marketing and use will occur during the first 20–30 hr after baking (12–16). Taste-panel appraisals correlate well with firmness measurements in showing the effect of age of bread at time of freezing on the quality of the bread after defrosting (17,18).

Several factors affect the rate at which bread can be frozen, and economic considerations impose additional limits with respect to practical feasibility. For example, sanitary and customer preference considerations require that bread be sliced and wrapped as soon after baking as possible. This means that the very rapid freezing attainable with high air velocities on unwrapped bread is infeasible for the average baker. The gain in quality retention must be weighed against the extra costs involved for the deviations from normal handling procedures. Fortunately, good results can be obtained with equipment and conditions common in the frozen food industry.

Extensive experimental work on freezing bread in air blast freezers (19) indicates that freezer temperature is the most important variable affecting the freezing rate of wrapped and sliced bread. Air velocity is without appreciable effect, provided the air moves at at least 500–600 linear ft. min. and individual loaves are well exposed to the moving air. Wrapper composition is of little significance, although the wrapper should fit the loaf fairly snugly. Loose wraps will delay freezing significantly (19,20).

As bread cools in the freezer, the temperature at the center of the loaf at first falls fairly rapidly. At about 18–20°F. (−8 to −7°C.) the curve begins to level off, but a sharp inflection due to a distinct phase transition is seldom seen. Somewhat later, the curve again falls rather rapidly until the temperature at the center of the loaf nears that of the freezer. These changes are illustrated in Fig. 7.2. Even though the deflection in the cooling curve near 20°F. (−7°C.) indicates a freezing zone, the crumb at this temperature is still somewhat deformable. Rigidity of the frozen loaf is not well developed until the crumb temperature falls to about 0°F. (−18°C.). Enthalpy measurements (21) show that water in bread is only about 70% frozen, even at 0°F. (−18°C.).

In an airblast freezer at −20°F. (−29°C.) with an air velocity of about 500 linear ft./min., the temperature of the center crumb of typical commercial white bread will fall from 70 to 20°F. (21 to −7°C.) in about 90 min. (19). This will add about 30 units of firmness (g./4 mm. compression, Baker Compressimeter at 72°F. (22°C.)) to the defrosted crumb (18). An additional increment of about 15 units of

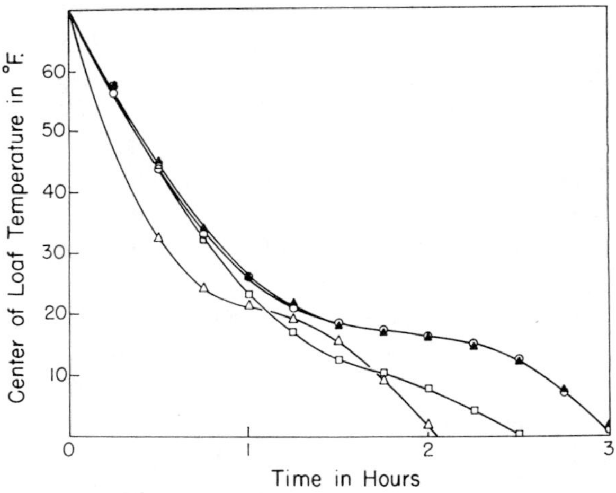

Fig. 7.2. Cooling curves for unwrapped variety breads at intermediate freezer temperature and air velocity (19). Reproduced by permission from *Food Technol., 9,* 342 (1955), copyright Institute of Food Technologists. Freezer temperature — 20°F.; air velocity 700 ft./min.; 1-lb. loaves, unwrapped; (○) white; (▲) whole wheat; (□) raisin; (△) sour French.

firmness will be added to the defrosted bread as a result of defrosting at room temperature, which will require about 2.5 hr. (18,22). The total change incurred in the freezing and defrosting is about the same as that for freshly baked bread during holding at room temperature for the first 24 hr. after baking. Thus, freezing and defrosting under these moderate and quite practical conditions add a day of age to the bread, with respect to firmness.

A firmness change of this magnitude would be of considerable concern for marketing of the bread were it not offset by a compensating change in rate of firming of the defrosted bread. Measurements of crumb firmness and of crumbliness (14) show that within 48 hr. after defrosting, frozen-defrosted bread and unfrozen bread are equivalent in these factors. This means that frozen bread can have essentially the same shelf life and use life after defrosting as unfrozen bread. The frozen bread, however, can be stored for many weeks before being defrosted for use or sale.

Although temperature changes during defrosting of bread are essentially the reverse of those during freezing, effects on firmness are not as marked. Fig. 7.3 shows some defrosting curves and Table 7.1 shows increments of firming due to different freezing and defrosting rates.

The values for comparable time periods in Table 7.1 are quite clearly different for freezing and for defrosting, but no clear exposition has ever been put forward to explain the results. Further experimental

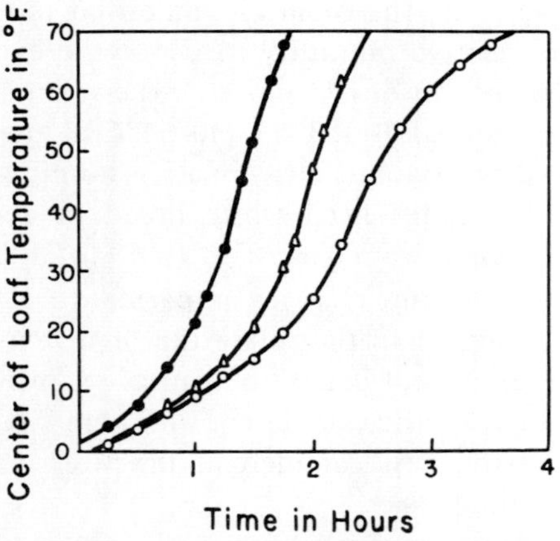

Fig. 7.3. Warming curves for wrapped white bread at intermediate relative humidity and several air temperatures (22). Reproduced by permission from *Food Technol., 10,* 492 (1956), copyright Institute of Food Technologists. Wrapped bread, 58–62% relative humidity; (●) 120°; (△) 100°; (○) 80°.

work would be necessary to determine whether the initial freezing of fresh crumb somehow imparts extra tolerance to crumb firming, as is suggested in the work cited above (13). Similar experiments would need to be run with day-old or two-day-old breads to establish this point.

For practical purposes, quality of frozen bread clearly is much more tolerant to defrosting rates than to freezing rates. This is quite fortunate because of limitations imposed by the necessity to protect wrappers and to avoid undesirable moisture changes in the breads during defrosting (22,23). The rather dry air used for freezing protects against moisture changes, but during defrosting serious gains or losses of moisture can be induced (22). In fact, control of air humidity during defrosting is as important as control of air temperature. How-

Table 7.1. Average Increase in Firmness of White Bread Crumb Caused by Freezing and Defrosting within Various Time Periods (18)

Freezing time, hr.	Firmness increase g./4 mm. compression	Defrosting time, hr.	Firmness increase, g./4 mm. compression
0.5	16	2.3	13
1.5	32	5.5	32
3.0	57	10.8	31
7.0	71	—	—

ever, the step of warming the room air somewhat to speed up defrosting also lowers its relative humidity to about the extent necessary to obtain the optimal conditions found in experimental work (22–24). These are air at about 120–130°F. (49–54°C.), moving at 200–500 linear ft./min., and with about 50% relative humidity. These conditions will defrost ordinary sliced white bread in about 4 hr. (loaf-center temperature raised from 0 to 70°F. (−18 to 21°C.)).

The tolerance to firming changes imparted to bread by freezing can be taken advantage of in the refreezing of frozen-defrosted bread. Each cycle of refreezing and defrosting causes an increase in firmness, but each increment is smaller, as shown in Table 7.2. Such an operation would not be the best commercial practice, but it is certainly feasible for use in emergencies.

Very little experimental work has been conducted with rolls and variety breads, despite the fact that these items are frozen to a much larger extent than ordinary white bread. Informal reports from industry representatives suggest that results are as good, or better, than they are with white breads. Freezing times for whole wheat and sour French breads (19) are at least as short as with white bread, so that stabilization of firming changes should likewise be as good. Studies with plain dinner rolls and cinnamon rolls (25,26) showed their general behavior to be very like that of white bread. Their tolerance to differences in freezing and defrosting rates was greater than that found for breads, however, presumably because of richer formulas. Even with the slower freezing and defrosting rates used in the study, the frozen-defrosted products were rated superior to unfrozen products 24 hr. after baking (26). Moisture changes were also generally similar to those found for breads. These products, therefore, are even more advantageously frozen than breads, as is consistent with the good results obtained in industry practice (2).

III. CHANGES IN QUALITY FACTORS DURING STORAGE

Although changes in crumb firmness of breads or rolls can be measured objectively with reasonable accuracy, the results are probably less meaningful than results obtained organoleptically, despite the difficulties of interpreting the latter. Evidence is clear, however, that temperatures of 0°F. (−18°C.), or lower, are necessary for effective retention of high quality during frozen storage of bread and roll products.

Reports of successful industry practice using 0°F. (−18°C.), or lower for storage, are reinforced by laboratory reports showing good

Table 7.2. Effect of Repeated Freezing and Defrosting on Firmness of White Bread Crumb (g./4 mm. Compression) (18)

Number of cycles	Firmness	Difference
0 (fresh)	57	—
1	85	28
2	102	17
3	114	12

correlation between firmness measurements and taste panel results (18). White bread frozen and defrosted at good practical rates acquired a firmness equivalent to that of unfrozen bread about 24 hr. after baking. Within 7 days of storage at 10 and 15°F. (−12 and −9°C.), however, the bread after defrosting was quite significantly firmer, and the bread stored at 15°F. (−9°) was rated organoleptically inferior to a day-old unfrozen control. The differences became more clearly marked with increasing storage time, whereas bread stored at 0°F. (−18°C.) remained approximately equal to the day-old unfrozen control for the 7-week duration of the tests.

Similar tests with plain dinner rolls and cinnamon rolls showed no significant advantage of storage at −20°F. (−29°C.) over 0°F. (−18°C.). Dinner rolls remained in very good condition for 8 weeks at 0°F. (−18°C.), but the cinnamon rolls deteriorated noticeably in freshness rating, although not in firmness. At 10 and 20°F. (−12 and −9°C.), deterioration was more rapid and discernible enough at 2 and 3 weeks as to preclude use of these storage temperatures with plain dinner rolls. The cinnamon rolls were less tolerant, and storage at 10 or 20°F. (−12 or −7°C.) for even 1 week was contraindicated. At 0°F. (−18°C.), 3 weeks of storage was about all that could be tolerated by the cinnamon rolls before the freshness rating declined enough to be of concern (26).

A. White Rings in Bread

During the studies of bread stored at 15°F. (−9°C.) mentioned above (18) a peculiar phenomenon was noticed that subsequently was found to be related to movement of moisture within the frozen loaf of bread (27). The defect, called "white-rings" had been known to occur occasionally in commercial freezing of bread (28) and was seen occasionally in bread stored at lower temperatures than 15°F. (−9°C.) (18). A study of possible causes of the phenomenon (27) revealed that the whitish opaque spots and rings in the frozen crumb appeared to result from migration of moisture within the frozen crumb.

During the baking of bread, the moisture content of most of the interior of the loaf remains close to that of the original dough (5,7,29). Substantial loss of moisture occurs only from the outer inch or so of the loaf. The crust, of course, becomes quite dry because of its direct exposure to the oven heat. Within the first inch or two inward from the crust a rather steep moisture gradient forms. Diffusion of moisture outward from the moist crumb of fresh bread soon lessens this gradient, but freezing fairly effectively stabilizes the distribution of moisture within the loaf (29). As storage of the frozen loaf progresses, however, a slow but steady movement of moisture continues through sublimation and redeposit of ice crystals. If the crumb moisture is 35–37% or above, the subliming moisture leaves behind tiny voids which cause the whitened appearance by way of the increased reflectance of light from the new air–solid interfaces (27).

The voids appear to be associated chiefly with the starch granules in the crumb. The contents of the swollen granules either shrink during the loss of moisture, or the initial ice crystal pushed aside the contents and then evaporated to leave a void within the granule (27).

White rings become noticeable within 2 weeks in fresh breads stored at 15°F. (−9°C.) (27). At 10°F. (−12°C.) about 5 weeks are required, and at 0°F. (−18°C.), 10 weeks or more of storage are necessary. In bread frozen shortly after baking, the rings form within the first half inch or so of the crumb just beneath the crust. In bread that is a day or two old before freezing, the clear zone outside the ring is considerably wider. Indeed if the bread is too old before freezing the rings may never form or may require several times longer to form. If loss of moisture from the loaf occurs because of loose or defective wrappings, for example, the white ring will grow in width as dehydration of the affected sufficiently moist area progresses. At the ends of the loaves, the entire faces of slices adjacent to the heel can become affected.

Upon defrosting, the whitish areas of the somewhat dehydrated crumb will sometimes revert to a normal semitranslucent appearance. If too dry, however, the ring will remain visible until the affected area is moistened slightly. Toasting of the slice effectively erases the ring, but mere warming to 140°F. (60°C.) will not always do so. Restoration of at least some of the missing moisture, as well as the heat, appears necessary to restore the shrunken or displaced contents of the starch granules to the turgidlike condition that existed in the freshly baked loaf (27).

Although not necessarily a serious defect in themselves, the "white-rings" in frozen bread provide a convenient index to the condition of bread at the time of freezing and a suggestion of its time–temperature history after freezing. White rings or whitened areas are sometimes seen in other frozen bakery products, such as rolls, biscuits, and cakes, although rarely in the latter.

IV. YEAST-RAISED SWEET GOODS

There appears to be substance to the widespread belief that bakery foods containing higher amounts of sugar and shortening than breads or rolls, and sometimes containing egg, withstand variations in freezing practices better than the products with leaner formulas. Scattered and fragmentary reports of industry experience are supported by experimental studies (30).

A. Coffee Cakes

The term coffee cakes is widely used both for chemically leavened quick-bread or cake products and for yeast-raised products which, as a class, include Danish pastries. The latter are available in a wide variety of size and shape, but all are yeast leavened and resemble bread in crumb texture and appearance more than they do cakes.

Rather extensive and systematic experimental studies (30) showed that Danish pastries were relatively insensitive to variations in freezing and defrosting times. Although measurable increases in firmness occurred as a result of freezing and defrosting, taste-panel results indicated no significant loss in freshness rating. The frozen-defrosted products were rated superior to unfrozen products 24 hours after baking, although the difference was not quite statistically significant.

This tolerance to differences in freezing and defrosting rates means that Danish pastries and rich sweet rolls may be packaged before freezing and may be confidently handled under average good commercial conditions. Recommended are freezer temperatures of about −15 to −20°F. (−26 to −29°C.) and air velocity of 200–500 linear ft./min. Defrosting may be accomplished at room temperature, although warmed and moving air is recommended to avoid excessive condensation on packages, with resulting bad effects on appearance. Undesirable net losses of moisture from unwrapped products can occur before defrosting is completed (30).

Once Danish pastries and rich sweet rolls have been frozen, use of proper storage temperatures is quite important. At 20°F. (−7°C), such products were found to become slightly stale within 7 days and clearly stale within 14 days, according to taste panel evaluations (30). A rapid change in crumb firmness measurements paralleled the organoleptic observations. Staling and increase in firmness were slower at 10°F. (−12°C), but still great enough to show that this temperature is also inadequate (30). At 0°F. (−18°C), in contrast, no significant loss of freshness or increase in firmness occurred in an 8-week experiment, and no advantage was found for storage at temperatures below 0°F. (−18°C.). The slow but steady loss of moisture during storage of frozen products requires that these rich sweet roll products be protected by adequate packaging if storage is to extend beyond 3 or 4 days.

B. Yeast-Raised Doughnuts

Although doughs for yeast-raised doughnuts often contain as much shortening, milk, and egg as Danish pastries, they seldom contain as much sugar, and they do not have the rolled-in extra shortening characteristic of the pastries. The doughnuts are therefore a less rich product and react somewhat less tolerantly to rates of freezing and defrosting. Slow freezing, whether or not accompanied by slow defrosting, caused a significant loss in freshness rating by a panel (31). The frozen-defrosted products, however, were considerably superior to unfrozen products held at room temperature for 24 hr. As with all other bakery products, the freezing of unwrapped doughnuts caused no significant loss of moisture. During defrosting, however, care must be taken to prevent losses of moisture, especially when warmed air is used (31). Packaging of the product prior to freezing effectively prevents undesirable moisture changes during freezing or defrosting.

Storage of yeast-raised doughnuts at 0°F. (−18°C.) appears to be adequate for preservation of freshness and prevention of undue firming. Above this temperature, significant losses in freshness occur within 7 days at 10 and 20°F. (−12 and −7°C.). During this period, quality will decline as much as in unfrozen doughnuts held at room temperature for 24 hours to give products unacceptable to a taste panel. In contrast, at 0 or −20°F. (−18 or −29°C.) little evidence of staling was evident after 8 weeks of observation (31).

V. CAKES AND CAKE-TYPE DOUGHNUTS

Among bakery products that are successfully frozen on a commercial scale for delayed marketing, cakes probably rank second only to frozen pies in tonnage. The tolerance shown by cakes to variations in freezing and defrosting practices and the good stability of cakes during frozen storage contribute heavily to this success. Although tolerant, cakes are not immune to poor freezing practices, so that careful attention must be given to all aspects of a frozen cake operation. The following information provides guidance for freezing cakes and cake-type doughnuts to preserve as much as possible of their original, freshly baked quality.

A. Freezing and Defrosting Considerations

In contrast to the behavior of the crumb of bread and rolls, cake crumb becomes firm and harsh more slowly at cool temperatures than at room temperature (32). The firming changes in cake therefore slow up during the freezing process, whereas those in bread accelerate. As a result, wide differences in freezing and defrosting times produce such small differences in freshness ratings that they are difficult to detect reliably.

In a study conducted specifically to determine effects of variations in freezing and defrosting on common types of cakes, taste panels could not distinguish between angel food cakes that had been frozen within 2 hr. and those frozen in 11 to 13 hr. (33). Similar results were obtained for chiffon cake. Ratings for slowly and rapidly frozen pound cakes or layer cakes differed somewhat, but not to a statistically significant extent. The rapidly and slowly frozen-defrosted layer cakes could not be distinguished from one another or from freshly baked cake, but were clearly superior to day-old unfrozen cake. Pound cakes were more sensitive to freezing in that the frozen-defrosted samples could be reliably distinguished from both freshly baked and day-old unfrozen ones, although rapidly-frozen and slowly-frozen samples could not be distinguished from one another (33).

Rates of freezing and defrosting appear to have no significant effect on freshness ratings or firmness measurements on cake-type doughnuts. In fact, frozen-defrosted doughnuts (frozen within 2 hr. after frying) are scarcely distinguishable from freshly fried unfrozen doughnuts (31).

Because rate of freezing has so little effect upon the quality of cakes of all common types, economic factors become dominant for selection of proper freezing conditions. These include such items as the size of installation necessary to handle the desired quantity of merchandise most efficiently, the refrigeration capacity necessary to freeze the desired quantity at the necessary rates, and the type of packaging most convenient to handle and best for subsequent display purposes.

Unwrapped cakes may be frozen without significant loss of moisture, but during subsequent storage a slow but steady loss of moisture can become serious within a week or two. Loss of even a few grams a day from exposed surfaces can soon change the appearance and texture of the products undesirably and unnecessarily. During defrosting, moisture may either be gained or lost. Use of warmed air or air of fairly low relative humidity will minimize condensation on the cold product (23).

As defrosting proceeds any condensed moisture will reevaporate, and a serious net loss of moisture will occur unless careful attention is given the process at this time. Packaging of products in paperboard boxes or other containers with free air space will minimize, but not completely eliminate, movement of moisture out of the products. Snug-fitting wraps are best in this regard, but do not completely solve the problem because of the normal shrinking of the product as it is frozen.

Cooling curves for cakes resemble those for bread in most respects, except that inflections due to liberation of latent heat are seldom discernible. The high content of sugar and other soluble materials in cake crumb lowers the ostensible freezing point and induces supercooling rather generally. That the latter actually occurs is shown by the cooling curves in Fig. 7.4. Most curves, however, show little or no evidence of a zone of liberation of latent heat due to crystallization. Cooling curves therefore drop rather promptly to temperatures near that of the freezing chamber (33).

Cakes to be frozen commercially differ so very widely in formula, size, shape, packaging, icing, and so forth, that attempts to summarize effects of such specific conditions as freezer temperature and air velocity tend to be somewhat futile. A number of generalizations of interest can be mentioned with regard to other factors, however. For example, freezing times for layer cakes increase in proportion to the square of their diameter. Rapid air movement reduces freezing time for unwrapped cakes more than for wrapped or boxed cakes. Snug packages or wraps have less effect on freezing or defrosting rates than typical

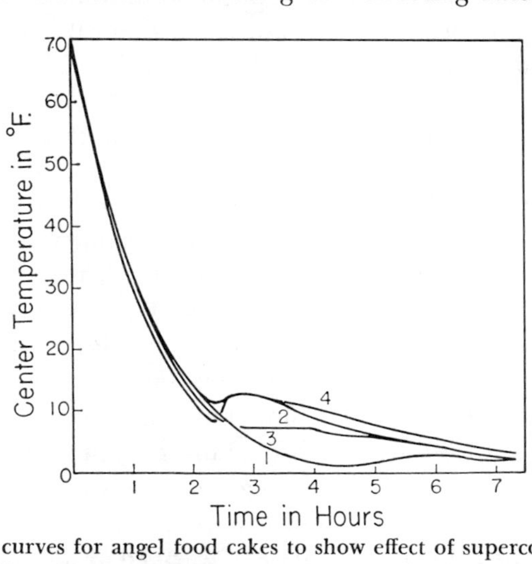

Fig. 7.4. Cooling curves for angel food cakes to show effect of supercooling on rate of temperature decline (33). Reproduced by permission from *Food Technol., 13,* 99 (1959) copyright by Institute of Food Technologists.

paperboard cake boxes allowing air space around the product. Icings on layer cakes prolong freezing or defrosting nearly as much as paperboard cake boxes. Cream-type icings insulate cake more effectively than fudge-type icings. Chiffon cakes freeze and defrost more rapidly than angel food cakes of the same size (33).

As with other frozen bakery products, defrosting with warmed and gently moving air is helpful to speed up defrosting and to minimize problems that may arise from excessive condensation of moisture on the cold surfaces (23). The extent to which the defrosting air can be warmed, however, is limited. Air temperatures in excess of 80°F. (27°C.), can cause damage to several of the major types of icings (33). Icings must also be protected from moisture condensation to avoid loss of gloss and other appearance features.

B. Changes During Storage

Although cakes are quite tolerant to variations in freezing and defrosting conditions, they are affected significantly by storage temperature. Commercial experience has been good under a wide variety of conditions (34,35), but careful studies have shown the desirability of storage temperatures of 0°F. (−18°C.) or lower (36). Changes during storage are somewhat characteristic for different types of cake.

C. Batter Cakes

Layer cakes are reasonably stable at 10°F. (−12°C.), but within 1–2 weeks a steady increase in crumb firmness and harshness makes them distinguishable from fresh unfrozen cakes (36). At 0°F. (−18°C.) and below, in contrast, the crumb became progressively softer and more moist-seeming. Panel judges commented favorably on this type of change during early weeks of experimental storage, but eventually the crumb became excessively gummy and somewhat pasty. Taste panels could detect this type of difference from freshly baked cake crumb within 3–4 weeks, but the eating quality of the cakes remained high throughout the 10-week duration of the tests (36). At −30°F. (−34°C.) the cakes could not be distinguished from freshly baked cake at the end of the 10-week study. Although definitive data are not available, there seems little doubt that layer cakes held at 0°F. (−18°C.) will remain in good saleable condition for 3 or 4 months. Layer cakes can be judged as about the most stable of all types of soft cakes for frozen storage.

Pound cakes, unlike the other cakes studied, immediately after freezing and defrosting could be distinguished from fresh unfrozen controls on the basis of greater firmness (32,36). A rapidly increasing crumbliness and mealiness of texture, however, led to an overall tenderization apparently well accepted by taste panel members. For example, cakes held for 5 weeks at 0 or −10°F. (−18 or −23°C.) could not be distinguished from freshly baked unfrozen cakes. Cakes held at 10°F. (−12°C.), on the other hand, did not show this improvement in acceptability. By making some allowances for possible personal variations in acceptability, pound cakes can be judged to be among the most stable during frozen storage for practical purposes (36).

D. Foam Cakes

Chiffon cakes, unlike batter cakes, show the same type of texture change at 10°F. (−12°C.) as at 0°F. (−18°C.) and lower, the major change being a slow loss of crumb resiliency and an increasing crumbliness and harshness of texture. A somewhat unique granular crumbliness was observed only with this type of cake. As soon as such changes became detectable, taste panels consistently preferred the unfrozen freshly baked cakes used as controls (36). The approximate time periods involved were 1 week at 10°F. (−12°C.) as at 0°F. (−18°C.), 4 weeks at −10°F. (−23°C.), and 5 weeks at −30°F. (−34°C.). Again, the still high quality of the cakes after the experimental storage periods at 0°F. (−18°C.) or lower should be emphasized. A commercial storage life at 0°F. (−18°C.) of at least 2 months can be confidently expected.

Changes in texture of angel food cakes during frozen storage were characterized by the relatively rapid loss of the unique toughness of the very soft crumb of the freshly baked cake (36). During early weeks of storage, panel judges commented favorably upon the increased tenderness of the crumb. Later, however, this change became distinctly undesirable, resulting in a crumb that tended to be breadlike. The change in texture was detectable within a week at 10°F. (−12°C.), 2 weeks at 0 and −10°F. (−18 and −23°C.), but required more than 8 weeks at −30°F. (−34°C.) (36). These times should in no way be construed as limits of storage life because of the good acceptance of the newly acquired tender bite of the frozen cakes. Acceptance of the bready texture that developed later was so variable and depended so much on personal preference that no significant decisions could be made regarding storage limits for angel food cakes. A limit of 3 months of 0°F (−18°C.) can, however, be suggested. Note should also be made, however, of the great advantage of storing angel food cakes at very low temperatures (−30°F.) (−34°C.); this advantage is greater than for any of the other cakes studied (36).

E. Cake-Type Doughnuts

Freshness ratings for cake-type doughnuts were significantly below those of controls after 14 days of storage at 10°F. (−12°C.), although firmness measurements had increased significantly within 7 days (31). At 0 and $-20°F.$ (-18 and $-29°C.$) freshness ratings required 21 days to become significantly different from those for controls, but no further decreases occurred during the remainder of the 8-week duration of the studies (31). Firmness measurements remained steady through the early weeks of storage, but were significantly higher at the end of the test.

As with other cake or cakelike frozen bakery products, cake-type doughnuts freeze and store very well, provided storage temperature is at 0°F. ($-18°C.$) or below. Good results may be expected for periods of at least 3 months for practical purposes. The products should be well packaged, however, to guard against moisture losses which can occur at a very low but steady rate.

VI. FRUIT PIES

Commercially produced frozen pies have met wide consumer acceptance for many years, and comprise by far the greatest proportion of frozen bakery products on either a weight or a sales basis (3). Fruit pies, especially, are well adapted to this manner of merchandising. Frozen, already baked pies freeze and keep well, but offer little advantage in convenience over unbaked pies. Easier handling and superior freshness at time of serving strongly favor the use of unbaked pies for commercial freezing operations. The tolerance of unbaked crusts to physical handling and the relatively good stability of fruit fillings permit the ready marketing of high-quality products.

A. Unbaked Pies

General industry experience with unbaked pies has been very good despite use of a wide range of formulations, handling practices, packaging, and size of product. Until recently, however, little published information was available for evaluating effects of product and processing variables.

1. Freezing Considerations

Because the unbaked frozen pies are generally not defrosted before baking, or are defrosted by the consumer just before baking, the commercial operator need concern himself only with possible effects of variations in freezing practices. Moreover, because baking takes place just before use of the product, most of the desirable final quality characteristics of the product are not directly subjected to effects of freezing variables. On the other hand, freezing rates are well known to affect retention of original quality of fruits, so that fast freezing would seem desirable to maintain best quality of fruit fillings. Quality and stability of other types of fillings could also be important.

Systematic study of the effect of freezing rates on unbaked cherry and apple pies revealed no significant effect on the quality of fillings or crusts under conditions covering the range used commercially (37). The pies were frozen in as short a time as 1–2 hr. and as long a time as 28 hr. to simulate the packing of pies in a shipping carton prior to freezing. The pies were then baked without defrosting. A slight soaking of the bottom crust was noted in some of the pies, but this was no more prevalent in the pies frozen at the slowest rate than in those frozen most rapidly (37).

Because fruit pie fillings are relatively high in moisture content, cooling curves show a substantial inflection due to release of latent heat. The freezing zone generally appears in the range of 15–20°F. (−9 to −7°C.), and evidence of supercooling is often seen when freezing rates are slow (37).

When selecting conditions, equipment, packages, and so forth for the commercial freezing of pies, the good tolerance of fruit pie fillings to variations in freezing rate allows primary emphasis on economic factors. Nevertheless, an operator would be wise to provide for reasonably rapid freezing rates in order to insure retention of best quality in the fillings.

2. *Storage Considerations*

Experimental study of factors concerning stability of frozen, unbaked fruit pies has focused attention on the crust and on the fillings, separately. One study showed that soaking of bottom crusts tended to be the limiting factor to satisfactory storage life (37). After 3 weeks at 20 or 10°F. (−7 or −12°C.) bottom crusts became very soggy and top crusts were soaked in irregular spots. During baking these spots became black and developed a bitter taste. Moreover, the fillings of these pies boiled out during baking because of the weakened structure of the crusts. This caused adhesion to the pan and difficulty in serving. As a result the pies were judged unsuitable for commercial distribution. At 0 and −20°F. (−18 and −29°C.), soaking of bottom crusts was slight for apple pies and moderate for cherry pies after 16 weeks. For boysenberry pies it was moderate after 8 weeks. No change in quality of fillings was noted during the 16-week course of the study (37).

In another study using commercially prepared unbaked fruit pies (38) a change in the flavor of the filling, rather than in the flavor or texture of the crust, was the primary factor affecting judgment of the stability of a pie. However, minor color and appearance changes in the crusts were detected before flavor changes were detectable. In general, flavor stability varied from about 6 to 18 months at 0°F. (−18°C.), depending upon the fruit and the method by which it was processed before being used for the pies. Variations in processing and packaging of the frozen fruit used to prepare the pies appeared to account for approximately a twofold variation in the stability of frozen unbaked apple and boysenberry pies. The best treatments yielded apple, boysenberry, peach, and cherry pies of excellent flavor stability. Blueberry pies were the least stable (38).

Relative flavor stabilities of apple pie fillings in this study were 2 weeks at 30°F. (−1°C.), 7 weeks at 20°F. (−7°C.), 24 weeks at 10°F. (−12°C.), and 36 weeks at 0°F. (−18°C.) (38). Figures for crust from the same mix but stored separately were 8 weeks at 30°F. (−1°C.), 40 weeks at 20°F. (−7°C.), 56 weeks at 10°F. (−12°C.), and greater than 104 weeks at 0°F. (−18°C.). Other studies have also shown the excellent stability of pie doughs, provided temperatures are low enough to suppress mold growth (39).

The above divergence of stability estimates for crusts of unbaked fruit pies emphasizes the rather wide range of properties such crusts may have at the outset of storage. Commercial pie doughs for unbaked frozen pies usually contain from 50 to 65% of shortening, on the basis of flour weight (37). At a midrange value of 60%, crusts can be rich enough to provide very high-quality crusts but the flour used and the manner in which the crusts are made can have a great effect on the subsequent performance of the crusts during storage of the unbaked frozen pie. A major factor, therefore, in maintaining high quality in the pies is the production of crusts sufficiently coherent to hold filling ingredients, yet have the eating qualities desired. Experience, alone, in day-to-day operations of each particular production plant with the ingredients available can guide successful operations in this regard. Published information is not readily available.

B. Baked Pies

Although fully baked pies freeze rather well and have been frozen in the home quite widely, several disadvantages in comparison to frozen unbaked pies have largely precluded their use commercially, except for specialty types, such as cream pies or custard pies, including pumpkin. Baked crusts, tender or flaky enough to justify a rating of high quality, are rather fragile and easily damaged during distribution. During storage their flavor and freshness decline steadily and are not readily restored by reheating after defrosting. Pies baked after freezing emerge from the oven at the peak of their attractiveness. Crusts of baked pies are more susceptible to sogginess and uptake of filling liquids than the crusts of unbaked pies.

The types of pies with fillings that do not lend themselves readily to reheating are customarily frozen in baked condition or in baked shells. These include cream pies, chiffon pies, and custard-types such as pumpkin. Practices used commercially for these types of pies have largely been developed by the companies marketing the products and no significant guiding information appears to have been published. Nevertheless, operators would be well advised to freeze baked pies rather rapidly and to store them at 0°F. (−18°C.), or below, for best retention of original freshness.

VII. BREAD AND ROLL DOUGHS

Commercial marketing of frozen doughs for home baking of bread and rolls achieved a healthy impetus during the early 60's (40), but subsequent growth has been seriously hampered by problems associated with product quality and its deterioration during frozen storage, especially in retail cabinets. On the other hand, the freezing of unbaked dough products in central locations for subsequent on-premise baking in supermarkets continues to grow in volume. The chief problems with frozen doughs relate to the stability of yeast performance upon which the final product quality depends very greatly.

The interaction of how the product is made initially and how it is handled while frozen affects frozen dough quality at time of use more, perhaps, than is true for any other bakery product. Frozen bread doughs are not of marked convenience for housewives, but the resultant freshly baked character exerts a considerable appeal which appears to be largely responsible for the success of this type of product. Improper treatment at any time prior to final use can result in complete failure or very serious impairment of the ability of the doughs to rise and to yield anything near a satisfactory product. Close attention to all phases of the operation are therefore necessary.

Retention of sufficient yeast viability and gassing power during frozen storage of doughs to avoid excessive or even impossible proofing times after thawing the frozen product is most important for doughs destined for home use. Frozen products intended for institutional use or in-store bake-off seldom are held for more than 2 or 3 weeks and cause no particular problem. Products for home use, however, must have a storage life of at least 10–12 weeks at 0°F. (−18°C.) for a successful commercial operation.

Several factors affect the ability of yeast to withstand frozen storage, but most workers agree that the amount of fermentation given doughs prior to freezing is the most important (41–44). Yeast spores are considerably more cold-resistant than vegetative cells, and the cells in unfermented frozen doughs are more resistant than those in fermented doughs (41). For this reason most doughs are frozen with little or no fermentation (40). Information on details of preparing doughs for freezing is available elsewhere (41,43,44). See also Chapter 3 of this volume.

The rate at which doughs are frozen has been reported to affect the preservation of the biological activity of the yeast (41). In fact, flash freezing with liquid nitrogen is used by at least one large frozen dough plant (45). In one of the few published studies on this topic one-pound doughs in a freezer at −10°F. (−23°C.) required 2½ hr. to reach 0°F. (−18°C.), those in a freezer at −40°F. (−40°C.) required only 1 hr. to reach 0°F. (−18°C.) (46).

Storage of frozen doughs at 0°F. (−18°C.), or slightly below, seems quite satisfactory for the minimum 10–12 weeks required for successful commercial practice, provided the doughs were properly prepared initially. Lower temperatures would be difficult to achieve in retail distribution. Maintaining a satisfactorily uniform low temperature is a very important factor and one that strains the tolerance of doughs, especially during the period after purchase by the consumer and transport to a home freezer. Refreezing of doughs that have softened is reputed to result in almost complete loss of yeast activity. Confirmation of this loss of activity has not been published, and there is no evidence that it has been investigated.

REFERENCES

(1) G. T. Carlin, *Baking Ind., 124,* No. 1575, 28 (1965).
(2) N. L. Rollag and R. V. Enochian, "The Freezing of Commercial Bakery Products: Current Practices, Problems, and Prospects," U.S. Dept. Agr. Marketing Research Report No. *674* (1964).
(3) Anon., *Quick Frozen Foods,* p. 180 (Nov. 1967).
(4) D. K. Tressler and C. F. Evers, "Bread and Rolls," in *The Freezing Preservation of Foods,* 3rd ed., Vol. 2, Avi Publishing, Westport, Connecticut, 1957, p. 312.
(5) C. W. Bice and W. F. Geddes, *Cereal Chem., 26,* 440 (1949).
(6) C. H. Bailey and E. Munz, *Cereal Chem., 15,* 413 (1938).
(7) W. B. Bradley and J. B. Thompson, *Cereal Chem., 27,* 331 (1950).
(8) W. G. Bechtel, D. F. Meisner, and W. B. Bradley, *Cereal Chem., 30,* 160 (1953).
(9) T. J. Schoch and D. French, *Cereal Chem., 24,* 231 (1947).
(10) J. R. Katz, *Bakers Weekly, 81,* No. 12, 34 (1936).
(11) F. R. Senti and R. J. Dimler, *Bakers Dig.,* 34, No. 1, 28 (1960).
(12) J. R. Katz, "Gelatinization and Retrogradation of Starch in Relation to the Problem of Bread Staling," in *A Comprehensive Survey of Starch Chemistry,* Vol. 1, R. P. Walton, Ed., Chemical Catalog, New York, 1928, p. 100.
(13) J. W. Pence and N. N. Standridge, *Cereal Chem., 32,* 519 (1955).

(14) E. C. Edelman, W. H. Cathcart, and C. B. Berquist, *Cereal Chem., 27,* 1 (1950).
(15) D. F. Meisner, *Baker's Dig., 27,* 109 (1953).
(16) W. R. Steller and C. H. Bailey, *Cereal Chem., 15,* 391 (1938).
(17) D. J. Kirk, *Baker's Dig., 27,* No. 1, 58 (1963).
(18) J. W. Pence, N. N. Standridge, T. M. Lubisich, D. K. Mecham, and H. S. Olcott, *Food Technol., 9,* 495 (1955).
(19) J. W. Pence, T. M. Lubisich, D. K. Mecham, and G. S. Smith, *Food Technol., 9,* 342 (1955).
(20) J. W. Pence, *Baker's Dig., 35,* No. 5, 64 (1961).
(21) H. C. Mannheim, M. P. Steinberg, A. I. Nelson, and T. W. Kendall, *Food Technol., 11,* 384 (1957).
(22) J. W. Pence, N. N. Standridge, and M. J. Copley, *Food Technol., 10,* 492 (1956).
(23) E. J. Barta, *Baker's Dig., 32,* No. 4, 50 (1958).
(24) B. Belderok and W. H. B. Wiebols, *Food Technol., 18,* No. 11, 155 (1964).
(25) K. Kulp and W. G. Bechtel, *Cereal Chem., 35,* 276 (1958).
(26) K. Kulp and W. G. Bechtel, *Cereal Chem., 37,* 170 (1960).
(27) J. W. Pence, N. N. Standridge, D. R. Black, and F. T. Jones, *Cereal Chem., 35,* 15 (1958).
(28) N. H. Walker, *Baking Ind., 107,* 54 (Feb. 23, 1967).
(29) J. W. Pence, N. N. Standridge, D. K. Mecham, T. M. Lubisich, and H. S. Olcott, *Food Technol., 10,* 76 (1956).
(30) K. Kulp and W. G. Bechtel, *Food Technol., 15,* 273 (1961).
(31) W. G. Bechtel and K. Kulp, *Food Technol., 14,* 391 (1960)
(32) J. W. Pence and N. N. Standridge, *Cereal Chem., 35,* 57 (1958).
(33) J. W. Pence and M. Hanamoto, *Food Technol., 13,* 99 (1959).
(34) W. H. Cathcart, *Cereal Sci. Today, 1,* 8 (1956).
(35) D. F. Copell, *Baker's Weekly, 170,* No. 8, 36 (1956).
(36) J. W. Pence and M. Heid, *Food Technol., 14,* 80 (1960).
(37) K. Kulp and W. G. Bechtel, *Food Technol., 16,* No. 7, 104 (1962).
(38) D. G. Guadagni, J. Harris, and S. Okano, *Food Technol., 17,* No. 7, 114 (1963).
(39) E. W. Cole and D. K. Mecham, *Food Technol., 14,* 115 (1960).
(40) N. L. Rollag and R. V. Enochian, "Market Potentials for Frozen Dough," U.S. Dept. Agr. Marketing Research Report No. 787 (1967).
(41) W. J. Godkin and W. H. Cathcart, *Food Technol., 3,* 139 (1949).
(42) B. Meyer, R. Moore, and R. Buckley, *Food Technol., 10,* 165 (1956).
(43) P. P. Merritt, *Baker's Dig., 34,* No. 4, 57 (1960).
(44) L. Kline and T. F. Sugihara, *Baker's Dig., 42,* No. 10, 44 (1968).
(45) P. Gelber, *Food Process. Market, 27,* No. 12, 32 (1966).
(46) K. Lorenz and W. G. Bechtel, *Baker's Dig., 38,* No. 6, 59 (1964).

Chapter 8

QUALITY AND STABILITY OF FROZEN MEATS

Mogens Jul
Danish Meat Research Institute, Roskilde, Denmark

INTRODUCTION

So far, comparatively few time–temperature tolerance data from so-called "T-TT" experiments have been carried out on frozen meat or meat products. Because of this, much in this chapter has had to be based on rather limited data. In addition, some of the information is derived from older data that have been recalculated to fit into the usual T-TT pattern of presentation. In doing so, however, a certain amount of inaccuracy may have been introduced.

As is the case for fish and poultry, the keeping time for frozen meat depends to a considerable extent on the degree of denaturation of the proteins. This denaturation takes place during freezing and more specifically during freezer storage. It may affect the water-binding capacity of the meat, and a high degree of denaturation generally will result in increased drip from the meat upon defrosting.

Like many fish products, many meats contain a considerable amount of lipids that may easily turn rancid. Thus, any method of packaging which will exclude oxygen completely may increase greatly the storage life of some of these products. Recent years have seen much improvement in packaging material and packaging techniques. This is probably one of the reasons why recent findings often show practical storage lifetimes for meat products that greatly exceed those indicated in the literature only a few years ago (*cf.* Table 8.1).

As is the case with most poultry, fruits, and vegetables, man has a considerable degree of control of the breeding and feeding of the animals from which the meat is derived. Improved feeding may increase considerably the storage life of the resulting frozen meat products (*cf.* Fig. 8.10).

Whereas most other foods are found to be best when frozen as fresh as possible, a certain amount of ripening improves the quality of some meats. This creates specific problems in the freezing preservation of meat products.

II. DETERMINING JUST NOTICEABLE DIFFERENCE AND PRACTICAL STORAGE LIFE

As mentioned in Chapter 10 (Sec. II.C.1) most control samples in the early T-TT experiments were held at −20°F. (−28.9°C.) and the time up to the "Just Noticeable Difference" (JND) as defined in Chapter 10 was determined as the earliest time when a difference between the product being tested and the controls stored at this temperature could be detected by trained taste panels. This is not quite satisfactory for meat products, and especially not for pork or mutton products, since the organoleptic quality of these products may deteriorate quite

Table 8.1. Traditionally Accepted Maximum Practical Storage Life in Days at 0°F. (−17.8°C.) Compared with Maximum Storage Life for Well-Packed and Well-Formulated Products

	Traditional practical storage life	Practical storage life from Table 10.1.[a]	Practical storage life for well-packed, well-formulated products
Beef	240 (IIR (1))	500	Over 700 (Dalhoff and Jul (2))
Beef, chopped	Not available	Not available	Over 700 (Dalhoff and Jul (2))
Beef, cooked with gravy	Not available	Not available	Over 700 (Dalhoff and Jul (2))
Pork	135 (IIR (1))	270	Over 610 (Dalhoff and Jul (2))
Pork, ground	Not available	Not available	Over 645 (Dalhoff and Jul (2))
Pork, cooked with gravy	Not available	Not available	Over 700 (Dalhoff and Jul (2))
Lamb	170 (IIR (1))	460	Not available
Salted pork	60 (Wiesman (3))	Not available	Over 140 (Dalhoff (4))
Pork sausage	60 (Wiesman (3))	150	Approx. 600 (Bøgh-Sørensen (5))
Goulash	Not available	Not available	Over 1200 (Bøgh-Sørensen (5))

[a] From Chapter 10, Section II. B.

markedly at control temperatures as low as −40°F. (−40°C.), *cf.* Fig. 8.1. Thus, it is generally not quite realistic to use comparisons with frozen control samples.

Dalhoff and Jul (2) have suggested that instead of a comparison between test and control samples, one might use a decrease of 1 point in an organoleptic scale of 0–10 (Copenhagen) or −5 to +5 (Roskilde) as an indication of the time when the just noticeable change has been reached. Similarly, a decrease down to a score of −2 in the latter scale has been used to indicate the time when the products are no longer fully acceptable. Dalhoff and Jul (2) have found these criteria to correspond quite closely with the terms "Just Noticeable Difference" and "Practical Storage Life" (PSL) as defined in Chapter 10. In most of the present chapter these or similar definitions have been used. Since this has necessitated recalculation of results from earlier experiments and certain interpretations of earlier organoleptic tests, inaccuracies will have been introduced. However, by presenting the data in this form comparisons between different findings are facilitated.

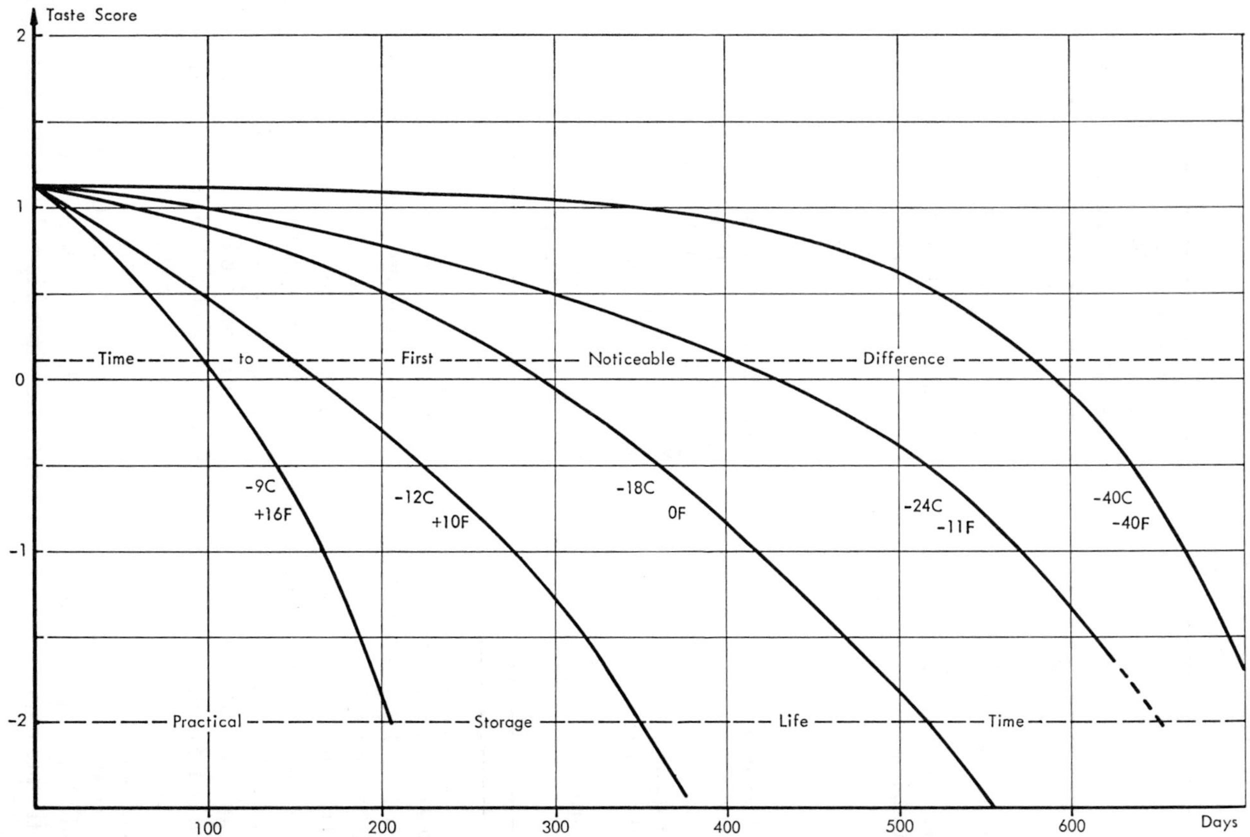

Fig. 8.1. Results of taste tests for pork chops at different storage temperatures. Criteria: taste scale of −5 to +5 (Roskilde). Packaging: wrapped in polyethylene in a freezer carton. Source: **Bøgh-Sørensen (5).**

It will be noticed that the use of a score of −2 in the −5 to +5 scale (Roskilde) as a limit for PSL causes difficulties for products of inferior quality. Similar reservations exist, however, for the use of triangle comparisons for such products.

From Fig. 8.1 it appears that the use of such definitions rather than comparisons with controls held at low temperatures is necessary. It may, however, also have some advantage in comparison with the usual use of control samples and triangle tests, since, as mentioned in Chapters 4 and 10, Guadagni required four replicates, *i.e.,* 12 samples for each time and temperature when he was using the triangle test technique. With a point scale system, fewer samples are required and more time–temperature data may, therefore, be determined, although with less analytical accuracy, with a given number of samples and taste-testing sessions.

It also may be worth noting that where control samples are used they should be kept at very low storage temperatures; in addition, they should be very well packaged, *e.g.,* vacuum-packed in tight-fitting bags made of film having very low water vapor and oxygen permeability. Such packaging may do more to preserve the control samples unchanged than merely maintaining a very low storage temperature for the controls; *cf.* curves *2* and *3* in Fig. 8.4.

In some T-TT experiments, criteria other than organoleptic tests have been used; *e.g.,* decrease in vitamin content. Results obtained with such objective tests are presently not available for meat products. Insofar as vitamin content is concerned, Lee *et al.* (7) found that freezer storage has no effect on the content of thiamine, riboflavin, niacin, pantothenic acid, or pyridoxine in meats, and analyses for these vitamins could, therefore, not be used to evaluate storage life.

The figures for keeping quality times for frozen meat products accompanying this chapter give a somewhat incoherent impression, *e.g.,* some curves are very flat, some others are steep, some are curved, others are straight lines, and some actually cross. This latter may be interpreted as indicating that there are cases where one product shows a longer storage life at, say −10°C. (14°F.), than another product, while the positions are reversed for the two products at −20°C. (−4°F.) However, most of this apparent lack of uniformity is likely to be due to the different conditions under which these products have been stored, packaged, and tested. A much clearer picture is likely to be the result if standard procedures are adopted. One of the more promising fields of research in the frozen meat industry is the determination of keeping quality curves, covering various environmental factors, using an experimental technique that would be suited for comparison of results between various laboratories and across national boundaries.

III. STORAGE LIFETIME DATA

Few data are available from exact experiments on the storage life of frozen meat and meat products; often, one has to rely on older, empirical data from observations in which full use of modern packaging

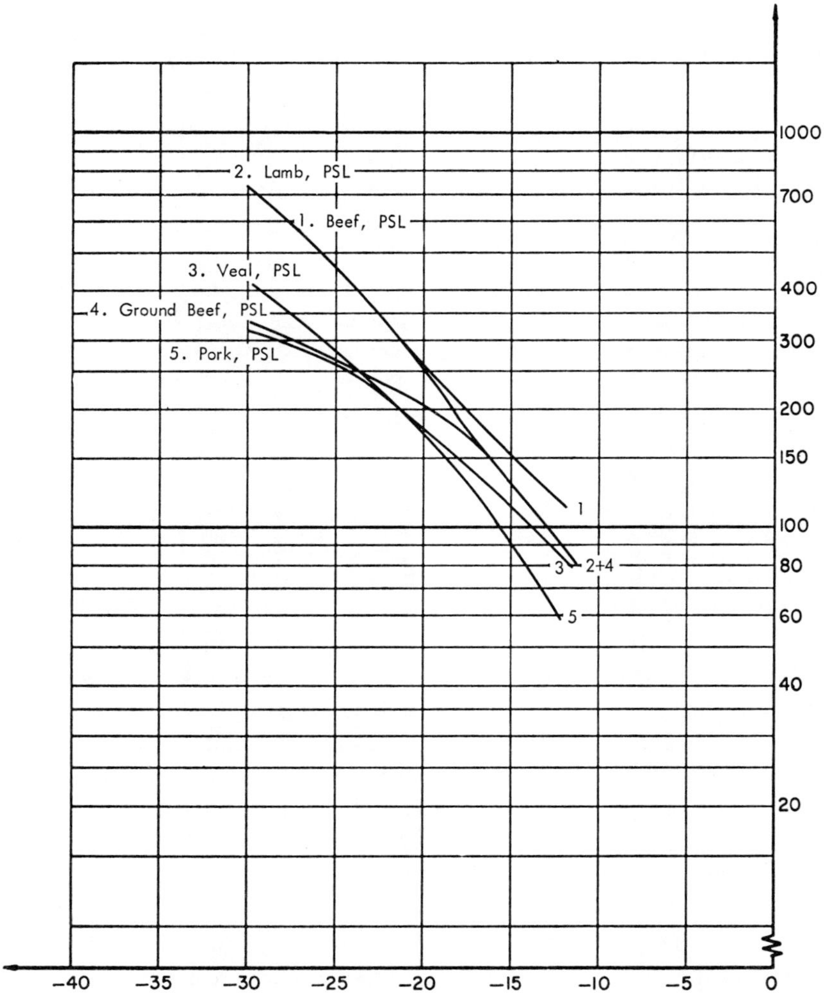

Fig. 8.2. Empirical practical storage life (PSL) for some frozen meats. Criteria: general acceptability. Products: no detailed information available. Packaging: no information available. Source: Wiesman (3).

materials and techniques was not made, and where often insufficient information is given on the nature of the product, the process it has undergone, the packaging, or the criteria used in determining storage lifetime.

Fig. 8.2 presents some older data on frozen meat often referred to in the literature. Fig. 8.3 presents somewhat similar data of more recent origin, but also based largely on practical experience.

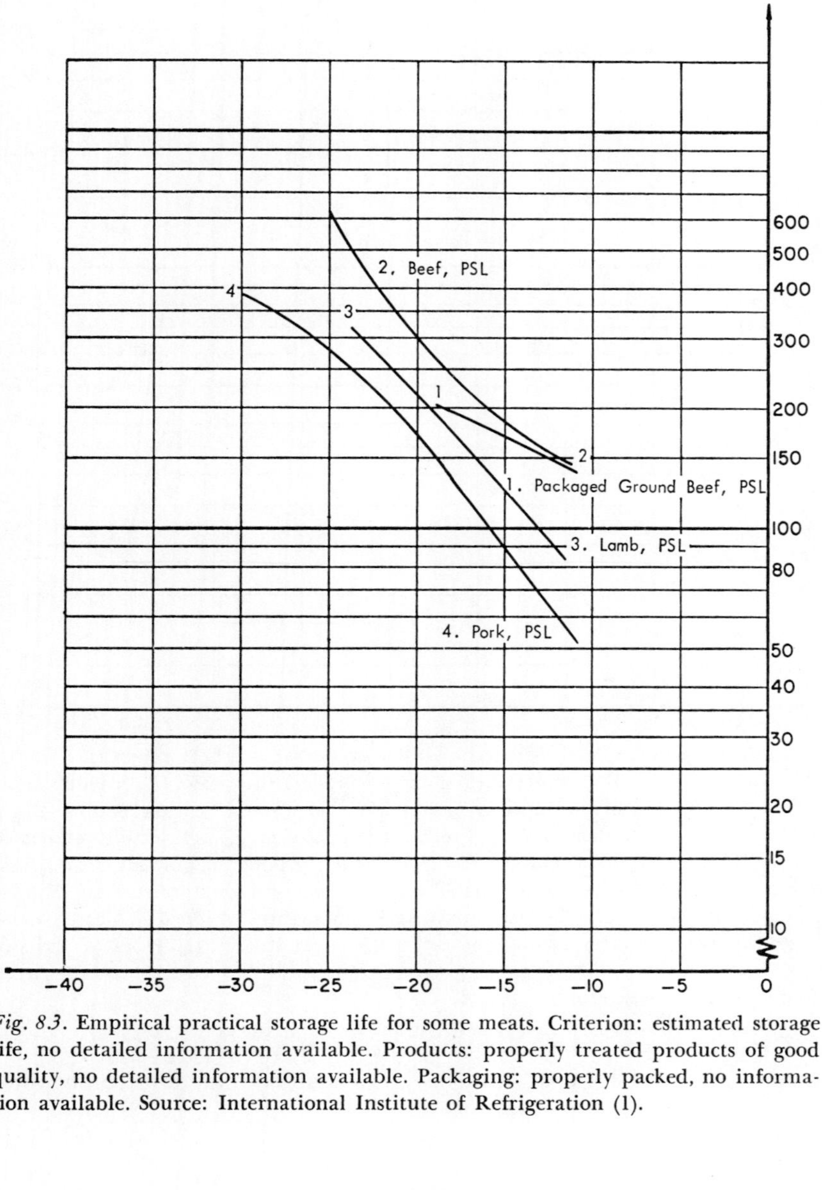

Fig. 8.3. Empirical practical storage life for some meats. Criterion: estimated storage life, no detailed information available. Products: properly treated products of good quality, no detailed information available. Packaging: properly packed, no information available. Source: International Institute of Refrigeration (1).

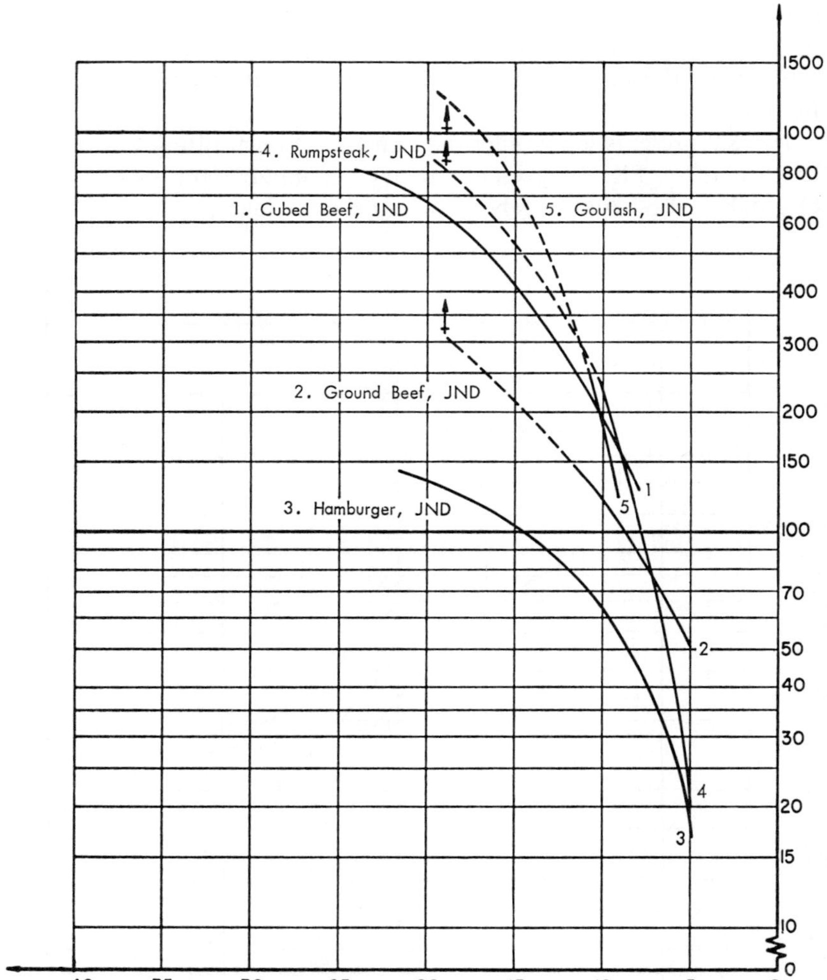

Fig. 8.4. Time up to "Just Noticeable Difference" for some beef products. Criteria: time for a decrease of 1 point in a − 5 to + 5 organoleptic scale. (*1*) Product: fresh, cubed beef. Packaging: wrapped in polyethylene in a freezer carton. Source: Bøgh-Sørensen (5). (*2*) Product: freshly ground beef. Packaging: tight-fitting, evacuated "Saran." Source: Dalhoff and Jul (2). (*3*) Product: ground hamburger meat. Packaging: wrapped in coated paper, placed in heat-sealed, coated freezer carton. Source: Dalhoff and Jul (2). (*4*) Product: rump steak. Packaging: wrapped in polyethylene, placed in coated freezer carton. Source: Dalhoff and Jul (2). (*5*) Product: goulash; see Fig. 8.9, curve *2*. Packaging: freezer carton with heat-sealed, coated paper. Source: Bøgh-Sørensen (5).

For comparison, Fig. 8.4 shows some newer storage-life data for frozen beef and Figs. 8.5 and 8.6 show similar data for frozen pork, based on exacting taste tests.

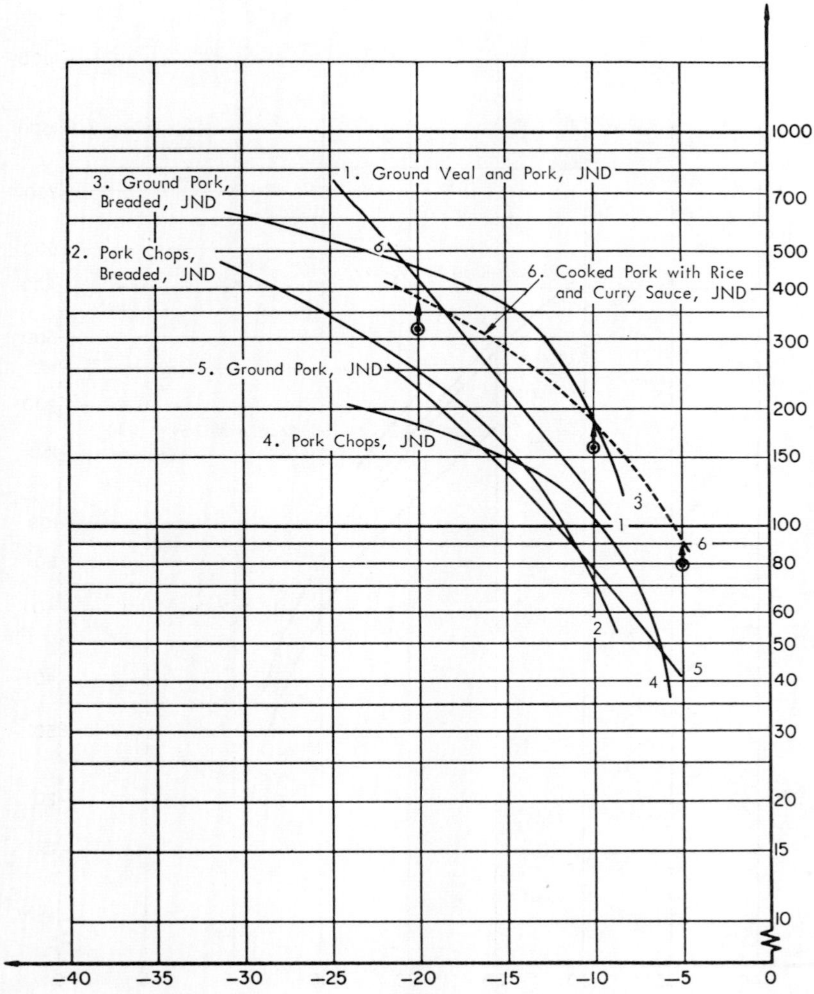

Fig. 8.5. Time for "Just Noticeable Difference" for some pork products. Criteria: time for a decrease of 1 point in a −5 to +5 organoleptic scale (Roskilde). (*1*) Product: fresh, ground veal and pork. Packaging: wrapped in polyethylene in a freezer carton. (*2*) Product: fresh pork chops, breaded. Packaging: wrapped in polyethylene in a freezer carton. (*3*) Product: breaded, ground pork. Packaging: wrapped in polyethylene in a freezer carton. (*1–3*) Source: Bøgh-Sørensen (5). (*4*) Product: fresh pork chops. Packaging: wrapped in wax paper, in heat-sealed carton. (*5*) Product: fresh ground pork. Packaging: as (*4*). (*6*) Cooked pork with rice and curry sauce. Product: as described. Packaging: freezer carton with heat-sealed, coated paper lining. Note: All samples used for curve *6* were found to be below the JND's state. Thus, keeping time may be still longer than indicated here. (*4–6*) Source: Dalhoff and Jul (2).

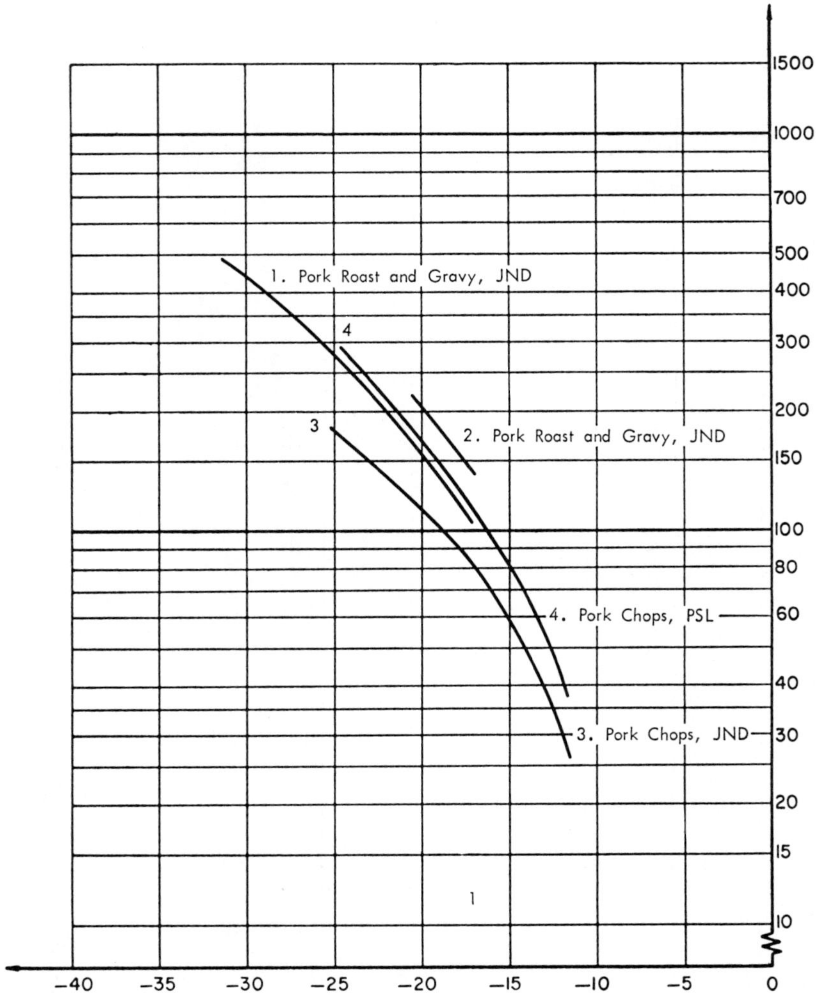

Fig. 8.6. Keeping times for some pork products. (*1* and *2*) Criteria: time for a decrease of 1 point in a 0–10 organoleptic scale (Karlsruhe). Product: pork roast (Schweinebraten) and gravy. Packaging: (*1*) aluminum container with mechanically tight-fitting closure; (*2*) aluminum container, hermetically sealed. Source: Ordynsky (8). (*3* and *4*) Criteria: (3) JND time for a decrease of 1 point in a 1–10 organoleptic scale (Karlsruhe); (*4*) PSL time for a decrease to a score of 7 in same scale. Product: middlefat pork chops. Packaging: Vapor proof (wetterfest) cellophane. Source: Gutschmidt (9).

For cured pork few exact data are available, and one may still have to refer to the empirical data given by Wiesman (3) and shown in Fig. 8.7. In this figure, however, more recent data for the time up to a JND for frozen, sliced bacon stored at 14°F. (−10°C.) are given,

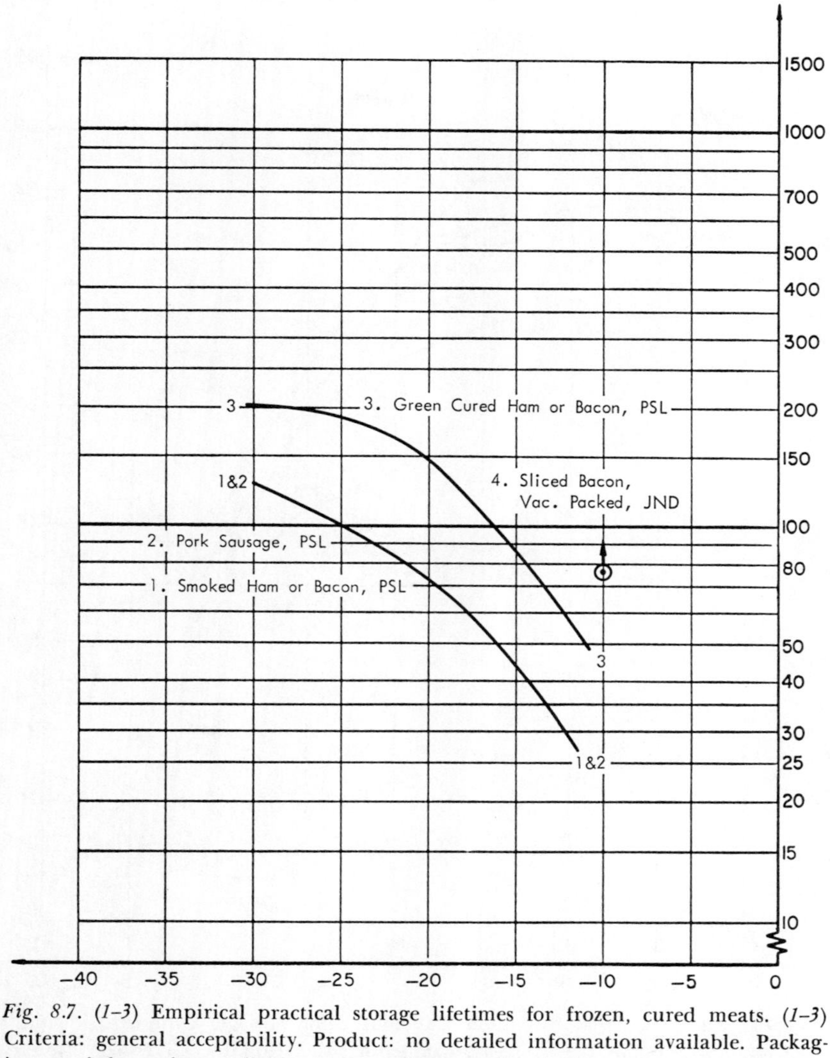

Fig. 8.7. (*1–3*) Empirical practical storage lifetimes for frozen, cured meats. (*1–3*) Criteria: general acceptability. Product: no detailed information available. Packaging: no information available. Source: Wiesman (3). (*4*) For comparison, the JND-time at 4°F. (−10°C.) for sliced, vacuum-packed bacon is indicated. Criteria: a decrease of 1 point in a −5 to +5 taste scale (Roskilde). Product: unsmoked, sliced bacon. Packaging: vacuum-packed in polyethylene/cellophane bags (Metathane X 320). Source: Dalhoff (4).

indicating that vacuum packaging this product in an oxygen-tight film may result in much prolonged storage life. More recent findings for one type of cured product, sausages, are given in Fig. 8.8, while Fig. 8.9 presents data for some prepared dishes, and Fig. 8.11 presents data for calves' liver.

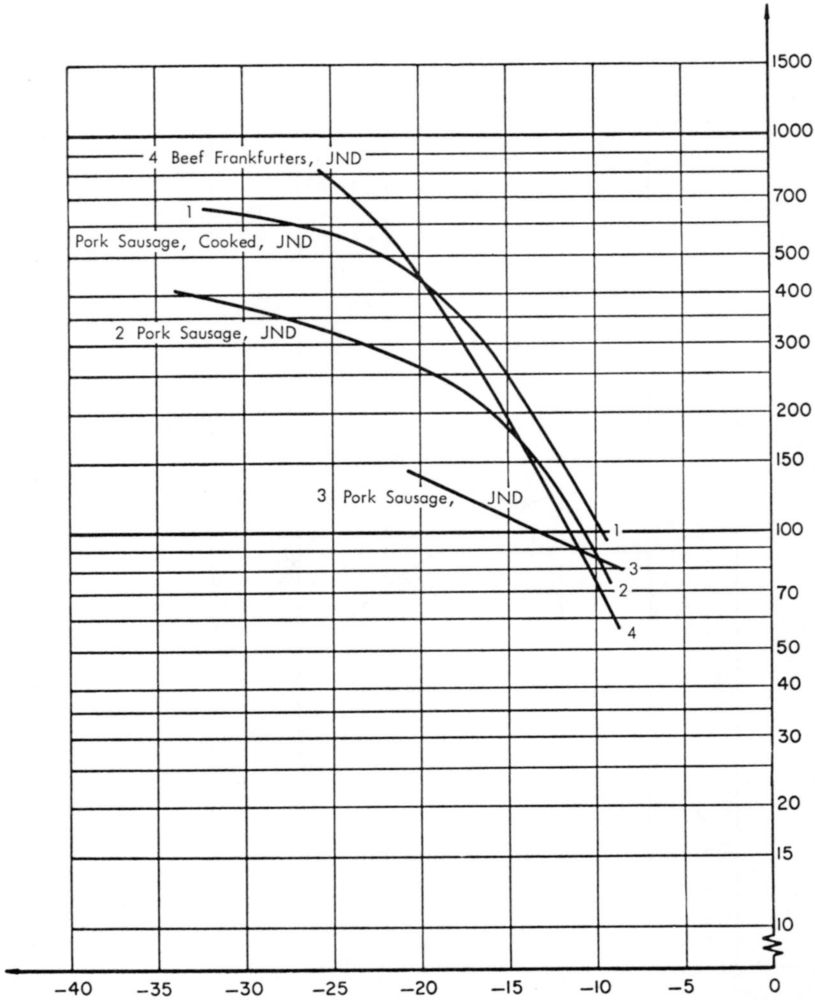

Fig. 8.8. Time for "Just Noticeable Difference" for sausage. Criteria: time for a decrease of 1 point in a −5 to +5 organoleptic scale (Roskilde). (*1* and *2*) Product: coarsely ground pork, containing about 2% salt (Medisterpølse). Process: (*1*) cooked; (*2*) uncooked. Packaging: freezer carton with heat-sealed, coated paper. Source: Bøgh-Sørensen (5). (*3*) Criteria: time for a decrease of 1 point in a 1–10 taste scale (Copenhagen). Product: similar to above. Packaging: aluminum foil. Source: Wismer-Pedersen and Sivesgaard (6). (*4*) Criteria: time for a decrease of 1 point in a −5 to +5 organoleptic scale (Roskilde). Product: frankfurter sausage, *i.e.,* beef, veal, and pork in natural casing. Packaging: freezer carton with heat-sealed, coated paper. Process: cooked, and smoked at 176°F. (80°C.). Source: Bøgh-Sørensen (5).

In Table 8.1, some of the traditionally quoted data for practical storage life of meat and meat products at 0°F. (−18°C.) are listed, together with the minimum storage life that may be expected where a

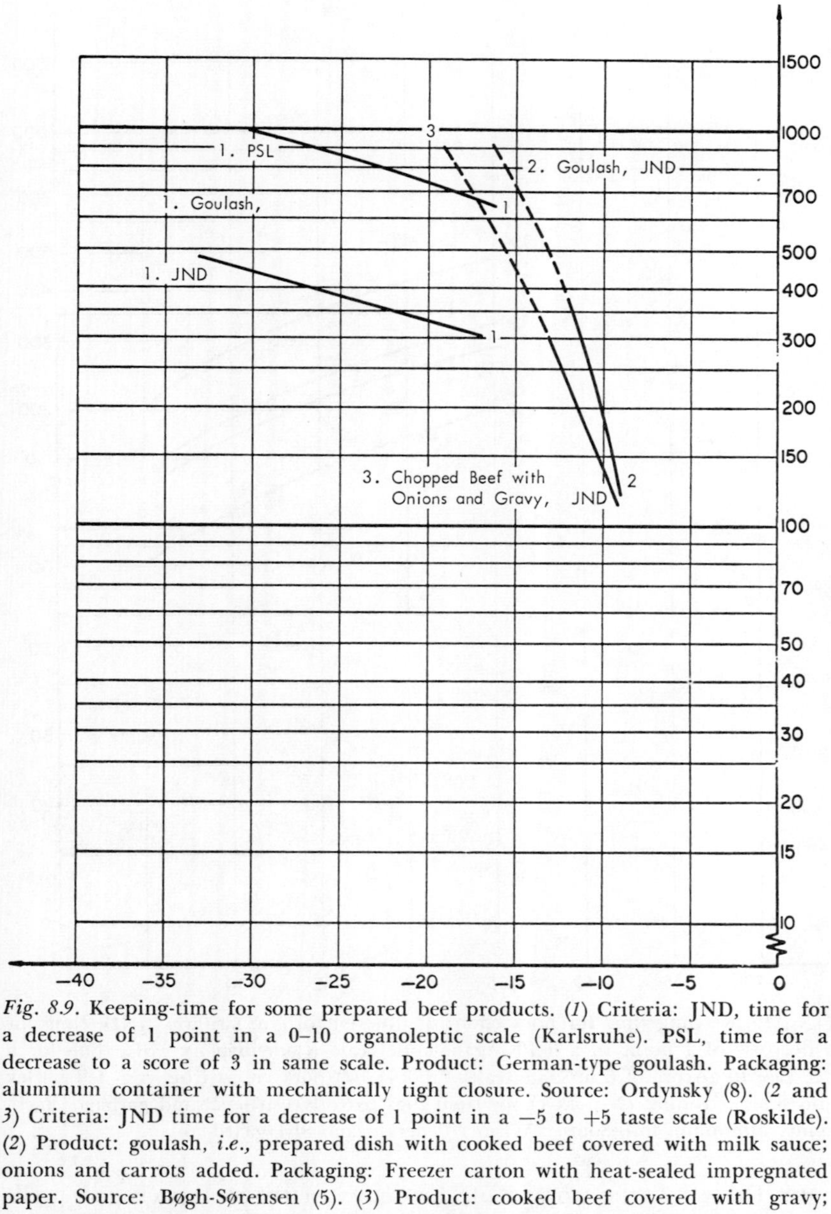

Fig. 8.9. Keeping-time for some prepared beef products. (*1*) Criteria: JND, time for a decrease of 1 point in a 0–10 organoleptic scale (Karlsruhe). PSL, time for a decrease to a score of 3 in same scale. Product: German-type goulash. Packaging: aluminum container with mechanically tight closure. Source: Ordynsky (8). (*2* and *3*) Criteria: JND time for a decrease of 1 point in a −5 to +5 taste scale (Roskilde). (*2*) Product: goulash, *i.e.*, prepared dish with cooked beef covered with milk sauce; onions and carrots added. Packaging: Freezer carton with heat-sealed impregnated paper. Source: Bøgh-Sørensen (5). (*3*) Product: cooked beef covered with gravy; onions added. Packaging: freezer carton with heat-sealed impregnated paper. Source: Bøgh-Sørensen (5).

good combination of product characteristics, processing method, and packaging precautions has been chosen. This table shows that measures of this nature may result in keeping times that exceed very considerably even those given in Table 10.1.

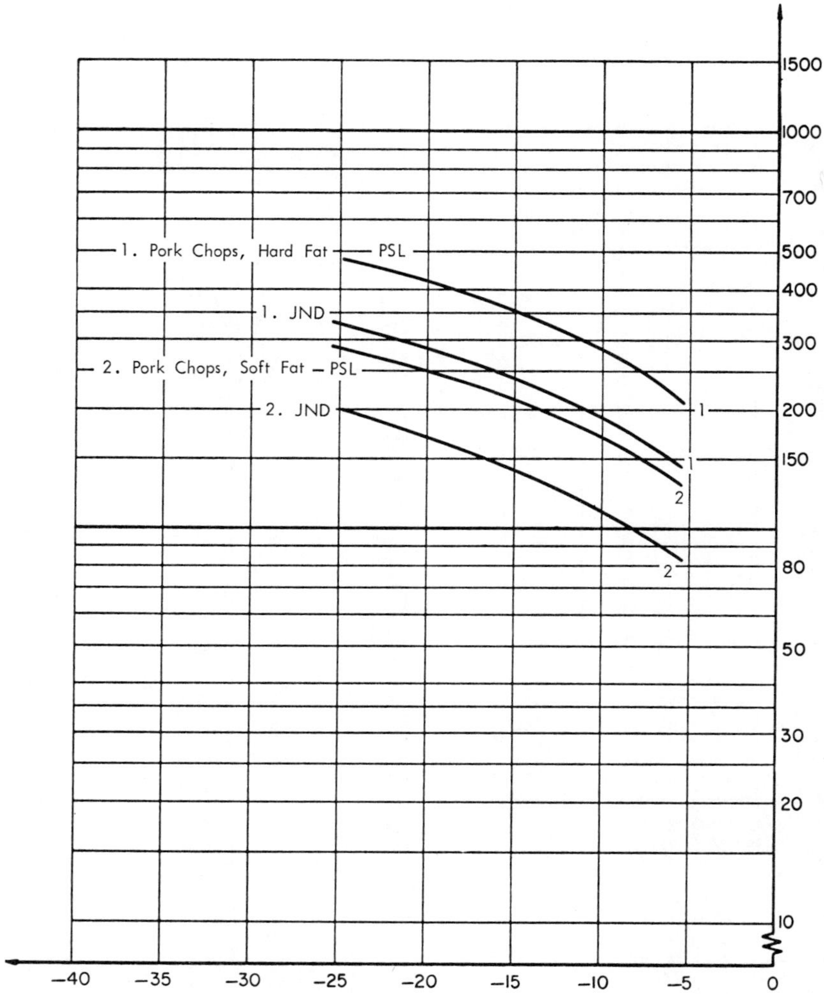

Fig. 8.10. Keeping time for pork chops of different origin. Criteria: JND, time for a decrease of 1 point in a 0–10 organoleptic scale (Copenhagen); PSL, time for a decrease to a score of 6 in same scale. Product: (*1*) pork chops from pigs fed barley and skimmed milk; (*2*) pork chops from pigs fed oily fish meal and garbage. Packaging: aluminum foil. Source: Wismer-Pedersen and Sivesgaard (6).

In Table 8.1, the storage life for many of the well-packaged, well-formulated products is not given except as: "Over 700 days." This is because, in planning the experiments on which the last column in the table is based, much shorter storage lives were expected than were actually found, *cf.* Fig. 8.3. Therefore, all samples were used up in the experiments before the actual end of the storage life was reached.

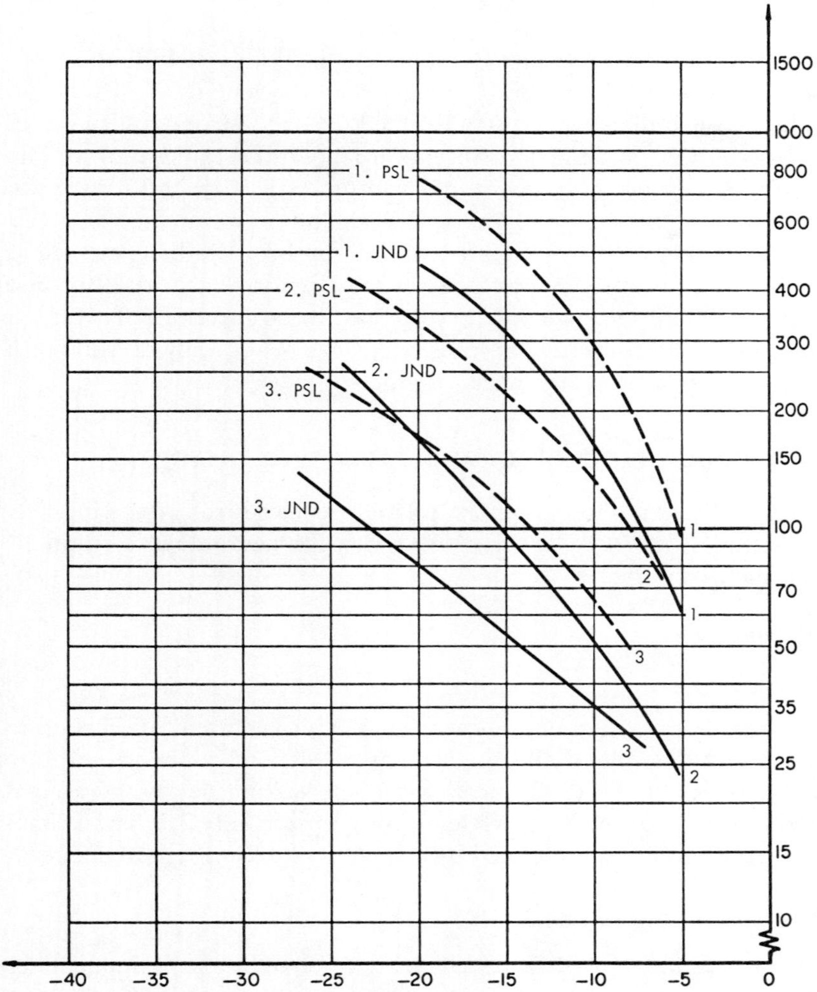

Fig. 8.11. Keeping times for calves liver. (*1* and *2*) Criteria: JND, time for a decrease of 1 point in a −5 to +5 organoleptic scale (Roskilde). PSL, time for a decrease to a score of −2 in same scale. Product: (*1*) calves livers of high degree of freshness, bacterial count of 30,000/g.; (*2*) calves liver after storage in the fresh state, bacterial count of 630,000/g. Packaging: coated sealed carton, individual pieces wrapped in wax paper. Source: Dalhoff and Jul (2). (*3*) Criteria: JND, time for a decrease of 0.5 points in a 0–5 organoleptic scale (Stockholm). PSL, time for a decrease to a score of 3 in same scale. Product: sliced liver. Packaging: two pieces in a sealed freezer carton, with polyethylene coating. Source: Löndahl and Danielson (10). Note: The criteria used in the last experiment are probably quite different from those used in the Danish experiment.

IV. ACCEPTABILITY FACTORS

As indicated in Chapter 10, the practical storage life (PSL) of a product is usually considerably longer than the time until "JND." The ratio between these two is often referred to as the "acceptability factor." This is an important factor, since most laboratory studies indicate time until JND, whereas industry and consumers are generally interested mainly in PSL. Andersen, Jul, and Riemann (11) estimate the acceptability factor to be 3 for pork and 4–5 for beef. Dalhoff and Jul (2) present data that suggest an acceptability factor of at least 2 for meat products. Data by Bøgh-Sørensen (5) and those presented in Figs. 8.6, 8.9, and 8.10 suggest that possibly a factor of only 2 should be employed for meat and meat products.

V. PRODUCT, PROCESSING, AND PACKAGING FACTORS

As indicated above, factors related to *p*roduct, *p*rocessing, and *p*ackaging (the "PPP" factors) have a very marked influence on quality and stability of meats.

A. Product

1. *Feeding*

Fig. 8.10 gives data determined by Wismer-Pedersen and Sivesgaard (6). They show that well-fed pigs may yield pork chops with PSL of 400 days at 0°F. (−18°C.), while pork chops from poorly fed animals under exactly the same conditions will keep for only 225 days. Almost identical results were obtained by Palmer *et al.* (12) as seen in Table 8.2.

2. *Condition of Animal*

The treatment of the animal prior to slaughter and immediately thereafter may have a marked influence on the quality of the meat,

Table 8.2. Practical Storage Life (PSL) for Pork of Different Iodine Numbers [a]

Iodine no.	PSL at 0°F. (−18°C.), days
68.44	240
58.75	400

[a] Criteria: Time for decrease to a score of 3 in a flavor rancidity scale of 1 to 5. Product: Ground pork. Packaging: Waxed paper or glassine laminated paper. Source: Palmer *et al.* (12).

e.g., on its pH. This has a considerable effect on such an important factor as drip. Thus, Sair and Cook (13) found that when beef, pork, and mutton with a pH of 5.0–5.2 are frozen, they give considerable drip, while at pH 6.4 little or no drip is formed. Similarly, Ramsbottom and Koonz (14) found 4.3% drip in beef with a pH of 5.7–5.9 and 0.7% drip in beef with a pH of 6.2–6.5.

The increase in pH may be obtained by holding the meat some days prior to freezing. When frozen too quickly after slaughter, meat, and especially beef and mutton also may show shortening of the muscle which is closely related to toughening of the meat. It appears that some ripening of the meat before freezing is necessary to avoid a considerable quality loss due to this factor. A long storage period for the frozen product may, however, partly offset this disadvantage in meat frozen before the onset of rigor mortis.

3. *Holding prior to Freezing*

Holding meat before freezing, however, also may have undesirable consequences. Thus, Harrison *et al.* (15) show that cooler storage of pork prior to freezing for one or three days has little influence on keeping quality, while a seven-days' holding period prior to freezing may reduce the practical storage life as follows:

Holding	Practical storage life, days
7 days at 40°F. (4.4°C.)	*ca.* 120
7 days at 30°F. (−1.1°C.)	*ca.* 133
3 days at 30°F. (−1.1°C.)	*ca.* 156
1 day at 30°F. (−1.1°C.)	*ca.* 212

In these figures, the time elapsed when the flavor score has fallen to 5 in a 10-step scale is indicated.

Fig. 8.11, curves *1* and *2*, give similar results for some calves' liver products. The PSL at 0°F. (−18°C.) is 660 days for a product of good bacteriological quality, while the corresponding PSL for a product of a somewhat inferior bacteriological quality is only 280 days.

Such effects are probably due to the formation of microbial enzymes that will be somewhat active even in the frozen product.

4. *Bacteriological Condition*

The bacteriological condition of the product may interfere somewhat with giving meat which is to be frozen the degree of ripening which is desirable. Thus, of the livers in Fig. 8.11, the product with a bacterial count of 630,000/g. received a higher score, +0.29 before freezing, than the untreated, very fresh calf liver with a bacterial count of 30,000/g., which received a score of −0.64, both on the −5 to +5 scale (Roskilde). However, when the products were frozen and stored in the frozen state the very fresh products soon showed the best organoleptic quality, as indicated in the curves for storage life in Fig. 8.11. Similarly for other meats, a certain maturation period is often desirable if the products are to be used immediately. If, on the other hand, they are to be frozen, this maturation may result in reduced storage life; thus, a product that is matured less may be preferred after freezer storage.

5. *Composition and Formulation*

Product composition is a factor of greater importance than is often realized. Compared with Fig. 8.2, Fig. 8.7 shows that the addition of curing ingredients is likely to reduce storage life. The influence of various curing ingredients and antioxidants was investigated by Zipser, Kwon, and Watts (16). The results are given in Table 8.3. This confirms the supposition that salt may reduce storage life very markedly, while the addition of an antioxidant may greatly increase it. It appears that when an ascorbate is used as an antioxidant, the effect is very pronounced up to a certain point in time. At that point, rancidity sets in and soon exceeds that of the product to which no antioxidants have been added. In this case, 0.1% ascorbate was below the amount that would give reasonable protection; however, this is likely to be highly dependent on the type of packaging material employed.

6. *Breading*

Breading a product may greatly improve its storage life; *cf.* Fig. 8.5, where breading appears to have lengthened the time up to the JND at 0°F. (−18°C.) for ground pork from 180 to 420 days. This may be due to the antioxidative effect of the flour in the breading preparation.

7. *Smoking*

It is a general feeling that smoking may increase the storage life of frozen meats. Fig. 8.8, curve *4*, may be taken to confirm this. The effect is likely to be due to the antioxidative properties of some of the constituents in wood smoke.

8. *Spices*

A protective effect may be found from the addition of spices, many of which show considerable antioxidative effect, *cf.* Fiskeriministeriets Forsøgslaboratorium (17) and Ostric-Matijasevic (18).

B. Processing

1. Heating

Heating a product to the boiling point appears to improve its storage life. Thus, in Fig. 8.8, curves *1* and *2,* Danish pork sausage (Medisterpølse) showed a length of time to the JND at 0°F. (−18°C.) of 230 days if it was uncooked and 360 if it was cooked. It is likely that this is due to heat inactivation of some autolytic enzymes. Grant and Alburn (19) found some enzymes to be even more active at −30.6°F. (−34.8°C.) than at 34°F. (1.1°C.). Thus enzymes may be responsible for changes during freezer storage.

2. Covering with Gravy

Covering a product with gravy may improve its storage life very markedly, as seen from Figs. 8.4 and 8.5. Here, covering with gravy increased the length of time before a JND at 0°F. (−18°C.) from 135 days to over 1,000 days for a beef product (*cf.* curves *3* and *5* in Fig. 8.4) and from 160 days to more than 350 days for a pork product (*cf.* curves *4* and *6* in Fig. 8.4). In both cases, however, some of the effect may be due to the heat treatment which the gravy-covered product had also undergone.

3. Freezing per se

For meat, freezing in itself has little effect on the quality of the end product. There have been many taste tests or other organoleptic tests

Table 8.3. Effects of Curing Ingredients and Antioxidants on Cured and Uncured Pork Stored at 0°F. (−18°C.) [a]

Product	Practical storage life, days
Uncured	30
Uncured, 0.1% ascorbate added	360
Cured (2% NaCl, 0.03% $NaNO_2$)	7–50
2% NaCl added	Very short
0.03% $NaNO_2$ added	360
4% NaCl added	Very short
5.1% KCl	30
Cured, 0.01% BHA added	370
Cured, 0.1% ascorbate added	200–360

[a] Criteria: Score of 4 in a 1–6 sensory score scale for odor. Product: Fresh ham, trimmed for fat and finely ground and cooked. Packaging: Polyethylene bags. Source: Zipser, Kwon, and Watts (16).

in which taste panels reported no or very little difference between an untreated product and the same product frozen and then immediately defrosted, *cf.* Dansk Køleinstitut (20).

4. Rate of Freezing

Rate of freezing, often measured in centimeters per hour for the crystalization front, or freezing time in hours, may have some influence on taste or quality. Thus, Du Bois *et al.* (21) found for cooked beef that when freezing times increased from 1 to 15 hr. the taste score decreased from 5.8 to 4.8 in a 0–6 point taste scale. For beef aged 14 days at 1°C. (34°F.) Bøgh-Sørensen (15) found an average tenderness score of 7.36 for the unfrozen product, 7.50 for a plate-frozen product, and 6.92 for bulk frozen, all in a 0–10 score tenderness scale (Copenhagen). Other investigators, *e.g.*, Pennsylvania Agricultural Experiment Station (22) found no influence of freezing rates varying from 10 to 0.1 cm./hr. on flavor.

Lee *et al.* (7) found that fast plate freezing gave no significant preference over slow freezing in a home freezer in respect to flavor and tenderness, and did not have any influence on thiamine, riboflavin, niacin, pantothenic acid, or pyridoxine content. Freezing and freezing rate may, however, have more influence on tenderness; thus, Hiner, Madsen, and Hankins (23) found for 1.5-in. thick samples of beef, aged five days before freezing, the following:

Freezing method	Decrease in shearing force, %	Weight loss during thawing, %
18°F. (−7.8°C.) still air	9.24	12.14
−10°F. (−23.3°C.) still air	12.86	7.54
−40°F. (−40°C.) still air	19.43	4.23
−40°F. (−40°C.) air blast	23.85	3.75
−114°F. (−81°C.) Dry Ice	28.55	1.91

Shearing strength was measured with a Warner-Bratzler machine. The samples were exposed to the freezing temperature for 24 hr.

Other investigators have failed to establish any organoleptically detectable difference in tenderness due to freezing rate.

Marsh (24) assumes that freezing may be so fast that the muscle is stiffened by the formation of ice crystals before cold contraction sets in, thereby preventing shortening and toughening. This, however, would be attainable only by freezing rates much higher than those now in commercial use. If it were done, this author would expect an improvement in tenderness.

Freezing rates appear to have some influence on drip after defrosting, as seen in the results by Hiner, Madsen, and Hankins (23) just quoted. Similarly, for cuts of 4–5 year old beef cattle, Cook, Love, Vickery, and Young (25) found the following relationship:

Freezing rate, cm./hr.	Per cent drip
1000	0
2.8	1.4
2.3	1.5
1.6	3.4
1.0	3.9

Nevertheless, Empey (26) failed to establish any correlation between freezing rates and drip.

A more detailed literature study of freezing rate as related to quality is given by Kondrup and Boldt (27).

5. *End Temperature in Freezing*

The end temperature to which a product is frozen may have some influence on quality and especially on drip, since it determines the percentage of the water content which is separated out as ice crystals, thereby conceivably affecting protein denaturation, etc. Thus, the results by Hiner, Madsen, and Hankins (23) quoted above may partly be attributed to the differences in end temperatures.

Hankins and Hiner (28) found that freezing at and to −10°F. (−23°C.) or −40°F. (−40°C.) makes beef more tender than freezing at and to +20°F. (−7°C.). There were no differences between −10°F. (−23°C.) and −40°F. (−40°C.). In all cases the meat was allowed to remain in the freezer until it attained the freezer's temperature. The tenderness was measured immediately after defrosting. The frozen beef showed a lower (12–18%) shearing strength than the unfrozen meat. All samples were aged 96 hr. at 34°F. (1°C.) before freezing.

Hiner *et al.* (23) state that intracellular ice formation may cause rupture of cell walls. They found more rupture by freezing at and to −40°F. (−40°C.) than at and to 0°F. (−18°C.), and suggest that cell-wall damage reduces the amount of drip from the thawed meat. This is generally explained by the improved diffusion conditions where cell walls are damaged.

Many such experiments are, however, somewhat difficult to interpret, since both rate of freezing and end temperature are involved. That this distinction is important is indicated by Moiseeva and Piskareva (29), who found that slow freezing down to −0.4°F. (−18°C.) gave 11–16%

drip. The same process, followed by quick freezing down to −319°F. (−195°C.), gave only 3–4% drip.

Much conflicting evidence exists as to the relative merits of slow versus fast and ultrarapid freezing, high versus low end temperatures, etc. There are advocates for many different types of freezing processes. This may be taken to indicate that the influence of the freezing process itself on the quality of the end product is of minor importance.

C. Packaging

Effective packaging may have a very considerable influence on the storage life of some frozen products. Ground beef and hamburger meat are fairly sensitive to oxidation and surface desiccation. As indicated in Fig. 8.4, curves *2* and *3*, Dalhoff and Jul (2) have shown that packaging ground beef in a tight-fitting, evacuated, heat-sealed film of low oxygen- and water-vapor permeability gives it a time up to a JND at 0°F. (−18°C.) of more than 270 days, whereas a standard commercial packaging method gave only 120 days.

Fig. 8.7, point *4*, and Table 8.3 show that frozen cured pork with no antioxidants added and without a highly protective packaging will keep for a few days or weeks only, while sliced bacon, which will normally turn rancid very quickly in the frozen state, may have a keeping time up to the JND of more than 70 days at 14°F. (−10°C.), if it is packed in a heat-sealed, evacuated film of low permeability for oxygen and water vapor.

On the other hand, Fig. 8.4, curve *5*, Fig. 8.5, curve *6*, and Fig. 8.9, curve *2* seem to indicate that for products covered with gravy the method of packaging is not important. The assumption is that such products are already well protected against oxidation and therefore show remarkably long practical storage lives even in freezer packages of only average quality.

This theory was confirmed by Bøgh-Sørensen (see Fig. 8.5, curves *2* and *3*), who packed breaded pork chops and breaded ground pork in two different types of packages, *i.e.*, one group individually wrapped in an ordinary plastic film and placed in a standard freezer carton and another group vacuum-packaged in a heat-sealed laminated plastic bag of very low oxygen permeability. No difference was found in either time up to the JND or in PSL.

The data given in the figures and tables accompanying this chapter may throw considerable light on the relative merits of various packaging materials and methods. A detailed discussion of this matter is, however, outside the scope of this chapter.

VI. FREEZER STORAGE CONDITION

A. Freezer Burn

It is well known and often experienced that freezer burn may sometimes seriously affect the quality of frozen meats. However, this defect is quite easily prevented by providing a water-vapor proof package if it can be made to fit the product tightly. If the shape of the meat makes it difficult to avoid air pockets in the package it may be necessary to provide protection by covering the meat with a film of antioxidant-containing lard, tallow, or glycerol, or by some breading process.

B. Fluctuating Temperatures

It is well known that freezer burn and other defects due to surface desiccation are especially likely to occur at highly fluctuating storage temperatures. If surface desiccation is prevented by effective packaging or by product composition, even considerably fluctuating storage temperatures are relatively unimportant.

Thus, Hustrulid, Winther, and Noble (30) showed that packaged, ground beef or ground pork were not markedly affected in quality by temperatures fluctuating from 0 to −10°F. (−18 to −23°C.) with 3 cycles per week for 6 months, compared to storage at a constant temperature of 0°F. (−18°C.). Where unsatisfactory packaging material was used, the effect of temperature fluctuation on quality was quite marked. Since most frozen meat products in consumer-size packages are well packaged, it appears that the strong emphasis generally put on keeping the storage temperatures constant with very small fluctuation is quite unnecessary for this type of freezer goods.

C. Time and Temperature Influence

The effects of various times and temperatures in storage conditions as given by different sources are indicated in Figs. 8.3–8.6, 8.8, and 8.9. These data need hardly be commented upon further but it may be noted that the slight tenderizing effect that freezing may have, in the view of Tressler, Birdseye, and Murray (31) may be continued during cold storage. Also, the tendency to form drip may increase during storage but Ramsbottom and Koons (14) found this effect only at temperatures above 10°F. (−12°C.).

As already mentioned, several meat products, especially some of those protected with a breading preparation, show very satisfactory storage lifetimes, even at temperatures considerably above those conventionally considered maximum for frozen food storage, *i.e.*, generally, −4°F. (−20°C.) or lower. This is true, for instance, for the products to which Fig. 8.4, curves *4* and *5*, and Fig. 8.5, curves *3* and *6* refer.

VII. DEFROSTING

In most considerations regarding the freezing of meat and meat products it is tacitly understood that the method of defrosting may have some influence on the quality of the end product, but few attempts have been made to determine the actual effect of various defrosting procedures.

Only one factor is well investigated. Many have demonstrated that cooking a product directly from the frozen state results in less drip but sometimes results in meat that appears to be less tender.

Insofar as vitamins are concerned, Lehrer *et al.* (32) gave the following average losses for pork wrapped in wax paper and stored frozen for 3 or 6 months at −5°F. (−26°C.) or 0°F. (−18°C.), compared with freshly cooked pork.

	Freshly cooked, % loss	Thawed and cooked, % loss	Cooked frozen, % loss
Thiamine	56	54	35
Riboflavin	34	27	11
Nicotinic acid	17	36	43

From this, freezing, frozen storage, and especially cooking directly from the frozen state appear to result in smaller losses in thiamine and riboflavin, but frozen storage and again especially cooking directly from the frozen state, resulted in considerably greater losses in nicotinic acid than when the fresh product was cooked.

VIII. METHOD OF COOKING

Little experimentation is available on the influence of the methods of cooking frozen meat upon its quality. It is known, especially from cooking frozen fish, *cf.* Fiskeriministeriets Forsøgslaboratorium (33) that efficient adaptation of recipe and methods of cooking, so as to fit the frozen product, may result in very considerable gains in taste and acceptability. This is an important field for investigation.

IX. SALES OF FROZEN MEATS

With the above studies in mind it is interesting to observe what is actually developing in the trade. Frozen cuts of beef, mutton, and pork are sold only to a limited extent *i.e.,* 2–4% of the total meat turnover in Denmark and Sweden, as well as in the U.S.A. The explanation generally offered is that frozen meats offer the consumer no particular advantage over fresh meats. Frozen chopped meats are sold to a considerably greater extent. As mentioned above, especially when they are vacuum-packed in high-quality film, these products will keep extremely well in the frozen state, while the fresh product normally must be sold within a day or a few days.

Breaded products and boil-in-the-bag products seem to be gaining ground. As mentioned above, they show very high acceptability even after long storage at relatively high freezer storage temperatures, or highly fluctuating temperatures.

Products covered with gravy are also gaining in popularity. As described above, they, too, stand up very well during freezer storage.

In Denmark, where the sale of not fully shelf-stable meats is limited by legal restrictions to butchers' shops and grocery stores, frozen cured meats have gained considerable acceptance because they may be sold from most categories of food stores. Only the use of very effective packaging makes this trade possible, as indicated in Fig. 8.7.

It is thus seen that those combinations of products, processes, and packages that give products a particularly high practical storage life, have also gained the widest consumer acceptance.

One remarkable product is frozen ham dinners in the U.S.A. They enjoy a good consumer acceptance and a relatively favorable rating, *cf. Consumer Reports* (34). Since they are not vacuum packed, this can only be explained by an unusually effective product formulation, *cf.* Table 8.3.

REFERENCES

(1) International Institute of Refrigeration, *Recommendations for the Processing and Handling of Frozen Foods,* Institut International du Froid, Paris, 1964, p. 112.

(2) E. Dalhoff and M. Jul, "Factors Affecting the Keeping Quality of Frozen Foods," in *Progress in Refrigeration Science and Technology,* Vol. 1, Pergamon Press, New York, 1965, pp. 57–66.

(3) C. K. Wiesman, *Ice Refrig. 112, No.* 4, 21 (1947).

(4) E. Dalhoff, *Holdbarhedsbestemmelser efter lagring ved frosttemperaturer af sliced bacon,* Report 140.19, Danish Meat Research Institute, Roskilde, 1963.

(5) L. Bøgh-Sørensen, *Undersøgelse af forskellige kødfrysevarers tid-temperatur-tolerance,* Report 6/65, Danish Meat Products Laboratory, Copenhagen, 1967.

(6) J. Wismer-Pedersen and A. Sivesgaard, *Kulde, 5,* 54 (1957).

(7) F. A. Lee, R. F. Brooks, A. M. Pearson, J. I. Miller, and F. Volz, *Food Res. 15,* 8 (1950).

(8) G. Ordynsky, *Tiefkühl-Praxis, 5,* 7 (1965).

(9) J. Gutschmidt, "Nährwert- and organoleptische Veränderungen beim Gefrieren und bei der Gefrierlagerung," in *Probleme der Ernährung durch Gefrietkost,* D. Steinkopff Verlag, Darmstadt, 1964.

(10) G. Löndahl and C. E. Danielson, *Livsmedelsteknik, 2,* 84 (1967).
(11) P. E. Andersen, M. Jul, and H. Riemann, *Industriel Levnedsmiddelkonservering,* Vol. 2, Teknisk Forlag, Copenhagen, 1965, p. 748.
(12) A. Z. Palmer, D. E. Brady, H. D. Naumann, and L. N. Tucker, *Food Technol., 7,* 90 (1953).
(13) L. Sair and W. H. Cook, *Can. J. Res., 16 D,* 255 (1938).
(14) J. M. Ramsbottom and C. H. Koons, *Food Res., 5,* 423 (1940).
(15) D. L. Harrison, J. L. Hall, D. L. Mackintosh, and G. E. Vail, *Food Technol., 10,* 104 (1956).
(16) M. W. Zipser, T. Kwon, and B. M. Watts, *J. Agr. Food Chem., 12,* 105 (1964).
(17) Fiskeriministeriets Forsøgslaboratorium, *Annual Report,* Fiskeriministeriets Forsøgslaboratorium, Copenhagen, 1966.
(18) B. Oštrić-Matijašević, *Technologija mesa, 7–8,* 195 (1963).
(19) N. H. Grant and H. E. Alburn, *Nature, 212,* No. 5058, 194 (1966).
(20) Dansk Køleinstitut, *Undersøgelse over frysning og påfølgende lagring af okse- og svinekød,* Meddelelse nr. 4, Akademiet for de tekniske Videnskaber, Copenhagen, 1944, p. 43.
(21) C. W. Du Bois, D. K. Tressler, and F. Fenton, "Influence of Rate of Freezing and Temperature of Storages on Quality of Frozen Meat," in *Proc. Conf. Inst. Food Technol.,* Chicago (June 1940) 167.
(22) Pennsylvania Agriculture Experiment Station, "Some Factors Affecting the Quality of Frozen Foods," in *Penn. Agr. Exptl. Sta. Bull.,* 565, 1953.
(23) R. L. Hiner, L. L. Madsen, and O. G. Hankins, *Food Res., 10,* 312 (1945).
(24) B. B. Marsh, "Meat Tenderness," in *Annual Research Report, 1965–66,* Meat Industry Research Institute of New Zealand, Hamilton, 1966, p. 14.
(25) G. A. Cook, E. F. J. Love, J. R. Vickery, and W. J. Young, *Australian J. Exptl. Biol. Med. Sci., 3,* 15 (1926).
(26) W. A. Empey, *J. Soc. Chem. Ind., 52,* 230T (1933).
(27) M. Kondrup and H. Boldt, "The Influence of the Freezing Rate upon the Quality of Frozen Meat and Poultry," in *Annexe 1960–3 au Bulletin de l' Institut International du Froid,* International Institute of Refrigeration, Paris, 1960, p. 309.
(28) O. G. Hankins and R. L. Hiner, *Food Ind. 1,* 49 (1940).
(29) N. Moiseeva and A. Piskareva, *Kholodil' naya tekhnika, 1,* 52 (1959).
(30) A. Hustrulid, J. D. Winter, and I. Noble, *Refrig. Eng. 57,* 38 (1949).
(31) D. K. Tressler, C. Birdseye, and W. T. Murray, *Ind. Eng. Chem., 24,* 242 (1932).
(32) W. P. Lehrer, A. C. Wiese, W. R. Harvey, and P. R. Moore, *Food Res., 16,* 485 (1951).
(33) Fiskeriministeriets Forsøgslaboratorium, *Annual Report,* Fiskeriministeriets Forsøgslaboratorium, Copenhagen, 1965, p. 21.
(34) Consumers Union, *Consumer Reports, 32,* 521 (1967).

Chapter 9

QUALITY AND STABILITY OF FROZEN SEAFOOD

F. Bramsnaes

Department of Technical Biochemistry, Food Technology Laboratory, Technical University of Denmark, Copenhagen, Denmark

INTRODUCTION

The production of frozen fish and fish products has been increasing steadily and considerably over the past several years. Canned fish and other preserved fish have not increased in usage nearly as much. Fatty fishes and shellfish sell equally well in canned or frozen state, but lean fish lend themselves much better to freezing than to any other kind of preservation.

Along with its commercial success frozen seafood has, however, given rise to a number of quality problems, mainly due to undesirable changes during storage and distribution of the products. More knowledge about the changes is now available, and frozen seafood today generally is of a much higher quality than it was a few years ago.

The factors contributing to the quality of frozen seafood may suitably be discussed under the following headings:

1. Quality of the raw material; handling and preparation prior to freezing.
2. Packaging.
3. Freezer storage conditions.
4. Defrosting and cooking methods.

1 and *2* combined are often called "PPP factors" (*p*roduct, *p*rocessing, and *p*ackaging).

Research related to these factors has been intensified in recent years and a great number of publications have appeared. It is possible here to mention only a few in order to illustrate the conclusions drawn. The reader is referred to several excellent reviews dealing with the subject in part or whole (Banks (1), Lane (2), Dyer and Dingle (3), Jones (4), Heen and Karsti (5), Jones (6), Dyer and Peters (7), and Slavin (8)).

II. JUST NOTICEABLE DIFFERENCE

Torry Research Station (Table 9.1) found for whole, glazed herring that the storage life for "good as fresh" was 3 months at $-21°$C. ($-6°$F.) and 6 months at $-29°$C. ($-20°$F.) Gutschmidt (9) reports a "high quality life" for lean fish fillets at $-18°$C. (0°F.) of 3–4 months. An international working party set up to draft a Frozen Fish Code recently agreed on the times in Table 9.3 (see Sect. III) for the occurrence of "slight but detectable changes." In all these three cases the times presumably correspond to JND, Just Noticeable Difference, as defined in Chapter 10, Section II.C. Edhborg (10), working with commercial packs, finds the results in Fig. 9.1. JND was in this case determined as a decrease of 1 point in a 7-point hedonic scale.

Table 9.1. Storage Life for Frozen Fish. Based on many experiments carried out by Torry Research Station, Aberdeen.[a]

Temperature	Gutted whitefish: Good as fresh[b]	Gutted whitefish: Inedible	Ungutted herring: Good as fresh[b]	Ungutted herring: Inedible
+15°F. or −9°C.	1 month	4 months	1 month	3 months
−5°F. or −21°C.	4 months	15 months	3 months	6 months
−20°F. or −29°C.	8 months	4 years	6 months	1½ years

[a] Source, ref. 10a, p. 9.

[b] "Good as fresh" indicates a quality still approximately as palatable as the raw material, and "inedible" means definitely distasteful to consumers accustomed to fresh fish.

III. STORAGE LIFE

The term "storage life" means, in most of the cases referred to in this article, the time required for various deteriorative changes in a freshly packed product to accumulate to such an extent that the test panel in question considered the quality to be at the limit of acceptability. This is a little different from the definition of storage life in Chapter 10, Section II.B.

Only a few exact data are available on storage life of frozen seafood.

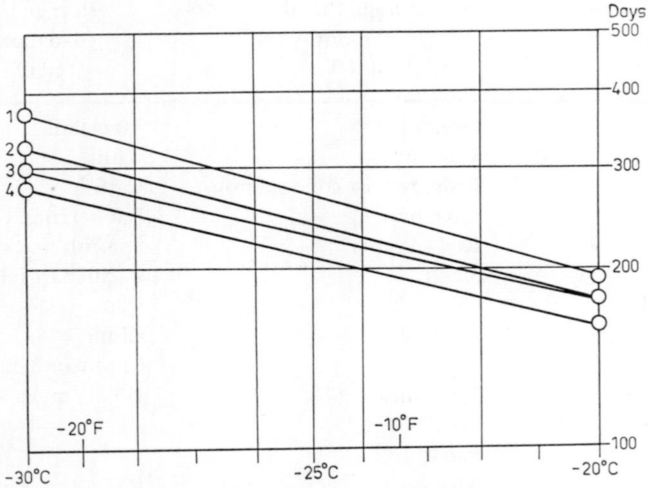

Fig. 9.1. Time after which an average decrease in the quality-determining factor (usually flavor) by one unit in a 7-point scale was noted by the test panel (10). (*1*) Plaice; (*2*) cod; (*3*) haddock; (*4*) whiting.

Most often one has to rely on publications in which insufficient information is given on the nature of the product, the process it has undergone, the packaging, or the criteria used for determining the storage life. Tables 9.1–9.3 and Fig. 9.2 give useful information of a general nature.

IV. CHANGES IN FROZEN SEAFOOD

For the most part there are three groups of changes (11): (*1*) denaturation of the myofibrillar proteins, (*2*) changes in components affecting color and flavor, and (*3*) breakdown of connective tissue and allied components.

A. Denaturation

Denaturation of the myofibrillar proteins is the scientific term for the fact that frozen fish during freezing and storage gradually become tougher, dryer, and more fibrous—changes which are particularly noticeable after cooking. In the later stages of this denaturation the raw fish muscle after thawing assumes a dull and opaque appearance and takes on a spongelike structure. The thawed raw fish readily lose

Table 9.2. Relative Degree of Suitability of Fishery Products for Freezing and Frozen Storage [a]

High: storage life of 7–12 months at 0°F.	Moderate: storage life of 5–9 months at 0°F.	Low: storage life of 4–6 months at 0°F.
Haddock	Ocean perch	Mackerel
Cod	Whiting	Tuna
Flounder	King, red, or coho salmon	Catfish
Hake	Lake herring	Sea herring
Shrimp	Red snapper	Spanish mackerel
Halibut	Smelt	Pacific sardine
King crab	Crawfish	Clam
Dungeness crab	Rockfish	Chub
Pollock	Carp	Chum or keta salmon
Sea scallop	Buffalofish	Whale meat
Blue pike	Swordfish	
Yellow perch	Pacific oyster	
	Alewives	
	White bass	

[a] Source, ref. 52, p. 288.

Table 9.3. Approximate Storage Times for Packaged or Glazed Whole or Filleted Fish and Shellfish

Product		Storage time in months at $-18°C$. ($0°F$.)[d]	
Fatty fish[c]	Mackerel Salmon Sea herring Smelt Sprat Trout	2–3[a]	4–6[b]
Lean and medium fatty fish[c]	Cod Haddock Flounder Ocean perch Plaice Pollock Sole	3–4[a]	7–10[b]
Shellfish	Shrimp Scallop	3–4[a]	6–8[b]
	Clam Lobster (cooked) Oyster	2–3[a]	4–6[b]

[a] Slight but detectable changes in quality occur; product is still of good acceptability.

[b] Very significant changes in quality occur, and product is of low acceptability.

[c] Values for fish up to three days in ice before freezing.

[d] The storage times will vary with the quality of the fish prior to freezing. Lower temperatures than $-18°C$ will give longer storage times. The storage life of fatty fish products may be extended beyond the stated times by the use of vacuum-packing. Precooked fish products, like fish sticks, have not been included in the table. The keeping quality of such products will, if they are properly processed and packed, normally be the same as for the raw material from which they are manufactured. However, in calculating the shelf-life one must remember to include the storage time of the half-finished product, *i.e.*, the fish blocks, for fish sticks and fish portions.

Source, (53).

fluid or drip, particularly if subjected to light pressure. The amount of drip, sometimes called thaw-drip, increases during the early part of the frozen storage period. Under good conditions, thaw-drip is usually less than 10% but may be considerably higher after long storage or under poor conditions.

All of these changes occur more rapidly in fish than in meat. No

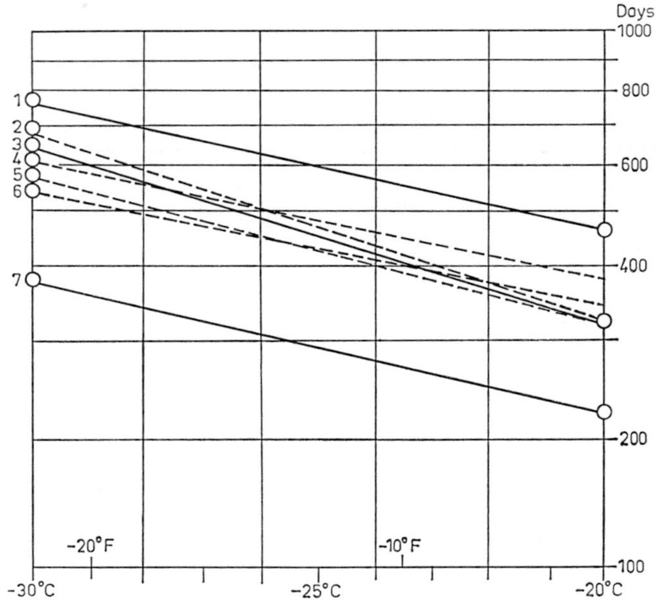

Fig. 9.2. Time after which half of the number of samples of frozen fish products were found to have passed the limit of acceptability. Dotted lines represent series in which only relatively few controls were made at —20°C. (—4°F.) (10.) (*1*) Plaice fillets; (*2*) haddock fillets; (*3*) cod fillets and cod portions; (*4*) fish cakes; (*5*) fish sticks (cod); (*6*) prawns (vacuum packed); (*7*) whiting fillets.

direct comparison between fish and meat seems to have been made. However, as the eating texture of a meat or fish product is influenced by these changes, there is ample evidence from commercial practice to show that the proteins of fish are more unstable during frozen storage and therefore require lower storage temperatures than those of beef, mutton, or poultry meat.

The main components of the fish muscle are the myofibrils (structural proteins), the sarcoplasma (muscle cell juice), and a small percentage of connective tissue and cell walls. Chemically the toughness and loss of juiciness have the result that the myofibrils during freezing and storage progessively lose their extractability by weak salt solutions (12,13) and some of their ability in thawing to reabsorb water from the melting ice. Dyer (13) found, as indicated in Fig. 9.3, a general correlation between loss of protein solubility and development of toughness. The panel scores shown in the figure are, as in all sensory assessments, the result of the particular scoring system used in Dyer's experiments, and would, therefore, have come out somewhat differently with another set of scoring instructions. However, the general trend is clearly visible

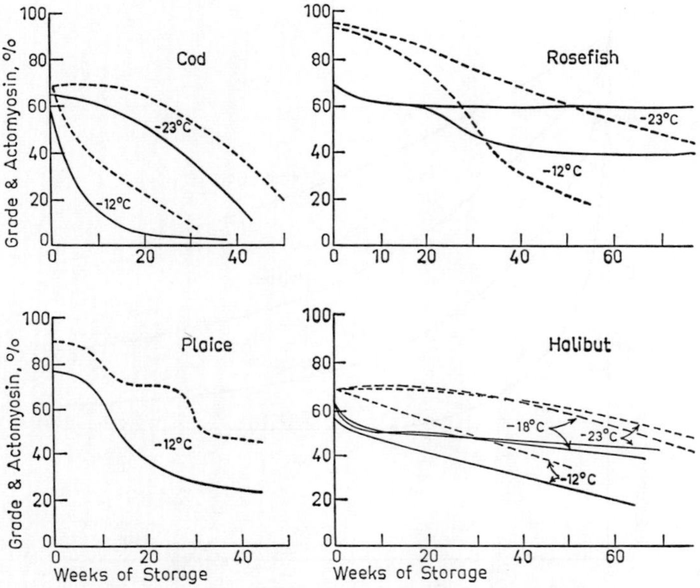

Fig. 9.3. Peptizable actomyosin nitrogen (solid lines) as per cent of total protein nitrogen, and taste panel grades (dotted lines), 0–100%, in frozen fillets of cod, American plaice, halibut, and rosefish, stored at −12, −18, and −23°C. (10, 0, and −10°F.) (3).

and, in addition, the experiments show that stability of frozen fish varies with the species. The myofibril denaturation thus plays a very important part in the texture changes. It is, however, not the whole picture. When the storage temperature is lowered to about −30°C. (−22°F.) no denaturation takes place. Nevertheless, slight texture changes may develop at this temperature and even at lower temperatutures (5).

Physically the reason for toughness and loss of waterholding capacity is an increased bonding between the myofibrillar proteins, sometimes called aggregation of the fibrils. The nature and cause of this bonding will not be dealt with here, but it should be mentioned that the subject is not at all clear and much work is going on at present to elucidate the field.

In contrast to the behavior of the myofibrillar groups of protein the extractability of the sarcoplasma proteins remains high during frozen storage (14). Consequently, the lower limit for protein solubility in salt solutions is roughly the same as the percentage of sarcoplasma in the total protein of the fish, namely 20–30%, as can be seen for cod in Fig. 9.4.

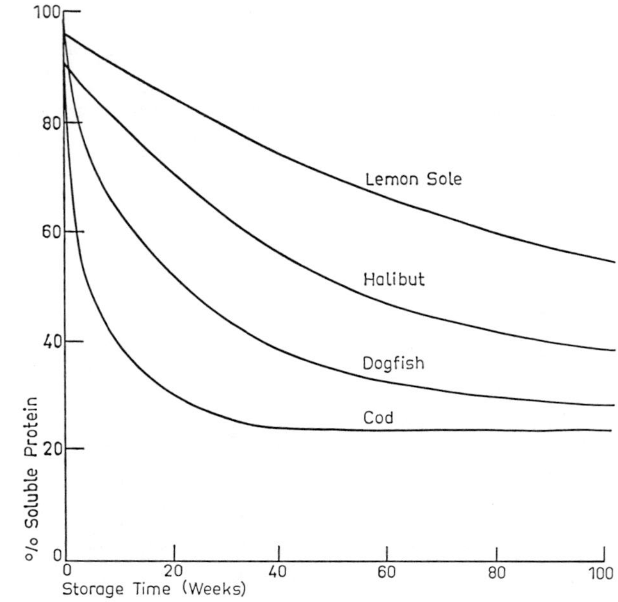

Fig. 9.4. Changes in extractability of proteins during frozen storage of fish muscle at −14°C (7°F.) (14).

B. Color Changes

1. Blood Pigments

Two heme pigments, hemoglobin in the blood and myoglobin in the cells, are the main coloring pigments in fish flesh. The dark parts of the muscles contain much more pigment than the light or white muscles.

In freshly caught fish the blood has the bright red color of oxy-hemoglobin. This color gradually changes to the red-brown or brown color of methemoglobin. Myoglobin suffers the same kind of changes.

If the fish are frozen in the round all the blood remains in the fish, and extensive discoloration can be expected from blood, mainly along the backbone and on the surface of the belly cavity. Even in eviscerated fish and in fillets considerable quantities of blood may remain if the fish is not sufficiently bled. The brown discoloration sometimes occurring on the surface of prerigor fillets has been shown to be due to the formation of a denatured form of methomoglobin which exudes to the surface (15).

In the muscle tissue the myoglobin is the dominating pigment. This is of importance in the darker flesh, of which such fishes as tuna have a considerable amount. In the presence of the low oxygen concentrations existing in the tissue after death the heme pigments are highly susceptible to oxidation (16). Practical experience confirms this. Japanese workers have shown that in frozen tuna meat, where discoloration of the meat is a problem, a more marked discoloration develops in the inner tissue than on the surface, possibly due to a lower partial oxygen pressure in the inner tissue (17).

2. *Carotenoid Pigments*

Some fishes like salmon, and some crustaceans, contain carotenoid pigments in the flesh. The red color astaxanthin tends to fade away during frozen storage. In crustaceans, like lobster, where the pigment is limited to the surface of the meat, the changes during frozen storage lead to yellow discolorations. Dyer and Horne (18) have shown that the red color pigment which is located in the fat of the tissue is oxidized, and this is accelerated by autoxidation of the fat itself.

3. *Fish Oils*

The oils and fats naturally occurring in fish vary with species from colorless, as in herring, through yellow to red, as in salmon. The color may also vary with the pigmentation of the fish food. While the natural pigments, the carotenoids, fade during frozen storage, the fish oil itself, being very unsaturated, will, unless heavily protected, gradually oxidize and turn yellow. Fat herring, with a layer of oil just under the skin, may thus develop a yellow color during poor storage conditions which can be seen through the skin. Such fish is said to have "rusted."

4. *Black Spot*

Shrimp, langustines, Norway lobster, etc., develop so-called black spot when stored fresh. Bruises and rough handling increase the occurrence of this discoloration, which is caused by an enzyme, acting on polyphenolic compounds. The blackening occurs first in the membranes which connect the ends of overlapping segments on the tail part (19). Because it is an enzymatic reaction, it is only slowed down, not stopped by frozen storage temperatures.

C. Flavor Changes

Systematic studies of flavor changes in frozen fish are lacking. Perfectly fresh fish, frozen and cold stored, is generally considered first to lose its characteristic fresh fish flavor, becoming bland and insipid; thereafter it develops flavors variously described as "cold stored," "salt fishy," etc. In practice the sequence of freezing and thawing alone leads to a loss of flavor (4). Superimposed on these changes comes the development of rancid flavors in fatty fish, and perhaps in lean fish too. Finally, the overall flavor impression is influenced by the freshness of the fish before freezing. Among the known off-flavors developing in unfrozen fish trimethylamine values remain on the prefreezing level during frozen storage, while hypoxantine concentrations seem to increase at higher storage temperatures but to remain fairly constant at about −30°C. (−22°F.) or lower (20).

D. Breakdown of Connective Tissue

On occasion, fish fillets are produced in which the muscle flakes (myotomes) will separate when the fish is filleted. This defect, often called "gaping," appears to result from a breakdown in the connective tissue and associated components which hold the muscle cells together (11). The problem is, by and large, connected with freezing at sea and fillet production from frozen whole fish. Under certain conditions, *e.g.*, if the whole fish have been in ice for about a week before freezing, this softening can be quite marked for cod and haddock. In such cases the desired firmness of the fish flesh can be secured if the storage in ice prior to freezing is limited to 2–5 days, depending on the species (1).

V. "PPP FACTORS": PRODUCT, PROCESSING AND PACKAGING

A. Product

1. *Fish Species*

It is well known from commercial experience that the stability of frozen fish varies with the species (see the group division in Tables 9.1–9.3). The experimental evidence mostly relates to texture and protein denaturation changes. The results from Canadian experiments, shown in Fig. 9.3, are typical. They show that in cod and haddock stored at about −23°C. (−10°F.), there is usually almost complete denaturation of myofibrillar protein in less than a year, but that in more fatty species, like halibut, rosefish (ocean perch), etc., little or no denaturation occurs for up to a year in storage. A similar difference between species can be seen in Fig. 9.4.

2. *Seasonal and Biological Variations*

The large number of factors contributing to the variability in biological material should be kept in mind. Fatty species of fish, *e.g.*, herring and mackerel, experience a yearly cycle as regards fat content. Even more important is the fact that the percentage of higher polyunsaturates varies, thereby varying the degree of stability against rancidity. With regard to lean fish it has become more evident lately that considerable fluctuations in quality occur, depending on nutritional status, sexual maturity, season, and fishing ground (4,21). Newfoundland trap cod, as one example, undergo changes during the summer season, which result in significant differences in texture stability during frozen storage (22). Danish experiments showed that autumn plaice are more resistant to double freezing and thawing than spring-caught fish (23). Various degrees of depletion and struggle before death result in Ph differences. Kelly (24) found that at low pH fish deteriorated in texture more quickly than fish at medium pH, which in turn were worse than fillets of high pH (Fig. 9.5.).

3. Holding Prior to Freezing

The initial quality of the raw material has a marked influence on the quality stability of frozen fish. Decreased freshness, measured by increased time in ice prior to freezing, strongly augments the amount of thaw-drip. Notevarp and Heen (25), using a quality scale which comprised flavor, texture, and "free water," found the differences in storage life shown in Fig. 9.6. In experiments where fresh whiting was used as a control Peters *et al.* (26) found that the storage lives at $-18°C$. ($0°F$.) of fish with prefreezing times in ice of 2,4, and 9 days were 12, 6–8, and 0–2 months, respectively. Haddock and American pollock behaved similarly (Fig. 9.7.) (27), but there was only small difference in storage stability of fillets from fish held between 3 and 9 days in ice. This is in line with commercial experience from fresh lean fish, where only small flavor changes become evident in the same period. In the case of fatty fish rancidity may set in quite early, and one must reckon with a decrease in the induction period for fat oxidation from the moment the fish is out of the water.

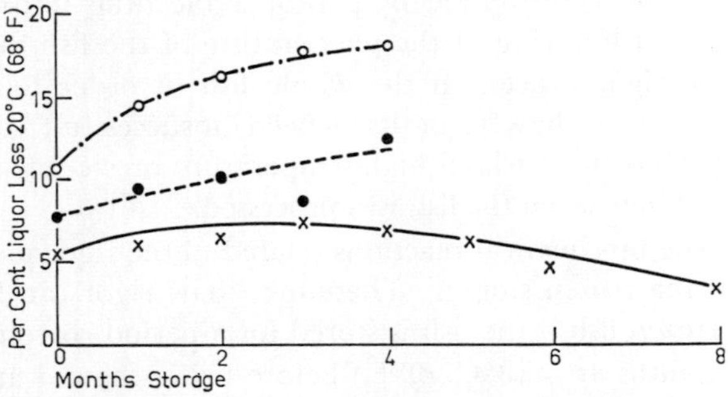

Fig. 9.5. Relationship between liquor loss, pH, and storage time at $-10°C$. ($+14°F$.) of frozen cod fillets (24). (X) pH 6.70; (●) pH 6.40-6.70; (○) pH 6.40.

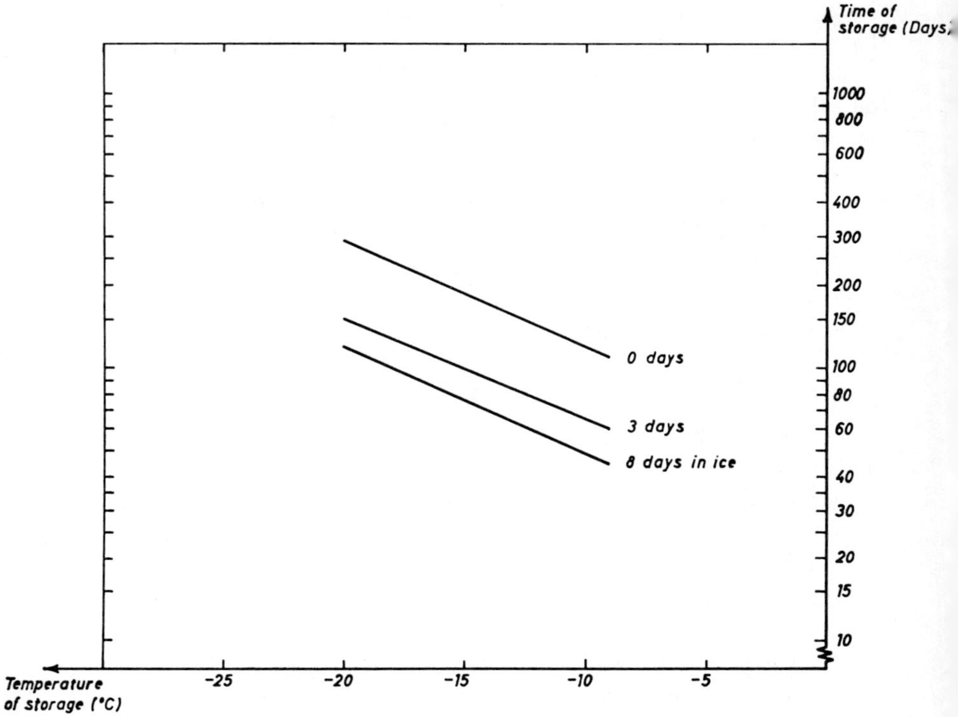

Fig. 9.6. The influence of freshness of cod on the storage life of the frozen product (25).

If one is dealing with very fresh fish, *i.e.*, fish frozen prerigor, whole or filleted, as on board freezer trawlers, three possible disadvantages are of interest to the later user of the fish, namely (4): (*1*) occurrence of blood discoloration, (*2*) fillet contraction, and (*3*) thaw rigor, which means contraction and extra drip loss in thawing. Discoloration (see Sect. IV.B.1) can be prevented by a proper bleeding technique combined with rapid lowering of the temperature of the fish. Attempts to accelerate the rigor process in the whole fish in order to counteract fillet contraction or thaw rigor have been unsuccessful; see Fig. 9.8, showing experiments in which high-temperature prestorage resulted in badly broken fillets when the fish was processed.

However, the biochemical reactions related to the rigor phenomenon continue during frozen storage. Therefore, thaw rigor can be avoided in prerigor frozen fish if the fish is stored for a period corresponding to about two months at -18°C. (0°F.) before being thawed and filleted.

Many experiments have been made in which the texture and flavor

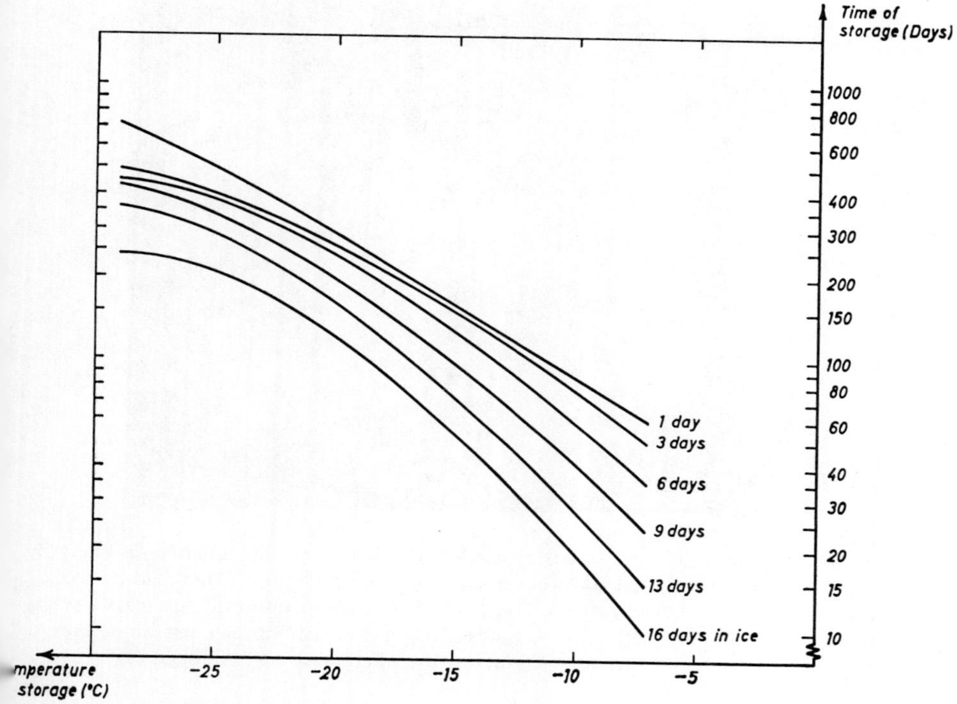

Fig. 9.7. The effect of iced storage time on frozen storage life of fillets of American pollock kept at various temperatures (27).

of fish and fillets produced from prerigor, in-rigor, and postrigor fish were compared. In the majority of cases an advantage has been derived from freezing fish prerigor rather than later (22,28). A granular or mealy texture which was found in prerigor frozen fillets (29) was shown to disappear after some months of storage. In some cases, however, no practical quality differences were noticed between prerigor and in-rigor fish. This was the case with North Atlantic cod and ocean perch (30) and with Pacific ling cod, ocean perch, red snapper, and rock sole (28). Slight preferences, if any, tended to disappear during storage. Fish sticks and fish portions, both products with a large sale today, are made from frozen blocks of fish fillets. Peters and Slavin (31) found that the storage life of frozen fish sticks varied inversely with the age of the parent fish block. The total storage life of the haddock fish block plus the storage life of the fish sticks was about 15 months at 0 to $-5°F$. (-18 to $-21°C$.).

Fig. 9.8. The effect of accelerated rigor "on the bone" on the quality of fillets frozen subsequently. Gutted cod were immersed in sea water until they had passed through rigor, then filleted, frozen, and thawed. The temperature of the sea water was from 0°C. (32°F.) (left) to 31°C. (88°F.) (right). In the warmest sea water the rigor period only lasted two hours; however, the fillets were badly broken (4).

B. Processing

1. Brining

The amount of thaw-drip can be greatly decreased if the fillets are lightly brined before freezing (32), and the practice of dipping fillets in sodium chloride brine has been used commercially for many years. Sodium chloride, however, has a tendency to accelerate rancidity on storage, resulting in off-odor and off-flavor development in many species, and for this reason its use is declining.

2. Polyphosphate Treatment

Various polyphosphates produce an effect similar to that obtained by brining, but without causing off-flavor and off-odor. A very slight "chemical" taste is sometimes noticed (33–36). One of the commercial uses is to dip fillets 2 minutes in a 12% tripolyphosphate solution before freezing. An increase in yield of frozen, thawed, and perhaps also cooked fish products is obtained because of a 2–5% absorption of dip solution, combined with the thaw-drip reduction obtained. The texture is also improved, although there does not seem to be any effect on the protein denaturation during storage.

In a few cases, depending on fish species and conditions, no e
polyphosphate treatment has been found, as compared with ur
controls (36).

3. Antioxidants

Rancidity, as measured by test panel, or high peroxide values and fading of red color, can be delayed considerably when salmon or trout are treated with various antioxidants in concentrations 0.01–0.05% in the fish meat (37). Ascorbic acid, being water soluble, seems to have certain advantages in application. Solutions of 0.25–0.5% are commercially used for dipping fish. Stronger solutions may confer an off-flavor. Storage life of frozen, eviscerated trout may be doubled by such treatment (38). Dyer (39) found delay of yellow discoloration in frozen ocean perch fillets by dipping in ascorbic acid solutions, but no effect on frozen flounder fillets. Banks (40) tried the most common antioxidants on frozen glazed blocks of herring and on herring fillets in small packages. He found very little, if any, improvement.

The sodium–calcium salts of EDTA (ethylenediamine tetraacetic acid) have been suggested as discoloration inhibitors for frozen fish sticks.

4. Breading, Covering with Gravy, Etc.

The experimental evidence published is too meager to determine whether breading, covering with gravy, sauce, etc., has a storage-prolonging effect on fish products like the effect recorded for meat (see Chapter 8).

5. Heating

In experiments with raw versus cooked frozen lobster Reay and House (41) found that it is preferable to cook the lobster before freezing and cold storage. The texture became softer and the flavor deteriorated more quickly in the raw frozen lobster, even at −30°C. (−22°F.). The same effect has been found for shrimp after extended cold storage. These experiments indicate that if enzyme activity can do damage to frozen seafood, it may be avoided by precooking.

6. Rate of Freezing

Microscopic examinations show that slow freezing results in greater destruction of the tissue than more rapid freezing. Variations in DNA, a nucleic acid, in the drip, depending on the rate of freezing, also indicate different degrees of cell damage (42).

In spite of these facts it is generally agreed that within the established commercial techniques in most countries the usual variation in freezing rates has no significant influence on product stability. Notevarp and Heen (25) find this to be true for cod fillets with rates down to 0.3 cm./hr. Certain species of fish like ocean perch and fatty fishes seem to be less sensitive to slow freezing than cod.

7. *Double Freezing*

The growing practice of freezing whole fish at sea for later thawing, filleting, and refreezing has brought up the question whether this "double freezing" damages the product stability. Some of the earlier experiments showed no essential damage (43). Later Dyer *et al.* (44), using inshore cod, found a very significant decrease in storage life after refreezing. The type of quality decrease was similar to that found to occur when frozen cod stored at 0°F. (−18°C.) is exposed to higher temperatures, say 15°F. (−10°C.), for a couple of weeks. Since then many experiments have been carried out using various kinds of raw material. The general picture which has come out is that certain species of fish, like ocean perch and flounder, suffer little or no damage by double freezing, while other species like cod seem to be more sensitive. It may even be that the extent of damage is dependent on the physiological condition of the fish. Thus Hansen (23) found that in plaice a reduction in the eating quality of 1 scale unit in a 10-point scale took place, except for autumn-caught fish, which was practically not affected by double freezing if the fish was less than 5 days in ice before being frozen.

C. Packaging

The importance of adequate packaging in relation to prevention of loss of moisture vapor and penetration of oxygen into the product is often overlooked. Sometimes packages believed to be vaportight are found to leak air, permitting dehydration to take place. This leads to underweight during prolonged storage. Dehydration also accelerates protein denaturation and of course oxidation of the lipid fraction results in flavor deterioration. Where possible the package must fit the product tightly. Where this is not possible moisture will migrate to the inside surface of the package, forming ice crystals. Sensitive products may suffer from this dehydration so that vacuum packaging may be preferable. Peeled shrimp in normal foil bags had only 6 months of storage life, while vacuum packing it more than doubled the storage life (48). Gutschmidt (9) found no differences after 6 months of storage of ocean perch fillets at −18°C. (0°F.) between vacuum and no vacuum

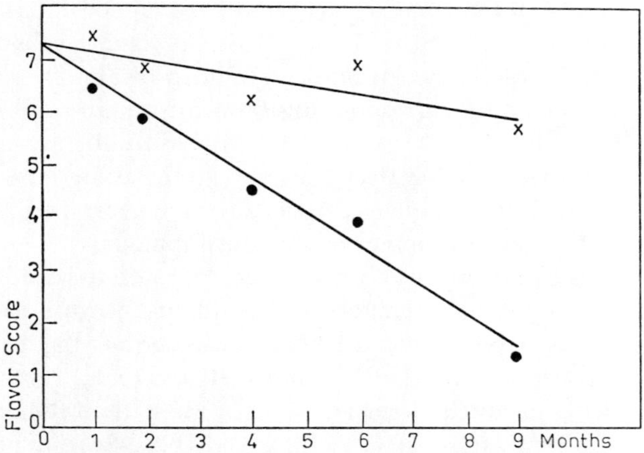

Fig. 9.9. Vacuum packed (x) and regular carton plus cellophane wrapper packed (●) eviscerated frozen trout stored at −20°C. (−4°F.). Flavor scores above 6 signified medium to good quality. Scores below 4 signified unacceptable quality due to rancidity (49).

if the fillets were tightly wrapped. Longer storage might have shown differences. Fatty fish benefit considerably by being vacuum packed. The storage life of retail-packed frozen eviscerated trout was extended from 5 to 15 months by vacuum packing (Fig. 9.9) (49). Glazes are, in the main, applied to whole fish or eviscerated fish. The storage life of fatty fish can be about doubled by this treatment (50). Many patents have been issued describing the addition of various chemicals to water to reduce the brittleness or slow the evaporation of the ice glaze. In many cases a plain water glaze, renewed when necessary, will give sufficient protection (7).

VI. FREEZER STORAGE CONDITIONS

A. Humidity

Where evaporation is possible seafood will lose weight at a significant rate when stored at low relative humidities. As an example, Slavin (45) found that whole tuna stored at 0°F. (−18°C.) and at a relative humidity of 70–80% for 10½ months lost 20 times as much weight as tuna stored similarly except at a relative humidity of 90–95%.

B. Fluctuating Temperatures

Regular and unavoidable small fluctuations (1–2°C.) in storage temperature are not important if the frozen seafood is properly packed. Larger fluctuations in the higher range of temperatures will affect the storage life.

Dyer *et al.* (46) found that if packages of frozen cod fillets stored at 0°F. (−18°C.) were placed at 15°F. (−10°C.) for 2 weeks and then returned to 0°F., there was an immediate drop of almost 10% in the palatability score and a corresponding drop in protein solubility. The storage life was decreased by about 25%. These results are typical for a series of experiments by Dyer and his group trying to assess the influence on quality of higher temperatures for short periods during refrigerator car transportation or other transshipment.

Experiments by Pottinger (47) with fillets of frozen bass and mackerel showed only a slight adverse effect on the quality because of temperature fluctuations. Dyer, aware of these results, points to the fact that fluctuations between −18°C. (0°F.) and −10°C. (15°F.) result in freezing and melting of much larger amounts of water than fluctuations in the interval −15°C. to −20°C. (5°F. to −5°F.), where many experiments, including the one reported by Pottinger, have been made.

C. Time and Temperature Influence

The effects of various time and temperature storage conditions given by different sources are indicated in Figs. 9.1, 9.2, 9.6, and 9.7, and Tables 9.1–9.3. While the tables are expressions of average experiences, the figures relate actual findings.

Variations will depend on PPP factors (Sect. V.). Partmann and Gutschmidt (30) find for ocean perch fillets, depending on quality of raw material, a storage life of 8–10 months at −28°C. (−18°F.), 5–7 months at −20°C. (−5°F.) and 2.5–5 months at −12°C. (10°F.). The figures given for fatty fish and crustaceans in Table 9.1 can be increased if vacuum packing is used. Q_{10} as a general rule is 2–3.* The bending curves in Fig. 9.7 indicate that Q_{10} is higher in the higher temperature intervals. Partmann and Gutschmidt (30) found Q_{10} 1.5–1.7 at −25°C±5°C. (−13°F. ±9°F.), 1.8–2.0 at −20°C. ±5°C. (−4°F. ±9°F.), and 2.1–2.5 at −15°C. ±5°C. (+4°F. ±9°F.) for ocean perch fillets. Edhborg (10) points out that in commercial loads the temperature histories of single packages will not be the same. He has found that the values representing the time at which an average loss of quality by one score unit has occurred also represents the time at which 5% of the samples may become unacceptable.

VII. COOKING METHODS

It is widely accepted that the method of preparing and serving samples to the panel may influence the marks scored by the panel. In testing frozen fish products the general experience has been that steam-

* Q_{10} is the "temperature quotient" for 10°C. See Chapter 10.

ing, baking, or cooking, with or without addition of a little salt, is the best method to measure differences in texture, flavor, and odor, while frying is an effective way of masking adverse texture and flavor. There is no work published where this experience is utilized in product development in relation to stability of the frozen product.

REFERENCES

(1) A. Banks, *Bull. Intern. Inst. Refrig.*, Annex 1962–1, 445, 1962.

(2) J. P. Lane, *Food Technol., 18,* No. 7, 156 (1964).

(3) W. J. Dyer and J. R. Dingle, in *Fish as Food,* Vol. 1, G. Borgstrom, Ed., Academic Press, New York, 1961, p. 275.

(4) N. R. Jones, in *Fish Handling and Preservation,* Organization for Economic Co-operation and Development, Paris, 1965, p. 31.

(5) E. Heen and O. Karsti, in *Fish as Food,* Vol. 4, Borgstrom, Ed., Academic Press, New York, 1965, p. 355.

(6) N. R. Jones, "Fish as Raw Material for Freezing," paper presented to FAO Technical Conference on Fish, Madrid, 1967.

(7) W. J. Dyer and J. Peters, "Factors Influencing Quality Changes During Frozen Storage," paper to FAO Technical Conference on Fish, Madrid, 1967).

(8) J. Slavin, "Frozen Fish," in *Freezing Preservation of Food,* Vol. 2, 4th ed., Avi Publishing Co., Westport, Connecticut, 1968, ch. 8.

(9) J. Gutschmidt, *Kältetechnik, 13,* 216 (1961).

(10) A. Edhborg, in *Fish Handling and Preservation,* Organization for Economic Co-operation and Development, Paris, 1965, p. 153.

(10a) Anon. OEEC Project No. 5/18, Paris (1960), p. 9.

(11) A. Banks, in *The Technology of Fish Utilization,* Fishing News, London, 1965, p. 89.

(12) G. A. Reay, *J. Soc. Chem. Ind., 52,* 265T (1933).

(13) W. J. Dyer, *Symposium on Cured and Frozen Fish Technology,* Research Publ. 100, Swedish Institute of Food Preservation, Gothenburg, 1954, ch. 7.

(14) J. J. Connell, in *Proteins and Their Reactions,* H. W. Schultz and A. F. Anglemeier, Eds., Avi Publishing Co., Westport, Connecticut, 1964, p. 255.

(15) T. R. Kelly, *Discoloration in Sea Frozen Fish Fillets,* paper presented to FAO Technical Conference, Madrid, 1967.

(16) W. D. Brown, personal communication, 1967.

(17) K. Amano, M. Bito, and N. Tsukuda, *Discoloration of Tuna Meat and Redskin Fish During Storage,* paper presented to FAO Technical Conference, Madrid, 1967.

(18) W. J. Dyer and D. C. Horne, *Fisheries Res. Board Can., Progr. Rept. Atlantic Coast Sta.,* circular, No. 2, 1963.

(19) J. A. Alford and E. A. Fieger, *Food Technol., 6,* 217 (1952).

(20) N. R. Jones, in *Progress in Refrigeration Science and Technology,* Vol. 2, Pergamon Press, Karlsruhe, 1965, p. 917.

(21) R. M. Love, "Condition of Fish and its Influence on Quality of Frozen Product," paper presented to FAO Technical Conference on Fish, Madrid, 1967.

(22) W. A. MacCallum, J. I. Jaffray, D. R. Idler, and D. N. Churchill, "Conditions of Newfoundland Fish, Inshore Cod, and its Influence on Quality After Freezing," paper presented to FAO Technical Conference on Fish, Madrid, 1967.

(23) P. Hansen, "Frozen Plaice Fillets from Fresh or Frozen Raw Materials," paper presented to FAO Technical Conference on Fish, Madrid, 1967.
(24) K. O. Kelly, "Factors Affecting Texture of Frozen Fish," paper presented to FAO Technical Conference on Fish, Madrid, 1967.
(25) O. Notevarp and E. Heen, *Z. Ges. Kälteindustrie, 47,* 122 (1940).
(26) J. A. Peters, E. H. Cohen, and F. J. King, *Food Technol., 17,* No. 6, 109 (1963).
(27) W. F. Dyer and F. A. Peters, "Factors Influencing Quality Changes during Frozen Storage and Distribution," paper presented to FAO Technical Conference on Fish, Madrid, 1967.
(28) N. Tomlinson, S. E. Geiger and S. W. Roach, "Quality of Sea Frozen Fish," paper presented to FAO Technical Conference on Fish, Madrid, 1967.
(29) D. R. Idler, W. A. MacCallum, D. Chalker, and I. T. Lander, in *The Technology of Fish Utilization,* R. Kreuzer, Ed., Fishing News, London, 1965, p. 98.
(30) W. Partmann and J. Gutschmidt, *Kältetechnik, 15,* 170 (1963).
(31) J. A. Peters and J. W. Slavin, *Bull. Intern. Inst. Refrig.,* Annex 1962–1, 533, 1962.
(32) G. A. Reay, *DSIR Food Invest. Board Rept. 1930,* 128, 1931.
(33) J. H. Mahon, U.S. Patent No. 3,036,923.
(34) J. W. Boyd and B. A. Southcott, *J. Fisheries Res. Board Can., 22,* 53 (1965).
(35) W. A. MacCallum, H. S. Shieh, D. A. Chalker, W. J. Dyer, and D. R. Idler, *J. Fisheries Res. Board Can., 21,* 539 (1964).
(36) W. J. Dyer, "Effect of Brining and Polyphosphate on Yield and Quality," paper presented to FAO Technical Conference, Madrid, 1967.
(37) H. L. A. Tarr, *J. Fisheries Res. Board Can., 7,* 137 (1947).
(38) F. Bramsnaes, H. Brennum, and H. C. Sørensen, in *Progress in Refrigeration Science and Technology,* Vol. 3, Pergamon Press, Oxford, 1960, p. 247.
(39) W. J. Dyer and M. L. Morton, *J. Fisheries Res. Board Can., 13,* 569 (1956).
(40) A. Banks, *J. Sci. Food Agri., 3,* 250 (1952).
(41) G. A. Reay and C. T. House, *The Fishing News,* London, Jan. 1951, p. 10.
(42) R. M. Love, *J. Sci. Food Agr., 6,* 30 (1955).
(43) M. E. Stansby and J. Dassow, *Pacific Fisherman, 46,* No. 4, p. 65.
(44) W. J. Dyer, D. I. Fraser, D. G. Ellis, D. R. Idler, W. A. MacCallum, and E. Laishley, *Bull. Intern. Inst. Refrig.,* Annex 1962-1, 515, 1962.
(45) J. W. Slavin, J. A. Peters, and S. R. Pottinger, *Food Technol., 12,* 602 (1958).
(46) W. J. Dyer, D. I. Fraser, D. G. Ellis, and W. A. MacCallum, *J. Fisheries Res. Board Can., 14,* 627 (1957).
(47) S. R. Pottinger, *Com. Fish. Rev., 13,* No. 2, 19 (1951).
(48) *Annual Report 1959,* Fiskeriministeriets Forsøgslaboratorium, Copenhagen, 1960, p. 44.
(49) F. Bramsnaes and H. C. Sørensen, *Bull. Intern. Inst. Refrig.,* Annex 1960–3, 281, 1960.
(50) A. I. Piskarev, A. K. Kaminarskaya, and L. G. Lukyanitsa, *Progress in Refrigeration Science and Technology,* Vol. 3, Pergamon Press, Oxford, 1960, p. 243.
(51) "Marketing and Consumption of Frozen Fish in OEEC Countries," OEEC Project No. 5/18, Paris 1960, p. 9.
(52) J. W. Slavin, "Freezing and Cold Storage," in *Industrial Fishery Technology,* M. E. Stansby, Ed., Reinhold, New York, 1963, p. 288.
(53) OECD, *Draft Code of Practice for Frozen Fish,* Paris, 1967.

Chapter 10

ESTIMATING QUALITY CHANGE FROM A KNOWN TEMPERATURE HISTORY

W. B. Van Arsdel

Western Utilization Research and Development Division, Agricultural Research Service, U.S. Department of Agriculture, Albany, California

If it were commercially practical to maintain every package of frozen food constantly at a temperature lower than 0°F. (−18°C.), from the moment of its original freezing to the moment of its final reheating for use, there would be no need for such a chapter as this. Although the processes of physical and chemical change in frozen foods, described in Chapter 2, are not slowed to a complete standstill at temperatures below zero, the extent of change within a year would generally be small.

However, real frozen foods in today's far-flung distribution and marketing system are not sheltered by an environment so nearly ideal as that. The writers of Chapters 12 and 13 present the results of extensive investigations into prevailing practices and performance, showing the extent of divergence from the wished-for ideal. Temperatures of product at successive stages in the long chain of manufacturing and distributing operations, involving many transfers of responsibility for the product, do not, in fact, remain constant; in the initial freezing operation and perhaps again at some stage of warehousing, the product temperature may be brought well below zero, but it may also rise substantially above zero at other stages of distribution and remain there for periods of hours, days, or weeks. Under some environmental conditions, temperatures will rise and fall in a periodic fluctuation for an extended period of time.

In addition, the hundreds of kinds of frozen foods that are now being produced and marketed differ greatly in their inherent *stability* —that is, their innate resistance to change under adverse conditions of temperature or access of air to package contents. Thus it has often been difficult to foresee the length of "life" a new product will attain in practical distribution.

I. STATEMENT OF THE PROBLEM

Every responsible person in the whole cold chain wishes to see the system perform so well that the frozen product reaching the consumer will be fully satisfactory; he also wishes to be sure that his own segment of the operation is above reproach. How can he be sure of that? The practical way is for him to monitor the product temperature at enough places and times to bracket critical phases of the operation, and then translate the resulting record of temperature history into the probable quantitative effect upon some selected attribute of product quality. This chapter is mainly a description of procedures for accomplishing the translation.

No great difficulty would exist if the temperature had remained constant during the entire period. It would only be necessary then to compare the duration of this period with the length of time that the product in question is known to remain "fully satisfactory" if held at this steady temperature. There are, of course, troublesome problems surrounding definition of such a term as "storage life"; they will be discussed in Section II.B.

If the temperature of the product is variable during the holding period several kinds of complication arise. In the first place, any rise in the temperature of a frozen food has much more than a simple linear effect upon the rates at which all of the deteriorative changes take place, as has been abundantly shown in the preceding chapters; consequently, substitution of an arithmetic mean temperature for the actual variable temperature is allowable only for quite small temperature changes.

Secondly, we cannot safely assume that the effects of chemical and physical changes in frozen foods simply add to one another, producing a total effect equal to the simple sum of all of them; nor can we assume that the combined result of several changes will be the same, regardless of the order in which the different changes occur. See Sections IV.A and B of this chapter for a discussion of these complications.

Finally, some kinds of quality loss in frozen foods are specific effects of temperature *change*—for example, repeated temperature fluctuation. See Section IV.D.

II. ESTIMATION OF "QUALITY CHANGE" IN A FROZEN FOOD

We might distinguish three quite different ways of defining and determining rates of quality change in a particular frozen food:

1. From objective measurements of a property or characteristic related to food quality,

2. From the consensus of opinions about the useful "storage life" reasonably to be expected of the product, and

3. From the measured "just noticeable differences" of a particular quality attribute.

A. From Objective Measurements

Computation of rate of change of an objectively measured property of a frozen food sample is, of course, straightforward. Thus, the ascorbic acid content of a sample of frozen sliced strawberries in composite cartons (2), 58 mg./100 g. initially, decreased linearly to 50 mg./100 g. in 40 days at 10°F. (−12°C.), so the rate of change of this valued "quality" constituent was −0.2 mg./100 g. per day at 10°.

In the foregoing example the rate of change of the property in question was found experimentally to remain constant. In most actual cases the rate at which a measured property changes with time increases or decreases as time goes on. For example, in an experiment with green beans at 10°F. (−12°C.) (3) the percentage of the original chlorophyll retained undegraded in the beans, 81% immediately after they were blanched and frozen, fell to 77% in 21 days at 10°, but only to 64% in 120 days at that temperature (Fig. 10.1). That is, the change became slower with increasing time at 10°. The data were well expressed by a linear decrease in the logarithm of the percentage retention.

B. From Empirically Determined "Storage Life"

Such expressions as "storage life" or "length of satisfactory storage" have come into rather wide use in the frozen food industry as rough indications of the stability of the different products. Tables 10.1 and 10.2 present two widely quoted estimates of this "life" for many common frozen foods. Now if we interpret a stated "storage life" as signifying the time required for various deteriorative changes in a freshly packed product to accumulate to such an extent that the market quality would be judged something lower than "excellent" or "highest quality," then the *reciprocal* of this length of time can be regarded as the *rate* at which the highest quality has disappeared. Thus if the "storage life" of freshly packed frozen lima beans is expected to be about 6 months at a steady temperature of 10°F. (−12°C.), the rate of change of quality during that period averages 1/6 per month (0.167

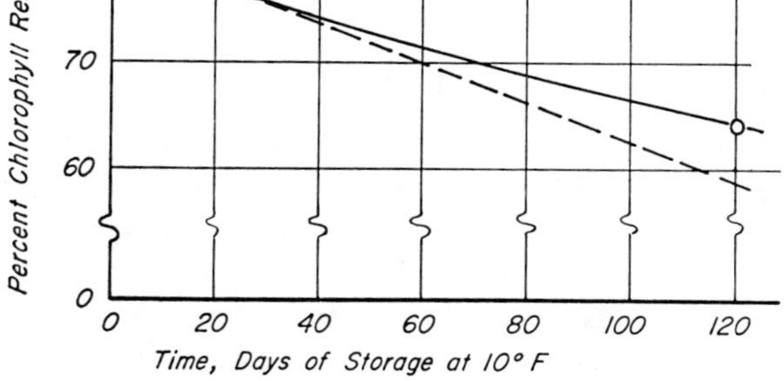

Fig. 10.1. The falling rate of loss of chlorophyll at a constant low temperature. Undegraded chlorophyll: (——) measured, (– – –) if rate of decrease had been constant.

Table 10.1.[a] Approximate Storage Life of Some Frozen Foods at Various Temperatures[b]

	+10°F. (−12°C.), months	0°F. (−18°C.), months	−10°F. (−23°C.), months
Fish			
fatty	4	6–8	10–12
lean	6	10–12	14–16
Fruit			
apricots	6–8	18–24	24
peaches, sliced	6–8	18–24	24
raspberries	8–10	18	24
strawberries, sliced	8–10	18	24
Meat			
beef, roast	6–8	16–18	18–24
lamb	5–7	14–16	16–18
pork, roast	4	8–10	12–15
sausage	2	4–6	8–10
Poultry, roasting	4	8–10	12–15
Shellfish			
lobsters	3–4	8–10	10–12
shrimp, raw	6	12	16–18
Vegetables			
asparagus	4–6	8–12	16–18
beans, snap	4–6	8–12	16–18
lima	6–8	14–16	24 or longer
broccoli	6–8	14–16	24 or longer
Brussels sprouts	4–6	8–12	16–18
cauliflower	6–8	14–16	24 or longer
corn, on the cob	4–6	8–10	12–14
cut	12	24	36 or longer
carrots	12	24	36 or longer
mushrooms	3–4	8–10	12–14
peas	6–8	14–16	24 or longer
pumpkin	12	24	36 or longer
spinach	6–8	14–16	24 or longer
squash	12	24	36 or longer

[a] Adapted from ref. 4, Table 100, p. 894.

[b] Average storage life for products that have been properly processed, packaged, and frozen, and warehoused under acceptable conditions.

Table 10.2.[a] Minimum Storage Life of Some Frozen Products (in months) at Different Temperatures[b] and Corresponding Temperature-Quotients

Product	Storage temperature, °F.						q_{10}[d]	
	+14	+5	−0.4	−4	−13	−22	+14 to −4°	−4 to −22°
Beef	3	8	15	20			2.9	
Pork	2	6	11	15			3.2	
Turkey	2.5	4	6	8	14		1.9	1.9
Chicken	2.5	6	10	13	26		2.4	2.3
Eggs	2	6	10	13			3.2	2.2
Butter	5	8	9	10			1.5	1.3
Lean fish	1.5	2.5	3	4	6	8	1.7	1.5
Fatty fish	0.8	1.5	2	2.5	4	6	1.9	1.6
Smoked fish	1	2		3.5	5	7	2.0	1.5
Kippers	0.7			2		4.5	1.8	1.6
Spinach[c]	7	12	15	20			1.8	
Asparagus[c]	5	8	10	11			1.6	
Peas[c]	7	9	10	11			1.3	

[a] Adapted from ref 5, Table 1.

[b] Time required at the given temperature to reduce quality from a rating of "excellent" to a rating of "very good," months.

[c] Much higher quotients were reported by others; see Table 10.3.

[d] q_{10}, the quotient of rate of change at some one temperature to the rate at a temperature 10°F. lower. See Section III.A.

month^{-1}). This amounts to adopting, as 1 unit of quality-change, whatever differences of all kinds combine into the first distinct step downward in a gradually decreasing practical market value.

Of course the "storage life," as thus understood, lacks nearly all the elements of a satisfactory quantitative measure, so its wide currency is rather surprising. Nevertheless, in the writer's opinion, it may be possible to derive from data on the relation of this "storage life" to storage temperature a meaningful relation between temperature and rate of quality decrease.

The data in Table 10.1 are adapted from ref. 4.* Table 10.2 is based upon one presented in an article by Kuprianoff (5). It lists a "minimum storage life" for several frozen foods at various low temperatures. Comparison of Tables 10.1 and 10.2 shows reasonably good agreement for the "lives" of meat and vegetables, but a wide divergence on fish; the Tressler and Evers estimate is more than three times as great

* Vol. 1, Table 100, p. 894.

as the Kuprianoff estimate. The reason for the wide difference is not immediately apparent.

C. From "Just Noticeable Difference" of Quality

Another approach to the problem, applied with marked success by U.S. Department of Agriculture workers in the "time–temperature tolerance" (T-TT) investigations, utilizes sensory tests, such as flavor or appearance, to provide a sensitive indicator of the first noticeable change in the property tested. The reciprocal of the time required to bring about this first change then serves as a measure of the mean rate of change under that set of conditions. Thus this procedure differs from the preceding one primarily in the selection of a different end-point to which the time will be measured.

1. The Detection of "Just Noticeable Difference"

A sensory difference between two samples can be detected most sensitively and certainly if the samples can be presented to the judge side-by-side and at the same time. But in a time series such as is in question here, side-by-side comparison of a sample taken now with one taken six months ago will be meaningful only if we can justifiably assume that all changes in the latter have been somehow brought to a halt during that six months. This can be accomplished satisfactorily by holding the "control" or comparison sample at a very low temperature. In most of the time–temperature tolerance experiments control samples were held at a temperature of −20°F. (−29°C.) (7). A still lower temperature should be employed for the holding of fish, shellfish, meat, or fried food samples.

Among the several possible procedures for detecting a small sensory change (6), Guadagni (8) selected a version of the well-known triangle test as most useful for long-time storage tests at several different storage temperatures. Five trained judges were employed, and four replications of the judgment, "which of these three samples is the odd one?" were made on each experimental material, making 20 judgments in all. The percentage of correct responses made by the panel was plotted against the duration of storage of the samples, and the time at which the resulting graph rose to and remained above the level of about 75% correct responses was taken as the time of first "real" difference from the −20° control sample.

It must always be borne in mind that this kind of procedure takes on definite meaning only when we define the particular sensory perception to be tested. A difference in the *odor* or *taste* of samples is to be detected, or a difference in visual *color,* not just an unspecified difference in "quality."

2. *Use of "Just Noticeable Difference" as a Reference Point*

The "just noticeable difference" (JND) is a psychophysical quantity which can provide the unit of quality change so greatly needed in this discussion. It has the very great advantage of good reproducibility at any lapse of time and by any team of investigators. In this respect it is far more firmly fixed and meaningful than the "storage life" we discussed earlier. The latter has its own set of advantages, but it has the grave inherent defect that it does not embody an operational kind of definition—especially a definition of those properties of a frozen food sample that would signify it has reached the *end* of that "life" period.

The JND can be used as a measure of quality *change,* but change from what? The base-line from which the change begins is left undefined, so formally it could as readily be the quality of a badly deteriorated sample as that of a perfectly fresh one. Thus the condition of the sample after the change has taken place must be at least perceptibly deteriorated from that of a fresh sample, but it may also be very much worse than that. Ordinarily, however, the JND has been used to establish the rate of change in fresh, high-quality samples (see, for example, the discussion in Chapter 4). A number of specific examples will be given in later sections of this chapter.

The interesting question whether a succession of JND's, each one starting with the termination of the preceding one, would create a uniformly graduated scale of the particular quality attribute (9) appears not to have been investigated. It is known, however, that sensory test panels have difficulty in achieving good sensitivity to flavor differences between severely degraded samples.

3. *Limitations on Significance of JND*

A change in flavor or color that is just surely detectable by a trained panel under laboratory conditions may, of course, be much too small to have any direct commercial significance. Whether this time would have to be multiplied by 2, or by 4 or even 10 to reach a degree of change obviously prejudicial to market acceptability cannot be determined from these data alone; that would require a merchandising decision, perhaps fortified by appropriate consumer surveys. For a further discussion, see Chapter 11, "Quality Management," especially Section III.B and C.

The length of storage time required to bring about a noticeable change in flavor or appearance in frozen strawberries, raspberries, or peaches held at 0°F. (−18°C.) or above could be determined reliably by the trained panel, according to Guadagni and his co-workers (2,8). Panels were particularly sensitive to flavor in the berry samples and color in the peaches. At the time when the flavor of an average sample of frozen strawberries, held at 20°F. (−6.7°C.), was just noticeably different from the control (9 days), ascorbic acid content had decreased only from an initial value of 58 mg./100 g. to about 45 mg., and redness of color had decreased by only about 2 units of the Hunter "A" value.

The kind of deterioration usually exhibited first by sliced peaches, on the other hand, results in the appearance of distinct brown color in a few slices (10). The JND may in this case be indistinguishable from a commercially significant change.

Limited surveys of the acceptability of the stored frozen strawberry samples referred to above (2) showed how difficult it may be to base an estimate of product "life" upon subjective judgment. Among a group of 20 quality-control people from the frozen food industry, only 15% judged that deterioration had gone far enough in 4 weeks at 20°F. (−6.7°C.) to cause consumer complaint (recall that JND for flavor at 20° appeared at about 9 days), whereas among 54 judges from other segments of the industry almost *half* thought the berries at 4 weeks would cause complaint. However, after 6 weeks of storage at 20° almost three-quarters of the quality control experts agreed there would be complaints, while among the other 54 judges there were still only half holding that opinion.

III. THE TEMPERATURE-DEPENDENCE OF RATES OF CHANGE OF QUALITY ATTRIBUTES

As the "storage life" data of Tables 10.1 and 10.2 show, the temperature of storage of the frozen food has a profound effect upon its stability. The several different types of physical change or chemical reaction which detract from the initial high quality are all accelerated by a rise in temperature, although not all by the same amount (5).

A. The Temperature Quotient of a Change

The effect of a difference of temperature is often stated in terms of a "temperature quotient," defined as the *ratio* of the rate of change of a property or attribute at some one temperature to the rate at a specified lower temperature (11,12). Because rates of change are increased by a rise of temperature, this ratio is greater than 1. An alternative definition, frequently simpler to apply, makes it the ratio of the *time* required to bring about a specified change at some one temperature to the time required at a specified *higher* temperature.

In physicochemical and biochemical investigations it has become customary to refer to a temperature quotient for a temperature difference of 10°C., and designate it by the symbol Q_{10}. However, most of the work done on frozen foods in the United States has been reported in terms of Fahrenheit temperatures. We have designated a temperature quotient for a temperature difference of 10°F. by the symbol q_{10}. The two symbols are related mathematically as follows:

$$Q_{10}=(q_{10})^{1.8} \tag{1}$$

Table 10.2 gives values of q_{10} for a number of frozen foods, computed by the relation of eq. (1) from the values reported by Kuprianoff. For certain meats and fish q_{10} is reported for the temperature range +14 to −4°F. (−10 to −20°C.) and also for the range −4 to −22°F. (−20 to −30°C.). The temperature quotient is in general a little smaller at the lower temperature—that is, the rates of various deteriorative changes are not lowered quite as much, proportionally, by a given decrease in temperature, at very low temperatures as at temperatures up closer to the freezing point (13).

It is well known (14,15) that frozen foods contain an appreciable amount of unfrozen liquid even at temperatures far below 0°F. (−18°C.), and even though the food may be frozen hard and apparently solid throughout. The proportion of liquid phase must have a great influence upon the rates of all chemical and physical changes, if only because of the effect on the mobility of reactants. Undoubtedly this proportion changes rapidly with small changes in temperature, especially in the temperature range between 0 and 30°F. (−18 and −1°C.). Thus it would not be surprising to find unusually high temperature quotients in this 0–30° range.

If the temperature quotient remains the same in different temperature ranges, however, as it substantially does for most frozen foods over not too wide a range, the rate of change of the property under consideration is a simple exponential function of the temperature:

$$\log r_t=\log b+at \tag{2}$$

Here r_t is the rate at which some property changes at temperature t°F., and a and b are constants to be determined. We can also write:

$$q_{10}=\frac{r_t+10}{r_t}=\frac{\theta_t}{\theta_{t+10}} \tag{3}$$

and we can readily show that:

$$a=0.1 \log q_{10} \tag{4}$$

Constant b in eq. (2) is the rate of change when the temperature of the product is 0°F., and θ_t in eq. (3) is the length of time required for the specified change to take place at a temperature t.

Table 10.3 gives average q_{10} values for a number of frozen foods not included in Kuprianoff's list (5), based upon a publication by Van Arsdel (16) which analyzed data from the U.S. Department of Agriculture's T-TT investigations. The procedure used throughout those studies was detection of the time at which 70–75% correct

Table 10.3.[a] Average Temperature Quotient and Stability at 0°F. (−18°C.) for Various Frozen Foods

Commodity	Pack	Property tested	Temperature quotient, q_{10}	Temperature range,[b] °F.	Stability at 0°F.,[c] days
Boysenberries	IQF[d]	Flavor	3.1	10–30	390
	Siruped	Flavor	5.1	10–30	650
	IQF[d]	Color	2.8	10–30	650
	Siruped	Color	3.6	10–30	310
Cauliflower	Carton	Flavor	3.4	−10–20	365
Chicken, cut up	Well packaged	Odor, raw	2.4	0–20	730
		Flavor, cooked	2.0	0–20	730
Chicken, fried	Fiber carton	Flavor	2.0	0–20	270
Green beans	Carton	Flavor	3.2	0–25	300
		Color	3.6	0–25	100
Peaches	Carton	Color	7.8	10–25	365
Peas	Carton	Flavor	3.4	0–25	320
		Color	4.2	0–25	210
Raspberries	Carton, siruped	Flavor	6.6	10–30	730
	Carton, siruped	Color	6.6	10–30	900
Spinach	Carton	Flavor	3.4	−10–20	140
Strawberries	Carton, sugar	Flavor	5.6	0–30	365
	Carton, sugar	Color	5.6	0–30	365
Turkey pies	Carton	Flavor	4.0	10–20	—

[a] Based on ref. 16.
[b] For averaging q_{10}.
[c] By triangle test of change.
[d] Individually quick frozen.

judgments of a real color or flavor difference from the control sample were obtained, and use of that occurrence to establish the relation between storage temperature and rate of change. The reason for the much higher temperature quotients for peas and spinach given in this table (3.4 for flavor change in both cases) than was found by Kuprianoff (5) for the same two vegetables (1.3 and 1.8, respectively) is not immediately apparent. The latter's samples were either much less susceptible to temperature damage than the samples examined by the U.S.D.A. workers or the method employed to detect a quality change was quite different.

Values of q_{10} given in Table 10.3 are averaged over the stated range of temperatures. The temperature quotient, q_{10}, did tend to be somewhat smaller at very low temperatures, just as Kuprianoff found it to be, but it was nearly enough constant so that the rate of these quality changes can fairly be said to vary exponentially with temperature, as expressed by eq. (2).

IV. EFFECTS OF CHANGES IN PRODUCT TEMPERATURE

Prediction of the effect that will be produced in a frozen food by exposure during a period of time to an environment whose temperature changes is complicated by several factors. These include possible special effects that might interfere with simple additivity and commutativity of successive experiences, the lag in establishment of uniform temperature within a body after its external temperature is changed, the formation of frost within packages and the accompanying desiccation or "freezer burn" of product, possible irreversible migration of water or other constituents in response to a temperature gradient, and the direct consequences of the dependence of all rates of change upon temperature.

A. Additivity of Effects

"Additivity" signifies that the total alteration in properties of a frozen food, produced by a succession of experiences at various temperatures, is the simple sum of the separate amounts of change and remains the same regardless of the number or spacing of breaks of continuity, provided all changes cease during the breaks (as by reduction of the temperature). Essentially, it means that defects of quality, once produced in the food, persist and are added to by all other defects that may be subsequently produced. Poor color or off-flavor is not healed by allowing a period for recovery after an adverse experience. The quality damage is irreversible.

Some observers have believed that in certain products a brief rise of the temperature almost to the thawing point might "sensitize" the product so that even a momentary subsequent adverse experience could trigger a disastrous breakdown. This would be an example of a *nonadditive* effect. So far as the writer knows, no case of this kind has ever been convincingly demonstrated.

One of the questions investigated in the "T-TT" studies was this additivity of effects on quality. No exceptions to the rule stated above were found in extensive experiments on the stability of strawberries (2), green snap beans (3), sliced peaches (10), orange juice concentrate (17), raspberries (18), turkey pies (19), fried chicken (20) raw cut-up chicken (21), peas (22), and cauliflower (23).

B. Commutativity of Effects

"Commutativity" means that the total alteration in properties of the frozen food is independent of the *order of presentation* of the various experiences at different temperatures. In many of the same series of experiments mentioned in the preceding paragraph, the order in which samples were exposed to temperatures of 0, 10, and 20°F. (−18, −12, and −6.7°C.) was purposely changed, or commutated, in various ways, leaving the total time at each temperature fixed. The end result was always the same within experimental error, so no exceptions to the rule were found.

C. Analytical Procedure for Summing the Effect of a Fluctuating Temperature

Hicks (24) and Schwimmer, Ingraham, and Hughes (25) have shown that if the rate of a reaction in a material varies exponentially with temperature and the temperature fluctuates with time according to some regular law (for example, in a sine wave) the extent of reaction after a period of time will be the same as it would have been if the material had been held at a certain steady "effective" temperature for the same length of time. If the temperature quotient is greater than 1, as it almost invariably is, this effective temperature, denoted by the symbol t_E, will be somewhat *higher* than the arithmetic mean temperature, t_M. For a sine-wave fluctuation (which can represent reasonably well most kinds of repetitious temperature change likely to occur in automatically refrigerated space) the difference can be expressed (25) by the curves of Figs. 10.2 and 10.3. In the former the difference between effective temperature, t_E, and mean temperature, t_M, expressed as a fraction of the amplitude of fluctuation, A (*i.e.*, one-half of the range from low to high), is plotted against the quotient,

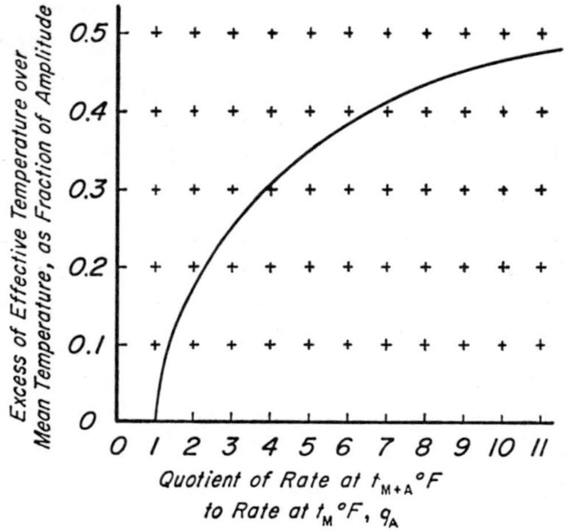

Fig. 10.2. Relation of the excess of effective temperature over mean temperature to the quotient of rates at t_{M+A} and at t_M, for sine-wave temperature fluctuation (25).

q_A, of the rate at the high temperature t_{M+A} to the rate at the mean temperature t_M. In the second figure this quotient is plotted against the standard temperature quotient, q_{10}, for various given values of the amplitude, A. Use of these relations may be illustrated by the following example:

The temperature of a lot of frozen strawberries ($q_{10}=5.6$) fluctuates between 0 and 15°F. ($A=7.5°$, $t_M=7.5°$). From Fig. 10.3, q_A, which is the quotient of rate at 7.5 +7.5°, or 15°, to rate at 7.5°, for a q_{10} of 5.6, is 3.65. Now from Fig. 10.2, the added fraction of the amplitude A is 0.29. Then the effective temperature is $0.29 \times 7.5°$, or 2.2°, above the mean temperature, and $t_E=9.7$°F.

1. *Experimental Test of Theory*

Tests of changes in properties after various storage periods under fluctuating temperature conditions were made on many of the commodities examined in the T-TT investigations (2, 3, 10, 18–23), employing repeated oscillations of as much as ±10°F. (±5.6°C.) in the range of −10 to 30°F. (−23 to −1°C.), and cycle periods of 24 or 72 hr. In addition, strawberries and raspberries were subjected to an especially thorough study of the effects of different temperature cycles (26).

In all of the cases tested the observed effects were, within experimental error, the same as they would have been after storage for the same length of time at a steady temperature equal to the computed "effective temperature," t_E. Earlier and less conclusive experiments by

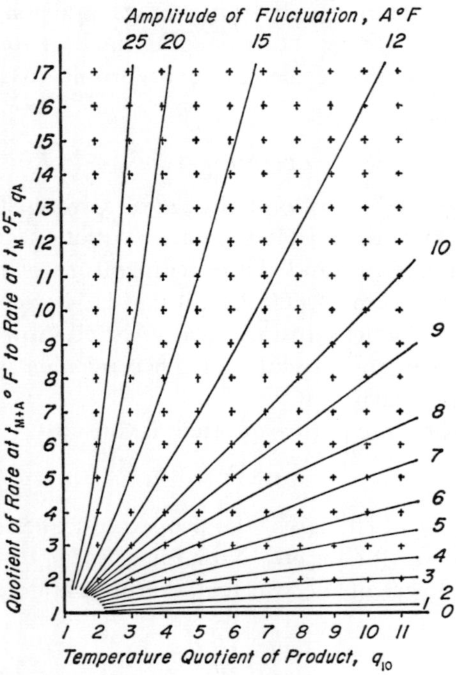

Fig. 10.3. Relation of rate at t_{M+A} to A, q_{10}, and rate at t_M (25).

other investigators (27–29) had led to essentially this same conclusion. In many of the cases in which t_E was only a degree or two above the arithmetic mean temperature, t_M, the latter would also have represented the result satisfactorily within the precision of the work. Thus the predictions of the Schwimmer *et al.* theory (25) are well confirmed by numerous experiments. One conclusion must be that temperature fluctuation in itself has no specific adverse effect upon those properties of frozen foods that are commonly judged or measured, such as color, flavor, and vitamin content. The often-heard caution that the temperature of storage environment should not be allowed to vary materially (27) is sound, but primarily for the reason that if it does fluctuate, it is higher than it should be for a part of the time. Also, as will be discussed in the next section, certain kinds of undesirable change are, in fact, specifically accelerated by repeated temperature change.

There are other limitations to the theory. Schwimmer *et al.* (25) pointed out that it assumes a single kind of deterioration reaction, with a constant temperature quotient. Probably neither of these assumptions is quite true. If two or more reactions with different temperature quotients go on simultaneously we should expect to find a qualitatively different result from storage under a fluctuating temperature than from storage at any one intermediate steady temperature. Perhaps sufficiently precise measurements would disclose this kind of effect.

2. *Special Solution if Temperature Change is Linear with Time*

If the "storage life" or rate of change of a frozen food at some one temperature is known, if its temperature quotient remains constant in the range of interest, and if the temperature of the food changes *linearly* with time, as may often be substantially true during a limited segment of the product's history, Van Arsdel showed (16) that the extent of change in the product can be conveniently computed with the aid of a graphic chart.

By eq. (2), if the temperature quotient is constant,

$$r_t = b\ (10^{at}) \tag{5}$$

where b is the rate of change of the property in question at a temperature of 0°F. and $a = 0.1 \log q_{10}$. Now for a linear temperature change between t_1, the lower of the two temperatures, and t_2, the higher one,

$$t = t_1 + \frac{\theta}{\theta_s}(t_2 - t_1) \tag{6}$$

where θ is the time when the temperature is t and θ_s is time at the conclusion of the change. If we write $dP = r_t d\theta$, where dP represents a differential change in the property, and substitute the value of t from eq. (6) in eq. (5), integration leads to the following expression:

$$f = 0.434\ (10^A)\left(\frac{10^B - 1}{B}\right) \tag{7}$$

Here f is the equivalent length of storage at a constant temperature of 0°F. (−18°C.) in days at 0° per day of actual experience;

$$A = 0.10\ t_1 \log q_{10}$$
$$B = 0.10\ (t_2 - t_1) \log q_{10}$$

Equation (7) can be evaluated by use of an alignment chart with parallel scales for A, B, and f.

D. Additional or Specific Effects of Temperature Change

Reports by Gortner *et al.* (27), Klose, Pool, and Lineweaver (30), Hanson, Fletcher, and Campbell (31), and Dykstra (32) describe the most convincing evidence for certain real effects of temperature fluctuation *per se*.

Gortner *et al.* (27) studied effects of 12-month storage of pork roast and commercially frozen strawberries, snap beans, and peas at steady and fluctuating temperatures between 0 and 20°F. (−18 and −6.7°C.). There was marked accumulation of frost within the packages of pork, beans, and peas stored at a steady 10°F. (−12°C.) and considerably more in the packages whose temperature had fluctuated between 0 and 20°.

Klose *et al.* (30) studied the storage of eviscerated turkeys in various types of wrap and under both steady and fluctuating (−10 to +10°) (−23 to −12°C.) temperatures. Far more frost formed in the body cavity and in small pockets between the flesh and the wrapping film of the birds held at the fluctuating temperature than in those held at a steady 10°F. The difference was most pronounced when the film was a water-vapor impermeable polyethylene. Klose referred to the effect as a one-way "pumping" of water vapor from place to place within the package.

Hanson *et al.* (31), seeking to improve the resistance of thickened puddings and white sauces to separation and curdling during freezing and storage, found that a temperature fluctuating from −10 to +10° was more damaging to the texture than a constant temperature of 0°. However, in view of the extremely high temperature quotient of the effect they did not believe their experiments were precise enough to prove the complicity of fluctuation *per se* beyond a reasonable doubt.

Dykstra (32) found the avoidance of temperature fluctuation to be important mainly to decrease the occurrence and seriousness of frost or "cavity ice" in frozen food packages. Other factors, such as the ratio of interstice volume to total volume and the insulating value of the package wall, also influence the rate of frost formation.

1. Cavity Ice Formation

The reality of the effect of temperature fluctuation upon the rate of formation of in-package frost and cavity ice is undoubted. However, the mechanism of this accelerating effect is still obscure. As the temperature of the air in the storage room or cabinet starts to fall from the high pont of its cycle, the temperature of the surface of the frozen load lags a little behind and remains warmer, and temperatures deeper within the body of the load remain warmer yet. A temperature *gradient* then exists between different points within the load and between the surface of a package and its interior. This gradient causes a corresponding gradient of water vapor pressure. Ice sublimes from the warmer locations within a package and the water vapor moves toward the cooler places near the package surfaces. The gradients, the difference in vapor pressure, and the mass transfer of water vapor are all reversed in the other half of the temperature cycle. However, it is quite possible that reabsorption of water vapor into areas near the package surface that have been partly desiccated in the cooling phase of a cycle is very much slower than was the rate of evaporation in the preceding phase. Then if the temperature fluctuation is even approximately symmetrical with time, like a sine wave, there may never be time enough during each rising temperature phase to redistribute the water back to the places from which it sublimed. Consequently, frost will accumulate in the open spaces within each package.

While the foregoing seems a reasonable explanation of the phenomenon, there is an unexplored region here which should be investigated.

a. Effect of Temperature Gradient. A persistent temperature gradient in one direction, even if small, could produce in the course of time substantial sublimation of water from product on the warmer side of each loosely filled package in a load, and deposition of the water as frost on the cooler sides of the same packages. In view of the notorious difficulty of eliminating all temperature differences from refrigerated spaces, this seems a probable source of much frost difficulty. A clear-cut example of this effect was reported by Woodroof and Malcom (33); the commodity was frozen bread packed hermetically in cans. The deposition of frost inside packages can produce enough product desiccation to be a serious cause of quality loss (34).

2. *Other Suggested Effects of Fluctuation*

The suspicion of Hanson *et al.* (31) that temperature fluctuation accelerates curdling or separation in puddings and white sauces has been mentioned. One might speculate that some such physical mechanism as the following would produce an effect of that kind:

These tender gels or emulsions owe their stiffness and smoothness in large part to a high and uniform degree of hydration of a starch or protein gel. An appreciable proportion of the material remains as an unfrozen liquid phase at temperatures down to well below 0°F., but if the temperature fluctuates, some part of the water in this phase must freeze out at each downward excursion of temperature. This partial removal of liquid water from the system, even though only temporary, may be accompanied by partially irreversible dehydration or denaturation of the hydrated molecules and consequent progressive breakdown of emulsions.

Several of the early investigators of frozen foods (35–37) noted the gradual growth of large ice crystals in a frozen substance at the expense of the smaller crystals, and its acceleration by temperature fluctuation.

The phenomenon has been confirmed many times (14,38). It was natural to expect a destruction of the normal food texture to occur. However, no adverse effect upon the texture of the thawed food appears ever to have been unequivocally demonstrated; in fact, the extensive tests on stored frozen fruits, vegetables, and poultry meat made by the T-TT investigators failed to bring out any instance of a down-grading for a texture defect clearly traceable to temperature fluctuation.

E. Graphical Procedure for Summing the Effect of an Irregular Temperature History

The accumulated effect of a known temperature history upon the measured quality characteristics of a frozen food can readily be estimated, provided that reliable data on the stability (or *per contra,* the rate of deterioration) of that kind of product are available for several steady temperatures in the required temperature range; and provided, further, that quality changes in that product are known to be additive and commutative (1).

From the available data on rate of change of the property or quality in question (such as flavor or ascorbic acid content) a smooth curve showing this rate of change as a function of temperature is laid out. From the curve, values of the rate at equally spaced increments of temperature are determined. Then a blank rectangular coordinate chart is drawn, with a uniform scale of time for the axis of abscissae, and for the ordinates a scale taken from the tabulation of rates at equal increments of temperature. The ordinate scale will be labeled with the *temperatures* corresponding to these rates.

The data on temperature history of the frozen product are then plotted in this new coordinate net which shows temperature versus time. Now the *area* under the resulting curve, representing the total accumulated change in the product as a result of this temperature history, is to be measured, for example with a planimeter. The zero baseline for this measurement will, of course, be the ordinate for zero rate of change, not the line for zero temperature. Ways of expressing and reporting the result, with a numerical example of the graphical construction and its use, will be presented in Section VI.

1. Experimental Test of the Procedure

Experimental data on rates of various objective quality changes in frozen strawberries, sliced peaches, and raspberries, as functions of temperature, were used as a basis for constructing curves relating rate of change to temperature (1). These were then applied as just described

Table 10.4.[a] Comparison of Calculated and Actual Values of Various Factors in Frozen Strawberries, Raspberries, and Peaches, as Affected by Temperature Experience

Temperature experience [b]	Ascorbic acid loss, mg./100 g.		Color units [e]	
	Calculated [c]	Found [d]	Calculated [c]	Found [d]
		Strawberries		
Least severe	5–8	6.9	1.2– 1.6	1.5
Intermediate	7–11	9.0	1.6– 2.4	2.1
Most severe	18–23	21.1	4.0– 5.2	4.8
		Raspberries		
Least severe	1.0–1.2	1.1	2.1– 2.8	2.3
Intermediate	1.4–1.8	1.6	3.0– 4.0	3.4
Most severe	2.8–3.6	3.3	6.0– 8.0	6.8
		Peaches		
Least severe	8–12	10	2.0– 2.6	2.1
Intermediate	14–19	16	2.8– 3.6	3.0
Most severe	26–32	30	9.0–10.0	9.4

[a] Adapted from ref. 1, Table 2.
[b] See ref. 1 for description.
[c] Ninety-five per cent of the lots tested fall within the range of values shown.
[d] Average of several typical lots.
[e] Color units: Strawberries, Hunter "a"; raspberries, in syrup; peaches, Hunter "b."

to calculate the probable total change in each of these objective measurements following three different prescribed temperature experiences. Table 10.4 summarizes the results and compares them with observed changes which occurred in samples subjected to these same prescribed experiences. The agreement with the calculation was, it will be seen, entirely satisfactory.

V. WAYS OF EXPRESSING THE INTEGRATED EFFECT OF A TEMPERATURE HISTORY

Total area under the rate-temperature curve will signify the accumulation of changes in the product during the time covered by the curve. The units in which it will be expressed are consistent with units used to express the rate of change. For example, if the rate variable is milligrams of decrease in ascorbic acid per 100 g. of product per day and the time variable is days of elapsed time, integration under the history curve will give the total decrease in ascorbic acid, in milligrams per 100/g. of product. On the other hand, if the rate variable is the reciprocal of an estimated "storage life" (see Sect. II.B) in months and elapsed time is expressed in months, integration up to a certain point of time will give the *fraction* of that estimated "life" that will then have gone by. This fraction may, of course, be either less than or greater than 1; if a result is greater than 1, the product will have undergone a greater amount of exposure than that supposed to bring that kind of product to the end of its storage "life."

If the reciprocal of the time at which the "just noticeable difference" is detected is used as the rate of change (see Sect. II.C), integration of the curve up to a certain time will give the total elapsed fraction of the exposure required to produce a perceptible change in the product. Here again, the result may be either less than or greater than unity.

One should be careful of his semantics in interpreting these results. If rate has been determined from data on storage life or just noticeable differences, a fraction of a unit in the answer has significance only with respect to *time,* not to any measurable characteristic of the product. If the integrated result is 0.75, the total exposure has been equivalent to ¾ of that at which the just noticeable difference would occur, or which would bring the product to the end of its expected life—whichever of those terms is applicable. Of course if the rate of change has been expressed in objective terms, the integrated answer will also be in terms of an objective amount of change.

The results of a computation of this kind may be expressed in terms of equivalence to the effect of exposure to some selected *steady* temperature in either of two useful ways:

1. The length of time at a selected steady temperature, such as 0°F. (−18°C.) which would produce the same total change;

2. The "effective" steady temperature which would produce the same total change in the same length of time (24,25).

A. The "High-Quality Life" Concept

A "just noticeable difference" of the quality of a sample of frozen food from the quality of a freshly packed sample of the same kind of food possesses the unique capability, as pointed out previously (Sect. II.C.2), when taken in conjunction with the length of time required for it to occur, of being a fairly reproducible unarbitrary unit of the rate of deterioration of the food in question. This fact led Van Arsdel and Guadagni (1) to propose that the length of time at some standard temperature, such as 0°F. (−18°C.), required to bring about this unit change in the freshly packed product be denoted the "high-quality life" of that product at 0°. Then by a simple extension of the concept, the effect of any specific temperature history subsequent to packing could be described by determining the integrated effect of the history in terms of equivalent length of time at a steady 0° temperature, and by difference from the normal value of "life at 0°," stating how many days or months of "high-quality life" remained for a product which had experienced that history.

When fully understood and correctly used, the phrase "high-quality life" has been found convenient and it is sufficiently striking to catch and hold the attention. However, it unfortunately has not always been correctly interpreted. It seems to imply that a frozen food which has changed in flavor or appearance to even a slightly greater extent than the "just noticeable difference" from freshly packed material is no longer high in quality, because it has gone beyond the "high-quality life" of that product. In fact, of course, it might still be completely acceptable in the market, and distinguishable from freshly packed material only by a trained panel under laboratory conditions (see Sect. II.C.3). The term also ignores the undeniable fact that "freshly packed" frozen foods themselves may vary widely in quality because of differences in raw material used and handling and packing practices. The phrase is, of course, completely out of place in description of any frozen food which was of poor quality even when freshly packed.

VI. EXAMPLES OF ESTIMATION OF THE EFFECT OF TEMPERATURE EXPERIENCES

The following example illustrates the procedure. We shall use data from the tabulation published by Kuprianoff (5) because they are presented in terms of "minimum storage life" and there is no question that they are intended to represent commercial conditions. The product is "chicken" (presumably eviscerated, well packaged). Let us suppose that the temperature of packaged product rises linearly with time from an initial −10 to +2°F. in 0.5 month, is held at +2° for 2 months, rises linearly to +14° in 0.1 month, is held at +14° for 0.1 month, falls to −6° linearly in 0.3 month, and is held at −6° for an additional 3 months (see Fig. 10.4 for a diagram of this history). What would be the equivalent duration of storage at a steady temperature of 0°F.? How much longer could the product be held at the final temperature of −6° before it reached the end of its expected "minimum life"?

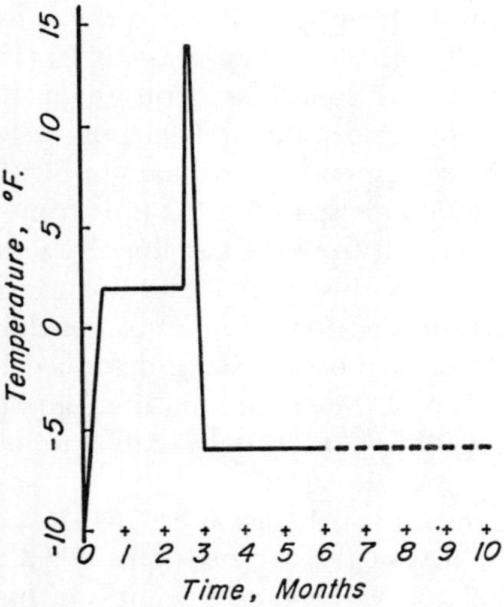

Fig. 10.4. Temperature history assumed for example.

The data for minimum storage life are given in Kuprianoff's tabulation as follows:

Temperature	°C.	−10	−15	−18	−20	−25
	°F.	+14	+ 5	− 0.4	− 4	−13
Time, months		2.5	6	10	13	26

The *reciprocals* of these times, in (months)$^{-1}$, plotted against temperature, give the smooth curve of Fig. 10.5. The data given indicate a

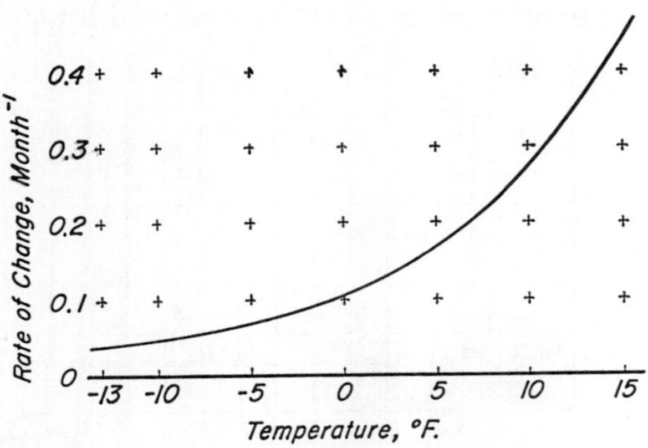

Fig. 10.5. Rate of change of acceptability in frozen chicken (5).

mean value of q_{10} as 1.54, but this is not quite constant; it is about 1.48 at −15°F. (−26°C.), about 1.83 at +15°F. (−9.4°C.).

In Fig. 10.6 the special coordinate grid (light lines) has equally spaced verticals for the time scale, and horizontals which are spaced proportionally to the corresponding ordinates in Fig. 10.5; for example, the horizontal for +15°F. is spaced 0.442 unit from the baseline, the horizontal for 0° 0.105 unit from the baseline. Now the given temperature history is plotted on this special coordinate grid; it appears as the cross-hatched area in Fig. 10.6. Application of the planimeter gives this area as 0.61. That is, the experience described has gone 61% of the way toward the end of the "minimum life," and only 39% of that expected satisfactory "life" remains available to absorb additional adverse experiences.

Figure 10.5 gives the rate of change at 0° (−18°C.) as 0.105 month^{-1}, so the answer to the first of the two questions asked at the end of the first paragraph of Section VI is that 5.8 months of holding at a steady temperature of 0°F. would give the same extent of change as the history actually experienced. The rate of change at −6°F. (−21°C.) is 0.063 month^{-1}; in answer to the second question, then, at this temperature the remaining 39% of product "life" would be used in 6.2 additional months.

The procedure just described was applied quite extensively by Jul (39,40) in combination with data obtained in 1952 by Munter, Byrne, and Dykstra (41) by means of surveys of times and temperatures experienced by frozen foods in channels of distribution. Munter *et al.* reported statistical distributions which made it possible to estimate both the average history of products in the market at the time of the survey, and the history that had been experienced by the top 5% of all cases so far as temperature damage would be concerned. In the very great

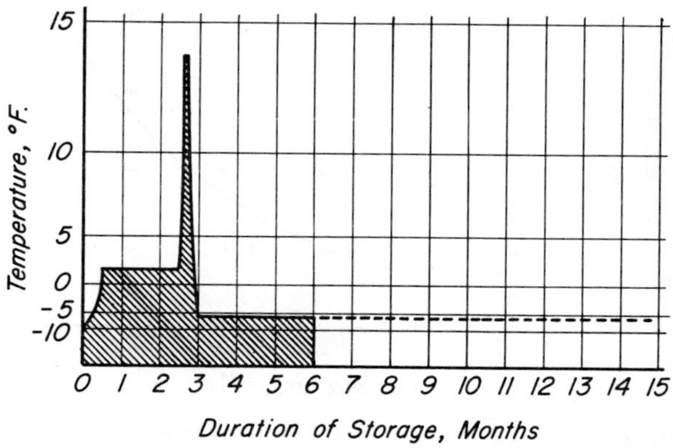

Fig. 10.6. Transformed coordinate grid and specified history shown plotted in it.

majority of cases examined, Jul deduced that the total equivalent exposure of the frozen materials would have been insufficient to produce a noticeable change in quality, but in the worst 5% of some frozen products severe deterioration would have been evident. This method of analysis therefore provided a fairly quantitative way of evaluating the performance of the system and of disclosing the most troublesome areas.

REFERENCES

(1) W. B. Van Arsdel and D. G. Guadagni, *Food Technol., 13,* 14 (1959).
(2) D. G. Guadagni, C. C. Nimmo, and E. F. Jansen, *Food Technol., 11,* 389 (1957).
(3) W. C. Dietrich, M.-D. Nutting, R. L. Olson, F. E. Lindquist, M. M. Boggs, G. S. Bohart, H. J. Neumann, and H. J. Morris, *Food Technol., 13,* 136 (1959).
(4) D. K. Tressler and C. F. Evers, *The Freezing Preservation of Foods,* 3rd ed., Avi Publishing, Westport, Connecticut, 1957.
(5) J. Kuprianoff, *Kältetechnik, 8,* 102 (1956).
(6) M. A. Amerine, R. M. Pangborn, and E. B. Roessler, *Principles of Sensory Evaluation of Food,* Academic Press, Inc., New York, 1965.
(7) W. B. Van Arsdel, *Food Technol., 11,* 28 (1957).
(8) D. G. Guadagni, *Food Technol,. 11,* 471 (1957).
(9) C. S. Ough and G. A. Baker, *J. Food Sci., 29,* 499 (1964).
(10) D. G. Guadagni, C. C. Nimmo, and E. F. Jansen, *Food Technol., 11,* 33 (1957).
(11) R. Plank, *Intern. Congr. Refrig., 10th, Copenhagen, 4.50,* 45 (1959).
(12) L. Riedel, *Pflügers Archiv, 271,* 157 (1960).
(13) E. Dalhoff and M. Jul, *Intern. Congr. Refrig., 11th, Munich,* P–9, *1,* 57 (1963).
(14) O. Fennema and W. D. Powrie, *Advan. Food Res.,* Vol. XIII, *13,* 220 (1964).
(15) J. Kuprianoff, *Bull. Intern. Inst. Refrig., 40,* 775 (1960).
(16) W. B. Van Arsdel, *Food Process., 22,* No. 12, 40 (1961).
(17) R. J. McColloch, R. G. Rice, M. B. Bandurski, and B. Gentili, *Food Technol., 11,* 444 (1957).
(18) D. G. Guadagni, C. C. Nimmo, and E. F. Jansen, *Food Technol., 11,* 633 (1957).
(19) H. L. Hanson and L. R. Fletcher, *Food Technol., 12,* 40 (1958).
(20) H. L. Hanson, L. R. Fletcher, and H. Lineweaver, *Food Technol., 13,* 221 (1959).
(21) A. A. Klose, M. F. Pool, A. A. Campbell, and H. L. Hanson, *Food Technol., 13,* 477 (1959).
(22) M. M. Boggs, W. C. Dietrich, M.-D. Nutting, R. L. Olson, F. E. Lindquist, G. S. Bohart, H. J. Neumann, and H. J. Morris, *Food Technol., 14,* 181 (1960).
(23) W. C. Dietrich, M.-D. Nutting, M. M. Boggs, and N. E. Weinstein, *Food Technol., 16,* 123 (1962).
(24) E. W. Hicks, *J. Council Sci. Ind. Research (Australia), 17,* 111 (1944).
(25) S. Schwimmer, L. L. Ingraham, and H. M. Hughes, *Ind. Eng. Chem., 47,* 1149 (1955).
(26) D. G. Guadagni and C. C. Nimmo, *Food Technol., 12,* 306 (1958).
(27) W. A. Gortner, F. Fenton, F. E. Volz, and E. Gleim, *Ind. Eng. Chem., 40,* 1423 (1948).
(28) S. R. Pottinger, *Ice Refrig., 121,* 13 (Aug. 1951).
(29) J. D. Winter, A. Hustrulid, I. Noble, and E. S. Ross, *Food Technol., 6,* 311 (1952).

(30) A. A. Klose, M. F. Pool, and H. Lineweaver, *Food Technol., 9,* 372 (1955).
(31) H. L. Hanson, L. R. Fletcher, and A. A. Campbell, *Food Technol., 11,* 339 (1957).
(32) K. G. Dykstra, *Bull. Intern. Inst. Refrig., Annexe 1962–1,* 405.
(33) J. G. Woodroof and H. R. Malcom, *Food Technol., 12,* 268 (1958).
(34) L. van den Berg, *Food in Canada, May 1966,* 37–38.
(35) H. F. Taylor, *Food Inds., 3,* 205 (1931).
(36) M. A. Joslyn, *Ind. Eng. Chem., 24,* 665 (1932).
(37) J. G. Woodroof, *Georgia Agr. Exptl. Sta. Bull., 201* (1938).
(38) R. M. Love, *J. Sci. Food Agr., 13,* 269 (1962).
(39) M. Jul, *Fryseteori, MS6,* Aug. 23, 1960, Konserveslaboratoriet, Copenhagen.
(40) M. Jul, *Food Technol., 17,* No. 8, 46 (1963).
(41) A. M. Munter, C. H. Byrne, and K. G. Dykstra, *Food Technol., 7,* 256 (1953).

Chapter 11

QUALITY MANAGEMENT

Henry N. Couden

Safeway Brands Buying Division, Safeway Stores, Inc., Oakland, California

I. INTRODUCTION

Wherever a product is there is its quality also. Each economic group in the frozen food industry is involved with and, in a general sense at least, is concerned with quality. Thus it is that quality comes to mean different things to different people, particularly to the various functional specialists.

One should realize also that the qualities of performance by production process, freezing, storage, and distribution, while they may be "recorded" by the product itself and in its own way, are not always the same qualities which are apparent to or understood by the consumer. Consumer qualities are those which, because they are present, make for a more acceptable product. The National Association of Frozen Food Packers calls these the "Class" qualities. Only in part are they the finished-product qualities which appear in buyer specifications and in U.S.D.A. grade standards. Recent consumer surveys for frozen foods have indicated that convenience and out-of-season availability are more important qualities than any others (1), and these of course do not appear in any product or handling standards. In a previous (1963) survey of western states to determine at the time of purchase, why shoppers purchased frozen foods, convenience outranked product quality by 2½ to 1 and out-of-season availability ranked second (2).

Other chapters of this book will impart much information concerning the quality considerations important to specific economic groups and their specialists. This chapter addresses itself to the quality problem *per se*—how we recognize it, regard it, and act upon it.

II. QUALITY CONSIDERATIONS

A. Direction of Quality

Under our free-enterprise system the direction of the quality requirement tends to be from buyer to seller, from demand to supply. This is the opposite of the direction of production, which is from seller to buyer, from supply to demand. There are certain exceptions to this generality, such as limitations within supply and quality capabilities, such that an interdependence does exist. But generally speaking production and distribution must satisfy the demands of the retailers (or institutions), who in turn must satisfy the demands of their customers.

B. Retailer Quality Requirements

It is therefore appropriate to review at the outset some of the qualities a retailer must consider in order to accept an item from his supplier and continue to stock it. His merchandising decisions are intimately involved with product cost and profit concepts. We list those we feel are the most important, but we make no attempt to list them in order of importance, for this varies. A change in any variable of the marketplace affects profitability. The customer is regarded as the most perishable item, and in the last analysis customers must be offered what they prove by their patronage they want to purchase.

1. Cost

The resultant price, namely cost plus markup, must be at a level the customer will pay.

2. Turnover

The product must continue to move in order to pay its way. The more frequently it moves the greater its appeal as a traffic builder, and the more often its individual profit is realized.

3. Markup

This is usually between 19 and 30% for frozen foods. It must cover not only the laid-in product costs, but unlike other grocery items it must also cover costs of cabinet operation and depreciation, and backroom storage at −10°F. (−23°C.). Markup is strongly related to turnover, demand, competitive activity, and price of competitive items.

4. Demand and Sales Potential

This is related to product quality as well as to preselling influences such as advertising or brand-name acceptance. What is the use of the product and how often will it be used? Is there a season for heaviest and lightest sales?

5. Impulse Purchasing

Related to label appearance, the "silent salesman." Surveys show that approximately 82% of all decisions to buy frozen foods are in-store decisions, and of these 62% are impulse decisions (3).

6. Convenience

With 3 of 10 women working it is understandable that convenience of use is the factor rated highest by customers.

7. Availability

Dependability of supply. "Out-of-stock" items may lose a customer, and is the most frequent reason why a customer makes a substitute purchase (3).

8. Display Position

Shall the same brands be displayed together, or shall items be displayed together or with other items that complement each other, or that compete?

9. Cabinet Space

Acceptance of a new item usually means elimination of an old one, because space is limited. While the elegant technique of computing direct product profit per cubic foot (DPP/CF) is possible, though not often fully applied, this concept is always very meaningful (4).

10. Ease of Handling

Size of package unit and the number of units packed in a shipping case may or may not be compatible with customer needs or operating requirements.

11. Quality Level

This is related to customer demand and customer loyalty. The merchant must consider whether size, weight, fragility, and quality are right for his customers.

In addition, there are questions as to just how many frozen items a retailer is prepared to handle, as well as how many he should handle in terms of his customer demands. Backroom storage, refrigeration capacity, and display space are among his considerations. From the nature of the customers served, some stores can handle a "price line" (lower quality implied) while others cannot.

Customer response, as measured by sales and profit, soon determines whether the merchandise manager has been correct in his accepting and offering an item for sale. If correct, the item earns a place in the cabinet. If incorrect, the item will be replaced shortly in favor of some other, because cabinet space is limited and every item is called upon to pay its own way in terms of the valuable space it occupies. On the other hand, if the merchandise manager's decision was incorrect in rejecting a successful item, this fact becomes apparent from the item's success with the competitor down the street. News of success travels fast.

Therefore, there is considerable assurance that, in the long run, it is the consuming public which determines by its patronage which items a retailer shall offer and what shall be their basic quality characteristics. The retailer's success depends largely upon how sensitively he gauges his customer demands.

C. Quality Changes and Differences

The changes in quality level acceptable to the frozen food industry and its customers are different from those documented in the U.S. Department of Agriculture's "time–temperature tolerance" series. A criterion of quality change much used in the latter investigations was the "just noticeable difference" in flavor or appearance, as judged by experienced taste panelists capable of detecting minor changes—such as in most instances would pass by the average consumer undetected (see Chapters 4, 5, and 10). Quality control experts of the National Association of Frozen Food Packers have rated the reserve quality of numerous frozen foods to be considerably greater than the "T-TT" limits. Yet the degree of a brand's success will be due in large part to the extent that customers find the quality that they expect when they use the product.

The concept of Class quality, suggested by NAFFP, must be included when considering customer acceptability. This is the composite rating of a product based upon both its quality and its condition. Quality or grade might be high, but its degree of soundness or condition might be poor. Thus this concept of Class becomes more important in determining the acceptability of a frozen food lot as it becomes exposed to the varying hazards of time and temperature changes in the channels of marketing. Since the Class quality tends to be lowered by the time the product is in consumer hands, industry has for many years operated under what has been called a "margin concept."

This is simply the idea that the product must leave the processor's hands at quality and condition margins above what quality management believes is the lowest quality limit acceptable to the consumer. There is general belief that higher product quality attributes tend to compensate for poorer product condition. This belief has considerable basis in the chemical and nutritive facts documented by the T-TT work; and there is also considerable physical basis in the effect upon appearance qualities particularly. For instance, excess moisture makes for unsightly ice buildup within the package and changes the attractive appearance of most frozen vegetables. Therefore, less residual moisture and a strong margin of good product color are preferred at the packaging station during processing.

D. Sources of Quality Information

For certain products in particular there are discriminating customers who are quick to point out differences in quality which may have been considered only minor. In spite of the fact that the number of these customers continues to increase there are serious limitations in marketplace feedback. In many instances where quality differences seem to have significance to the expert there will be no discernable customer reaction. The French scientist, H. Cheftel may have this in mind when he states, "Unless the consumer becomes more discriminating in his choice of foods it is but a matter of a few generations until man will be merely ingesting meals and not enjoying them" (5). However, we must also understand that what people buy and what may be their favorite items are not necessarily identical, because of price, buying habits, and sometimes diet restrictions.

This boils down to the fact that decisions concerning quality are essentially decisions of individual or committee judgment, aided whenever possible by whatever skills are developed for making objective decisions. The following aids are frequently used.

1. Taste and flavor panel procedures and techniques.

2. Finished-product evaluations of competitive products.

3. Plant and product surveys and appraisals of plant and production facilities by trained food technologists.

4. Field surveys of raw product, and inquiry into new varieties.

5. Product development and research facilities, including those of suppliers, research centers, trade associations, and reports in scientific journals.

The main job is to cull from these many sources the practical and useful, then organize and hold or apply this information as needed. Practically all procedures and evaluations appearing in product specifications are obtained from one or another of these sources. It should be recognized also that much of the research to improve the stability and quality characteristics of frozen foods remains in the files of the packers themselves, has not been published, and is not available.

It is not our purpose to discuss or even list the important contributions to frozen food quality research from the various sources. But it is interesting to know of the many industry associations which have contributed in a responsible measure to frozen food research. Those we know of are: American Trucking Association, National Association of Food Chains, National Association of Frozen Food Packers, National

Association of Refrigerated Warehouses, National Association of Retail Grocers of the United States, National Fisheries Institute, National Frozen Food Association, and National Prepared Frozen Food Processors Association.

Members of these associations, together with the Association of Food and Drug Officials of the United States, have compiled a Frozen Food Handling Code (1962) as a reference of good handling practice. Some seven states to date have adopted regulations in line with its temperature recommendations (6).

Finally, a quality decision may be made inadvertently by some unique quality structure as between competitive items. For example, if all or most brands of lemonade contained practically no pulp, it would seem that the presence of pulp in one's own lemonade would be a desirable plus quality, likely to attract specifically the group of customers who would appreciate its presence.

E. Complaints

A customer complaint marks the extreme negative pole of quality management efforts, because it is evidence of extreme dissatisfaction, as opposed to satisfaction. The complaining customer did not find what he expected upon use of the product, and the situation was of such serious nature he took the trouble to write about it. Such a complaint may have great importance to management.

1. First, it is a positive indication of a breakdown somewhere in production or marketing. If it occurs, the reason should be discovered before it can result in a serious loss of product acceptance in the marketplace.

2. If the customer has suffered distress or damages of some sort an expensive lawsuit may result.

3. Complaints are a form of feedback which gives information concerning how the product is being used by customers, and whether management's concepts of its use are correct.

4. A complaint may provide information about the label and whether instructions for use are stated clearly. Thus, it gives opportunity to discover and correct unintentional errors.

Therefore, management pays serious respect to complaints. Each one is investigated, a fact which favors heavily the policy of coding every package so that its history can be obtained from quality-control records. In addition to this type of search and investigation, records of all complaints will be logged for periodic analysis as a possible basis for some major correction program.

The serious complaints, such as those which threaten or claim damages, will be channeled for insurance or legal action. In practically every instance the complaining customer will be contacted and offered reimbursement.

We ordinarily do not know how many additional customers are inconvenienced or dissatisfied for each one who complains. Most customers probably register their feelings by failing to repeat purchases of the product or the brand. The kinds and causes of complaints are too numerous for compiling a series of dependable figures. A single complaint may airse from a single instance, such as the return of a unique machine screw from a filler, or from thousands of equal instances, such as a faint foreign, or unwanted flavor which might remain in a faulty batch of product. We only know that the complaints arising from the somewhat faulty batch are in very small proportion to the amount of product which is later discovered to have reached the marketplace.

There is a widely held belief that the incidence of complaints increases during times of or in areas of general economic recession. This is usually explained from the fact that a written complaint frequently elicits either a monetary refund or some free merchandise, and sometimes both. In addition to some supporting data for this belief, there are instances where rashes of complaints against a single brand will progress through a neighborhood or area, as if through word of mouth broadcast, and with little or no evidence of legitimacy upon investigation. Accordingly the standard formula for reimbursement is no longer the highly rewarding experience for the complainant that it once was. However, it is good business and only right that the customer always receive satisfaction in both acknowledgment and reasonable reimbursement for his inconvenience and expense.

Possibly it is more often the loyal customer than the casual one who takes the trouble to register his complaint. At least a significant percentage do protest their loyalty, so these customers are especially worth contacting and retaining. Therefore, contact by management should be immediate, with appreciation and concern expressed. The customer will be reassured that all possible care is taken to prevent such accidents or errors, and opportunity can be taken to explain the company quality policy. The offer of reimbursement as one alternative will usually be accompanied by a more generous offer of product replacement as another, both to demonstrate confidence in one's own product and to tempt the customer himself to continue using it with confidence.

The logged record of the different kinds of complaints, and their sources, as discovered from the coded quality-control records, provides useful guides for correction. Complaints can be classified as:

1. Quantitative: fill, counts, weights, size.

2. Qualitative: "class" or condition, appearance, edibility, workmanship.

3. Extraneous: the presence of things that don't belong. Direct and immediate action for any one kind or group of complaints can often be taken, particularly where the source is ascertained. Periodic comparisons of the incidence of the various kinds of complaints indicate where trends are developing, so that corrective action may be taken before a serious situation develops.

F. Changing Specifications

Assume the marketer of a branded line of frozen foods is asked to accept a less flavorful variety of fruit or vegetable because it has higher yield, is therefore more competitive, and as a consequence is becoming more readily available. He might well consider this to be something of a threat to his product's quality. How does he decide whether to change his raw product specifications accordingly?

First we must realize that his decision has to be a dynamic one, and there is rarely sufficient time to develop objective comparisons or await the results of market research. Production and marketing sources will be consulted with respect to prospective changes, including price, competition, and the known results of market acceptance. Some facts may have to come from growers, some from seed or plant suppliers. Often samples will be especially prepared, either to demonstrate limits, to compare differences, or to make flavor panel evaluations. Shelf life experiments may be required. Published research as well as private will be searched and reviewed.

From samples a "trained" group—often those within management who are responsible for decision-making concerning the product—will judge for quality differences. One or more "experts" may make a flavor difference report based upon careful analysis (usually strictly subjective) of the flavor profile or upon some basis of flavor analysis. Or a panel preference test may be used. The final judgment whether the change will be significantly noted or adversely reacted to by his customers will be made in a decision by the top merchandise manager.

Most often there is no positive market reaction to be recognized concerning the effect of this decision, largely because of the dynamics of competition. The decision, once made, will stand without ever receiving positive objective assurance whether it was good or bad.

G. The Average-Value Pitfall

The dynamics of the marketplace create a constant downward pressure upon product qualities, since qualities usually bear a price tag.

Quality management must establish guidelines for wise reaction to these pressures.

Production management can readily assess the cost savings that would be realized by lowering the content of some important quality ingredient, say by 2 units. The problem of assessing the possible change in customer demand will be considerably more difficult, as we have just noted. But if the change is related to some significant provision of the finished product specification and this provision is properly stated, then quality management has a useful statistical tool which will help it decide upon the suggested change.

Let us assume the specification was originally set at an average of 30 units as a result of some satisfactory research, while experience with processing has revealed there is a normal spread of 4 units on both sides of this average. Management has learned that so long as the level goes no lower than 26 the product meets with consumer acceptance, and in fact when the level exceeds 32 the quality is so good as to build customer loyalty and helps to establish brand franchise. But in light of the possible savings in manufacturing costs it seems logical to argue for the 2-unit change, especially since 28 units would be midway between the lowest allowable and the specified target of 30.

It might seem that management's simplest approach to making this decision is to request samples for comparison. These would be carefully prepared at the 30 and 28 unit levels. Judgments will, of course, confirm that the 28 unit level is not significantly different than the 30 unit level, so a decision to drop to this level may well be made.

Reference to Fig. 11.1 illustrates the fallacy of decision-making on this basis. Quality levels are normally distributed about an average, or the specified target, and are not precise for every item produced. When the same normal spread of processing values is shifted 2 units lower, as from curve *A* to curve *B*, a significant portion of the product, at *b*, is manufactured below the 26 limit of acceptability. Furthermore, the exceptionally good quality, at *a*, is no longer produced. The product will actually experience significant quality losses from two directions, neither of which would be detected from the sample comparison analysis.

This example illustrates why it is good guidelines practice to specify values for the acceptable range as well as for the average target value.

H. The Industry Taste Panel

Taste panel work and much product testing, often with emphasis upon flavor, are generally conducted for some or all of the following purposes:

1. Comparison of products from different suppliers.

2. Comparisons of competitive products.

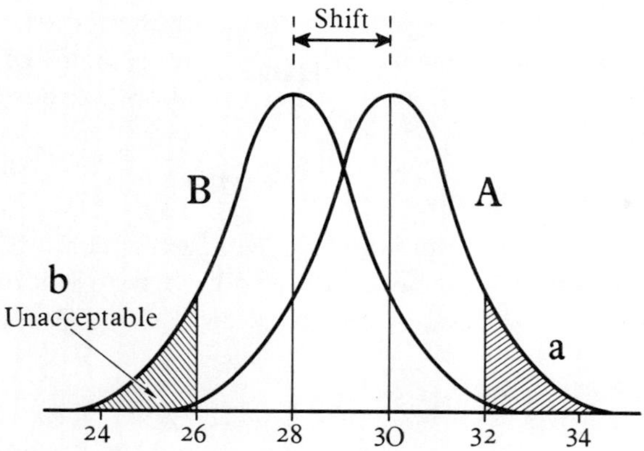

Fig. 11.1. Shift in distribution of lot values from A to B permits averages to remain within limits of acceptable quality while a significant proportion of individual units drops below acceptable limit at *b*.

3. New variety comparisons and evaluation.
4. Product improvement.
5. Acceptability evaluations.
6. Investigation of complaints.

There is recognition of different kinds of panelists, such as the experts, both trained and untrained, as well as those panelists who, from past results, have proved to be consistent in their reporting.

There is also recognition of different techniques and methods for presentation and analysis of panel results. The purpose of the test, of course, determines which kind shall be used. Probably the most frequent is the preference ranking test, though the triangle and the paired comparison tests are also used. These can be readily analyzed statistically, and each allows for panelists' remarks, which are sometimes the most important of the test results.

Very seldom is the brand supplier concerned with small differences. By the same token, it is also very seldom that the sophisticated taste test laboratory environment is offered the panel members. Instead, most taste tests are conducted in the regular production laboratory or in office surroundings. Certain precautions are taken with respect to sample preparation, temperatures, order of presentation, external noises, and the conversations and opinions of other respondents. The use of specific forms containing instructions and space for comments also seem to contribute to reasonably high validity of results.

I. New Product Development

Many factors come into play when a new product is being developed for the marketplace. Sequence of these factors will depend upon the type of business and its organization, the kind of product, availability of various resources, and the expertise of personnel. Some of the following stages may be bypassed, most notably number *3*, although all must be considered (7).

1. Management Approval

A new product must be suitable for a company's marketing program. Costs of introducing a new product are so high as to eliminate all but a very small percentage of products which may be submitted by research.

2. Concept Testing

From several dozens to several hundred consumers may be presented with the idea or concept embodied by the new product. Their answers determine whether there would be customer demand.

3. Product Testing

A limited consumer group, usually by families, tests the product and answers an accompanying questionnaire in which the product is also to be rated for acceptability. Sometimes a comparison ranking is also made with a competitive product.

4. Consumer Interviews

Person-to-person contact provides critical information for the selection of a name, type of package, art work, and design.

5. Home Consumption Testing

From 200 to 400 families are given a supply of the product to be tested, or a supply of the new product and its best competition. As in stage *3* records are made of consumer behavior toward the product and its consumption. There may also be a purchase phase at which time consumers are asked to purchase many of the products at a typical cabinet price.

6. Test Marketing

The more complex the new product the larger should be the test marketing. The market or markets should be representative in size and population characteristics, contain stores of all sizes, and include suburban and fringe areas. This test should last for at least one year, and it should include notations of competitor reactions.

A new product which has completed all these stages successfully is not necessarily assured a place in the grocer's cabinet (see Sect. II.B, above). The grocer's problem is not one of finding new items, but of finding items for which there will be a demand, which will show a profit, and have an advantage over competitive items which must be removed from his cabinets in order to provide space for the new item.

III. SPECIFICATIONS

Written specifications translate management's quality decisions into written form. In so doing they assign relative importance to raw products, manufacturing procedures, and quality characteristics. The talents of the entire business organization are brought to bear in accordance with quality policy. This concerns the following matters, most of which find their way into specifications of one kind or another. This is not to say that all specifications are in writing. Many times they are built into organizational structure and processing procedures.

A. Raw Materials

Raw fruits and vegetables may be specified according to U.S. raw product standards, by variety, and by conditions of maturity and soundness. In addition, raw product specifications may include such procedures as methods of harvest, restrictions for spray residue tolerances, harvest time and temperature maximums, and delay between harvest and processing.

The selection of sources of supply in the geographical sense is in effect a matter of raw product specifications. However, in some instances, this may be more a matter of production or marketing location than one of product quality selection.

In the area of prepared foods, specifications include the quality inspection and sometimes certification of incoming ingredients such as flour, shortening, salt, dairy products, oils, starches, colors, and flavorings. The certification may cover chemical composition, physical characteristics, and limitations of bacteria types and counts.

B. Preparation, Handling, and Processing

An abundance of considerations and procedures may be listed under these headings of a specification:

Sanitation	Line inspection and quality control
Size-grading	Blanching controls
Washing	Container handling and cleaning
Cooling	Weights or measure controls
Coding	Maturity control methods
Packaging	Freezing methods
Storage	Freezing time limits

Exceptions are sometimes taken to conventional production methods. A quality policy that wants the most possible flavor in its vegetables may see fit to restrict the amount of washing, pumping, and fluming, and specify more rapid handling and closer blanching controls. On the other hand, it may require more washing, screening and shaker-grading—all water flushing procedures which tend to leach flavors—to obtain a more attractive product and satisfy a different quality policy; or, closer attention must be given to size grading if size uniformity is the policy consideration rather than garden run; or, a container of fruit may have to be frozen in a specified position with respect to its label in order to present the most favorable appearance to the customer upon opening.

C. Finished Product

If U.S.D.A. grade standards are already in effect for the product at hand, the job of writing finished product specifications is made easier. Not only are the important quality and quantity levels already defined and established in U.S. standards, but so are language and terminology. The desired specification can be directly tied to these, such that this becomes a favored industry method. However, if no such standard is available it then becomes a practical necessity to establish one's own, or else run the much greater risk of nonuniform production and deliveries.

Other chapters in this book discuss at length the quality characteristics of various finished products. Finished product specifications should place limiting values and ranges, as objectively stated as possible, upon these characteristics. The selection of these values, while coming under the head of quality management, is nevertheless predominantly a matter of cost and supply-demand, or marketing economics. Yet there are many areas where production and marketing skills are able to enhance certain quality characteristics by changes in raw products, processing techniques, or ingredients such that similar products at the same economic price will differ considerably. Scientific research and the talents of food technologists and process engineers continually make for dynamic changes and improvements in this area.

D. Product Handling

Separate specifications are often issued to cover storage conditions, such as temperature, methods of stacking, aisle space, air circulation, and location with respect to doorways, stairs, elevators, and proximity to other stored products. Other specifications are concerned with shipment by rail or truck, and may cover handling, stacking, supporting of loads, temperature controls, and methods of reporting. Supplementary to these are instructions for handling in case of emergency; guides or charts may be furnished for use in deciding upon acceptance, follow-up inspection, or rejection of "hot" merchandise, based upon which product is involved, including its temperature history. Obviously such charts will often be based upon or interpolated from the time–temperature tolerance studies.

Finally, each distributor and each retailer specifies certain handling practices for the distribution, delivery, storage, and handling of frozen foods while they are in his care.

IV. QUALITY CONTROL

It should be apparent that quality is an all-pervasive business proposition. It is a noncompartmentalized, moving target. Its management is a dynamic process, including among its considerations a thoughtful analysis of production and distribution costs and risks, and market situations. It requires thorough understanding of the product and of the probable conditions which will exist for a period of time at least equal to its normal production cycle. While hopefully optimum, it will rarely be either minimum cost or highest priced. While it is not a level arrived at in some theoretical, sophisticated manner such as mathematically optimizing profits against a particular supply–demand situation, it is nevertheless a planned and consistent application of controls to establish and maintain quality at a profitable level (8). It includes knowing what is right about a product as well as what is wrong about it.

Because quality is everyone's business it must not be allowed to become nobody's. Management's quality decisions, once made, may then be assigned as a coordination function to a quality control department as its full-time activity. Thus quality control may operate at any or all levels of management, and the organizational structure which serves best will serve at all levels. While the technical functions involved with quality control procedures will be executed by the quality-control department, it continues to be the administrative function of management to audit them. The immediate goal of quality control is to assure customer satisfaction at minimum cost. It is therefore oriented to the quality problems of the business rather than technique oriented, and in essence it is an arm of the total marketing function.

A. The Quality Control Organization

The actual assignment which management makes to quality control and its resultant organization will be determined by the quality policy and the kind of answers given to such questions as the following:

1. Is the quality policy well defined, or is it relatively opportunistic and expedient?

2. Must every order be filled, regardless of quality?

3. Shall there be written quality specifications? Can they be compromised, and if so, under what circumstances?

4. What areas of quality responsibility and authority shall be assigned?

The assignment of quality control at some level of management is to make more effective the quality performance at that level. Since cost-saving techniques are most dramatic and effective on the production line, quality control activities are most often assigned to strictly production processes. Here its job is to direct and guide the production efforts pertinent to quality. The following list of duties shows this is a considerable assignment.

1. Identify the processes through which major quality characteristics are achieved.

2. Relate quality at key successive stages to cost and yields.

3. Relate costs to deviation from specified levels.

4. Provide consistent systems for the orderly continuous evaluation of the product and process.

5. Diagnose problems. Predict troubles before they happen.

6. Estimate the magnitude of drifts and shifts in production factors, and detect and localize deficiencies.

7. Assure management that the quality specified is the quality delivered.

8. Provide a system for determining how well the program is succeeding.

To accomplish these ends, quality control must know with what quality features customers are primarily concerned. It therefore addresses itself continually to basic and dynamic quality issues, such as the following:

1. What is the price range of the products?

2. How do our products compare with competition?

3. What are the complaints?

4. What quality is an added asset to sales? Is the cost justifiable?

5. Where is it necessary to apply basic research?

6. How are quality requirements best expressed in terms of specifications?

7. What are the needs for specification revisions?

8. What management reports are required?

The key costs of quality control are rejection of product, appraisal of product quality, and prevention of quality errors, in that descending order. Prevention costs probably represent only about 10% of the quality effort. Failure costs are perhaps six times this amount—one dramatic reason why the value of quality control can be demonstrated on the production line.

B. Procurement Quality Control

Inspection within reasonable time prior to purchase is a legal right. A quality-control department organized at the procurement level essentially monitors two management quality decisions: (*1*) that an adequate quality system exists among suppliers, and (*2*) that each supplier lives up to his own system. Procurement quality control involves:

1. Survey of vendor plants, product and process capabilities, and controls.

2. Audit of vendor quality system if it is complex, otherwise the complete test and survey of it.

3. Review of vendor's inspection and training programs.

4. Evaluate the manner in which he operates under other buyer specifications.

5. Evaluate his procedures for accepting incoming materials and ingredients.

6. Routinely inspect product and make quality audits of quality control and processing records.

7. Recommend procedures which will assist suppliers to correct their deficiencies and improve their operations as sources of supply.

8. Predict trouble before it happens, and help extricate the vendor from trouble if it does occur.

9. Review quality data, recommend specification changes when necessary, and analyze facts leading to these recommendations.

10. Rate the performance of different suppliers and supplier plant operations.

C. Sample Inspection Decision

Sampling plans must be developed to assure that the quality specified is indeed the quality produced or procured. Information obtained from the samples is used to make inferences about the lot or population being tested. This involves risks.

Quality control therefore addresses itself to the problem of determining when, where, and how much sample inspection shall be performed to ascertain whether the lot quality lies within or without specification limits. Too little sample inspection means higher risk and greater uncertainty, while too much is unnecessarily costly.

Two methods for determining lot quality from sample inspection might be identified as the subjective and the objective. Debate about the merits of each is the familiar one of art versus science.

The subjective method is often called intuitive. It is certainly a valuable method, requiring the inspection of fewer samples because it relies upon the inspector's experience with the product and process, and upon his talents of observation and interpretation. The number of samples required at any one time is largely based upon his personal judgment from cumulative results of previous samplings, or possibly from his current survey of the process itself. The inspector observes not only the magnitude of variations and the degree to which they are present, but he accounts for them in wider terms of his understanding their special significance and particular causes. His knowledge of raw product, handling procedures, and processing techniques gives more meaning to their presence or absence. He supplants the uncertainties and degrees of risk in purely objective methods by personal interpretation and analysis.

Of course this method is no better than the inspector's own ability, and there is no assurance of his success. He must have good perception abilities for all his senses, a strong food-processing background and education, and an analytical attitude. His ability to communicate is also important since the weakest link of the intuitive method is communication: from inspector to management, from management to inspector, and from inspector to inspector.

D. Statistical Quality Control

The objective method is based upon statistical mathematics and is called statistical quality control (SQC). Our purpose here is to discuss what it is and to illustrate why it works.

SQC may be used whenever results of any kind can be expressed quantitatively. This makes its use potential so universal that it has been hailed—perhaps a bit too enthusiastically, but nevertheless significantly—as the greatest scientific achievement of the last half century. The frozen food industry has been unaccountably slow to respond to it.

The advantage of SQC is that it provides a purely objective basis for decision making. It measures the degree of risk (or its opposite, the degree of confidence), based upon the statistical probability that the results obtained might have happened as a chance event. As a highly sophisticated technique it calls for planned, regulated, and uniform procedures. It possesses certain "elegant techniques" which objectively analyse and correlate several variable factors at a time. The planning of procedure to fit the exact specifications of the problem, so that results will express an answer with mathematical accuracy, is likewise highly sophisticated. Sometimes computations require calculation by electronic machine. However most SQC is being handled by relatively simple mathematical calculations.

SQC, as applied to the sample-inspection decision, calculates the sampling plan required to implement management's specification for acceptance. Such a specification might be to require 95% confidence that only lots containing not more than 6% deviant units shall be accepted. (A "deviant" is a unit which does not conform to the specified quality. Six per cent is the allowance under U.S.D.A. sampling plans.) The SQC answer, to insure meeting such a specification, will state the method of inspection, and the minimum number of samples which must be inspected including the allowable number of deviants.

We shall explain some of the basis for SQC problem solving by reference to Fig. 11.2. The quality value for some crucial acceptance factor (*e.g.,* net weight or total score) may be identified as acceptable

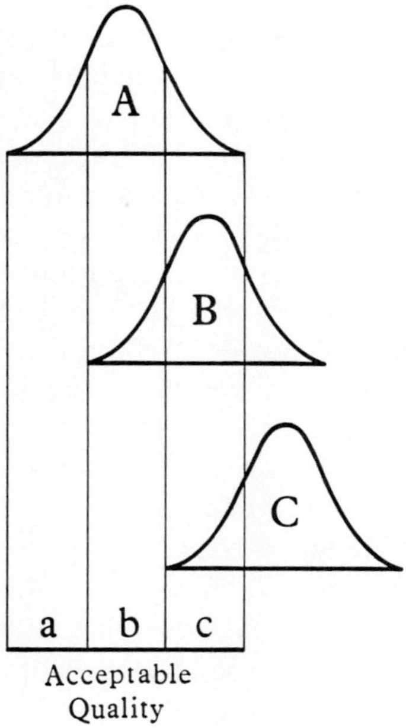

Fig. 11.2. The relationship of three possible distributions of lot-quality values to acceptable quality.

through a range of equal intervals as at *a, b,* or *c*. We assume the unit values within the lot are normally distributed, as is generally the case for repetitive processes under "good commercial practice."

It is apparent from Fig. 11.2 that if the individual unit values within a lot are represented by the probability distribution curve as at *A,* all of them will test within the specified limits. If they are as shown at *B,* there is a chance probability (about 1 in 6, or 17% as happens to be shown here) for a unit to test beyond *c* and thus fail as a deviant. At *C* this chance increases to 5 out of 6 deviants, with only the remaining 1/6 units within the specification limits. It is now evident that fewer samples should be required to provide acceptance criteria for the lot at *A,* or rejection criteria for the lot at *C,* than to determine the true status at *B*. The discovery and use of this kind of information are among the SQC tools.

Next, envision the probability distribution curve moving from the *A* position to *B, i.e.,* from 0 to 17% deviants. As this movement takes place the number of *a* and *b* unit values decreases while the number of *c* unit values increases. At some point during this movement the curve passes through the accept/reject boundary where exactly 6% of the distribution tail lies beyond *c*. Just as is the case at the locations *A, B,* and *C,* so also at this point (not shown) will there be a constant percentage of unit values lying within the *a, b,* and *c* intervals. Another SQC technique, then, is the discovery of what shall be the distribution of sample values which would give the desired confidence level that the curve has not passed beyond this 6% limit.

The nature of the normal curve is such that, upon locating its average value and computing the dispersion of sample values on both sides of this average, it is then possible to predict the number of sample values which should occur theoretically at any value distance away from it, either to the right or to the left. The SQC sampling plans and the SQC production control charts are based upon the statistical probability that the sample values, as checked during the inspection period, will fall, as predicted from the analyzed data, to within the computed risk limits. If they do, this is evidence the process is stable. If they do not, it is evidence of fundamental change, the cause for which should be established and corrected if possible.

SQC when used on the production line is basically a system of process control. Such charts give the exact amount of objective information computed into them for the quality or quantity characteristics the processing lines are imparting to the product at the time the test values are taken.

E. Conclusion

In a new sense we have come full circle. While quality always remains with the product, quality management and control procedures, and especially those of SQC, are still largely confined, in their sense of formal management, to the processing line. There are other areas of quality management and control to be sure; and, as with production, all of them reinforce the marketing requirement that the basic economic direction of quality management is from demand to supply. It is an essential part of the total marketing concept.

REFERENCES

(1) Agricultural and Technical Institute, Farmingale, N. Y., and State University of N.Y., *Quick Frozen Foods, 28,* No. 10, 43 (1966).
(2) Washington Agricultural Experiment Stations, "The Use of Fruits and Vegetables by Western Families," Bulletin 646, pp. 19–26.
(3) Ide Associates, Inc., *Quick Frozen Foods, 28,* No. 11, 108 (1966).
(4) R. D. Buzzell, *et al., Product Profitability Measurement and Merchandising Decisions,* Division of Research, Graduate School of Business Administration, Harvard University, Soldiers Field, Boston, Mass., 1965.
(5) H. Chaftel, *Food Technol., 20,* No. 4, 58 (1966).
(6) Association of Food and Drug Officials of United States, *AFDOUS Quart. Bull. 26,* 25 (1962).
(7) J. D. North, *Supermarket News,* May 2, 1966.
(8) D. L. Field, *Ind. Quality Control, 23,* No. 4, 178 (1966).

Chapter 12

DEVELOPMENTS IN THE HANDLING, STORING, AND DISTRIBUTION OF FROZEN FOODS

H. C. Diehl
Colorado Springs, Colorado

I. GENERAL DISCUSSION

During the USDA Time–Temperature Tolerance Investigations, it was clearly evident that unduly high temperatures, steady or fluctuating, would produce lasting, cumulative adverse effects on frozen food quality.

As a result, a meaningful phrase comes into general usage, "Frozen foods have a memory for adverse experiences."

However, any reference to frozen food behavior must equally emphasize the time periods involved as well as take account of the nature of the product and the ambient environmental temperatures, for both time and temperature are equally significant. Too often the emphasis has been placed on temperature only.

The actual temperature experienced by the frozen product during a stated time period determines the extent of the inevitable quality loss, as has been discussed in previous chapters. Assuming that the frozen food left the packer's establishment with the highest quality obtainable by good processing practice, this time–temperature experience also determines how good it will be when it reaches the consumer. However, this quality loss may proceed so imperceptibly in any given time period that it may escape notice until it has built up to the point at which more critical consumers may find the product unacceptable. Rejection by the latter may then be caused by the fact that the particular product sample, while not spoiled, simply does not have the desired appearance, flavor, or texture which the consumer has been led to expect.

It has been rightly said that more permanent harm is done to the consistent sale and public acceptance of frozen foods by the offering of products which "just do not look or taste right" than by the occasional and exceptional spoilage of an item, caused by gross mishandling.

The latter, at least, invariably results in a consumer complaint and this permits the vendor to explain and rectify the unpleasant experience. On the other hand, the marginally poor quality condition usually causes no indignant complaint; the customer simply does not again purchase the offending item or even products bearing the same brand. The potential consumer may even be so displeased that frozen foods, in general, are thereafter regarded unfavorably, especially if the purchaser has chosen frozen foods for the first time. The vendor may not be aware of the problem until diminishing sales reflect the public rejection, and then it is usually too late to remedy the matter effectively.

One must always remember that the consumer who buys frozen foods sees only the product and the brand name on the package and knows little or nothing, and also has no basic concern, about the processor's facilities or problems; nor does the consumer have any knowledge of all the services involved to bring the packaged product to its final resting place in the merchandising cabinet.

The product price and quality are the sale factors which influence the purchaser. It seems hardly necessary then to emphasize the fact that every person involved in the production and distribution of the product has a profound responsibility to bring that frozen item to the market place in the best possible condition in order to insure consistent, satisfied consumer acceptance. Upon this major fact rests the success of the industry, considered in the broadest sense, especially in the highly competitive food economy which prevails in the United States.

The moral of this situation is that even though satisfactory protective practices may prevail in every phase of the movement sequence of frozen foods from producer to consumer except one, successful marketing can be seriously and adversely affected or nullified by carelessness or neglect of essential temperature protection in that one phase.

There is, therefore, a joint and mutual responsibility resting on the persons concerned with every operation involved in bringing the frozen food from its origin to its exposure for sale in the retailer's establishment.

At the time the U.S. Department of Agriculture initiated the extensive series of studies which became known as the Time–Temperature Tolerance project in 1948 (1), many leaders in the frozen food industry had become convinced that continued healthy growth of the industry demanded the application of much more extensive knowledge of the actual behavior of commercially packed frozen foods than was then available. If major reconstruction or replacement of industrial facilities costing many millions of dollars were in question, then the persons who must decide what expenditures were justifiable needed quantitative information about the effects which all the circumstances of production, storage, and distribution would be likely to have on the acceptability and marketability of the frozen products. Such information was also indispensable for the realistic and practical administration of public health agencies' responsibility for monitoring the wholesomeness of foods.

The T-TT investigations, involving many simulated temperature histories characteristic of average transport, warehousing, distributing, and retailing conditions, created a great fund of significant data. This information has proved to be invaluable both in focusing on and identifying the insecurely based segments of the overall distribution pattern which were then prevailing, and for planning improvements in facilities, equipment, procedures, and employee training. The USDA investigations dealt primarily with frozen fruits, vegetables, and poultry. Frozen seafoods have more recently been studied by the Bureau of Commercial Fisheries of the U.S. Fish and Wildlife Service (2–4).

The concluding statement of the introductory T-TT article (1) is worthy of repeated emphasis:

> One generalization apparent both from the work of this project and from published reports by others is that the quality and acceptability of frozen foods

are eroded away by every experience, whether it be holding for a long time at even a "good" temperature, or experience of a higher temperature for even a short time, and the erosion is progressive and irreparable. The reader of articles in this series is therefore urged always to interpret the results of any one episode in terms of the integrated sequence of all the many steps between production of the commodity and delivery of the product to the consumer. Every one of these steps subtracts something—much or little, as may be—from the quality.

Certainly the T-TT investigations served to stimulate much activity by many groups and individuals seeking to improve all the conditions surrounding the production, storage, and distribution of frozen foods. Among these studies is a recent one, begun in 1966, by Better Management Services of Staten Island, N.Y., in cooperation with the U.S. Department of Agriculture, the U.S. Forest Service, and the Forest Products Marketing Laboratory of Princeton, W. Va. (5). The primary objective is the development of better methods in the movement of frozen foods in order to produce economies to the packer-processor, transport agencies, distribution centers, public refrigerated warehouses, and retail outlets. Protection of product quality will always be a major constituent of such studies.

II. HANDLING OF FROZEN FOODS

The general term *handling* refers to all those actions involved in the movement of merchandise after the product is assembled in shipping cartons at the manufacturing plant and thereafter moves in the channels of distribution and merchandising.

Modern developments in the mechanical or semimechanized handling of manufactured products, stimulated by military needs during World War II, have had a profound influence on the handling of frozen foods. Economic considerations were in these undertakings, due to the increasing costs and sometimes limited supply of hand labor as well as the urgency of expediting movement of goods.

Because frozen foods are not comparatively insensitive to their environment, as are nonperishable items, the considerable reduction of extent or severity of the nonrefrigerated conditions to which they are sometimes unavoidably exposed has been one of the significant advantages of these modern handling procedures.

The grouping of shipping cartons in palletized or unitized ladings has speeded up the movement of the frozen foods into a low-temperature environment as compared with the handling and transporting of a large number of individual shipping cartons. Even the bulking of such packages into larger masses has reduced the relative surface exposure to unfavorable temperature environments.

A. Palletizing and Unitizing of Shipments

In trade parlance, pallets are wooden or metal frames, or even heavy paper sheets, on which a group of shipping cartons is placed in such a way that mechanical equipment such as truck lifts, stackers, or towmotors can convey and lift the entire pallet load to a designated location. In refrigerated storages, these are usually electrical battery-powered in order to avoid introduction of combustion gases into storage rooms.

Since all persons responsibly concerned with the transport and warehousing of frozen foods are acquainted with these matters and types of equipment it is not necessary to discuss them in detail in this chapter. In fact, modifications and improvements of these devices occur in such rapid succession that it is not useful to attempt a detailed discussion.

As one might expect, many pallet types and dimensions have arisen. This condition has led in time to efforts to achieve some standardization of pallet sizes both in this country and abroad. The basic reason for this action has been that the relatively costly wooden or metal pallets have to be reused, so that in time there has been a mingling of pallets at warehouses, affecting the stacking and space patterns. As a result many pallets, even though marked by the original owner, have been dispersed, in the course of their use, to other than the originating establishment.

Furthermore, in order to permit utilization of stacking space in transport equipment and warehouses designed for palletized operation, some uniformity is essential. In transport equipment, with inside body widths determined by exterior dimensions in relation to standard highway requirements, economical payloads are influenced by pallet dimensions.

In order to achieve some uniformity of equipment and practice, many industrial conferences have taken place in recent years between users of pallets, representatives of rail transport operators and warehouse companies, and materials handling specialists. The constructive discussion of handling requirements and expenses has led to much better understanding of mutual problems and to improvements which are of interest to all concerned (6–9).

While complete uniformity in pallet types and dimensions cannot be achieved quickly for various important reasons, much progress has been made. Reduction of pallet dimensions to a few standard sizes, with only a few unavoidable special exceptions (such as pallets for handling drums of frozen citrus concentrates) has already been possible. It was found that many pallet types had one common dimension of 40 inches, which would permit two rows of pallets having one dimension of 40 inches to be placed in refrigerator trailer bodies with presently permissible inside widths.

While complete agreement on pallet dimensions may require some years for achievement, there has been an increasing number of individual agreements, based on two rather extensively used pallet sizes, A.P. 48×40 and 48×32 inches. Such agreements must precede the hoped-for eventual establishment of a national or industry-wide pallet pool and exchange agency, mutually agreed to and sustained.

It is interesting to note that some years ago, 13 european countries joined in an intereuropean pallet pool and as a preliminary, of course established a high degree of dimension standardization. At present, a National Pallet Pool organization with offices in Washington, D.C., has combined its offices with those of the National Wooden Pallet Manufacturers Association.

An extensive survey (10) of palletization and unitization of commodities, cooperatively supported by a number of trade associations, several private corporations, chain store groups, truck transport organizations, refrigerated warehouses, and others was begun in 1966 by the previously mentioned Better Management Services organization, under the direction of a nationally-known distribution specialist, Mr. John J. Strobel. Pertinent and comparative details of handling, storing, inventory control, and office procedures are being studied. The announcement stated that at present 8 different wood pallet sizes are in common use and further that the different sizes mean that space may remain unused during shipping or at warehouses, if pallets are too small. It means that certain pallets may not pass through warehouse doors or into truck-trailer bodies. It means that pallet exchange systems are rendered ineffectual because a packer cannot receive the size he needs in return for pallets delivered to the warehouse or store. It means that products loaded on pallets in haphazard stacking patterns may not fit into the space reserved for them or at the warehouse, necessitating the removal of items from the pallets by hand, one by one. This defeats the entire purpose of unitized loads.

These pallet and materials handling developments are briefly discussed in this chapter primarily because they make possible significantly improved protection of frozen foods from adverse temperature conditions during large-scale handling.

B. Unitization of Frozen Food Shipments

The terms unitization, containerization, and palletization are sometimes used loosely in referring to essentially similar practices. However, unitization refers properly to the use of insulated containers, boxes, or even demountable insulated and refrigerated truck-trailer bodies, denoted by the industrial term "piggy-back" (or, in the case of marine transport, "fishy-back") transport of frozen foods (11–13). Such relatively large containers are already in use for shipping nonperishable foods in land, air, and marine transport. It is likely that use of such unit equipment will increase in the future as distribution practices are further modernized.

Essentially only two types of such units exist. One has an attached source of mechanical refrigeration, while the unit is in use. The other relies on insulation primarily to hold a sufficient low temperature during transit time, after the box is loaded with frozen foods, often at a considerably lower than usual product temperature, in order to provide some margin of temperature safety en route. Supplementary refrigeration in the form of solid carbon dioxide or an initial liquid nitrogen treatment may also be employed.

There has been some discussion and limited use of module-type container units for transport on flat-bed, uninsulated rail or truck equipment provided with a lifting device for loading and unloading the units.

In this connection, it has also been suggested that orders of individual retail stores could be assembled at the originating warehouses, where retail price-stamping of items could be done as the mixed lots were assembled for delivery. Significant savings in time and expense at the retail store might be achieved.

These facts are mentioned to demonstrate that there is a continuing search for new patterns of distribution that will reduce the cost and time-span of handling and will employ the transport systems almost as moving warehouses, so that a continuous flow of the required frozen foods can be achieved. In all of these efforts to expedite the unitized movement of frozen foods, a paramount consideration is, of course, the continuous protection of the product by decreasing the number of necessary transfers of these foods through higher ambient air temperatures.

Whether weather-proofed module units, if delivered to the retail store at intervals, could function in lieu of the walk-in freezer chambers now provided for the frozen food inventory in some, but not in all, retail establishments, remains to be determined. Where there are now no such back-room freezer facilities, deliveries must be transferred directly to the merchandising cabinets. This, of course, entails exposure of cased products to ambient air temperatures, and unless close supervision is exercised by the store manager, undesirable delays during such transfers may, and have, occurred. Needless to say, such exposure is not likely to be favorable to the retention of highest quality in the frozen items, depending on the time involved and the ambient air temperature in and around the retail market.

While the export of frozen foods has had only limited growth, and applies especially to an off-shore state such as Hawaii, considerable study has been given to containerization of refrigerated cargo and has extended the range of experimentally validated procedures (14–16).

The Westling-Sabin Refrigerated Intermodal Container System (U.S. patents applied for), claims an advantage of practical flexibility, since on shipboard the refrigeration is supplied by resident ship refrigeration equipment, so that the containers can be stowed in ship holds below deck.

Present U.S. practice, using an individual refrigeration machine attached to each container, requires that the containers must be stowed above deck in order to permit heat dissipation. This method results in a multiplicity of small units with no reserve refrigeration capacity, as well as high initial and maintenance costs.

The W-S system applies, of course, to all refrigerated cargo, but is especially useful for frozen food shipments, with which this chapter is concerned.

There have been a number of large trial exports, shipments of insulated units, stored on deck, containing frozen foods refrigerated to very low temperatures by the use of liquid nitrogen. After the container unit has been sealed, the rate of heat penetration into the lading is influenced, unless precise controls are provided, by the ambient air temperatures en route and, of course, of the efficiency of the container insulation.

One of the interesting developments in marine transport of refrigerated cargo has been the Sea–Land Controlled Temperature Service, which at present has more than 800 Thermo-King units in service and the world's largest trailer-ship fleet, the largest and fastest of which, the C-45 type, has a ship capacity of 609 trailers.

Track-mounted dockside cranes or ship-based gantry cranes load or unload the trailer units either on or from the waiting truck-trailer chasis. LTL (less than truck-load) shipments are consolidated by the Sea–Land organization, or by carriers working with Sea–Land, into trailer load units, sealed and delivered to dockside for loading aboard the trailer ships. At the marine terminal nearest the destination, the consolidated loads are "broken out" for delivery to the consignee.

An LTL shipment, sealed in the trailer, is protected against weather and pilferage. Even when the products are shipped to foreign ports, export packaging is not necessary.

All Sea–Land ports have special reefer yards with spare parts, maintenance facilities, and mechanics available.

In principle, these marine containerization developments provide useful information for future land transport of refrigerated container (trailer) units, although some environmental conditions are different.

III. STORING OF FROZEN FOODS

A. Temperature and Time Periods of Storage

Because of the relatively long time periods involved in normal practice, storage of frozen foods in refrigerated warehouses is an appreciable segment of their entire time–temperature experience (Fig. 12.1).

Research and industrial experience have established, beyond any doubt, that in general practice storage room air temperatures of 0°F. (–18°C.) or lower should be consistently maintained almost universally. Public and private warehouses in which frozen foods are accumulated now provide such air temperatures. In fact, the trend has been to somewhat lower storage room temperatures, in the range of –5°F. (–20.5°C.) to –15°F. (–26°C.).

The choice of temperature often depends on the nature of the product and the estimated period of storage before the product moves into distribution and retail channels. Fish, pork, and some vegetables may require lower air temperatures. Practice in the United Kingdom and other European countries, where fish constitutes an important item in the average diet, tends toward storage of both lean and fatty species of fish at –10°F. (–23°C.) to –20°F. (–29°C.).

Tressler and Evers (17) have stated these generally recognized facts in suggesting that such products are best warehoused at air temperatures of –5°F. (–20.5°C.) or lower.

These authors have also emphasized several important factors in the successful warehousing of frozen products—air temperature, air humidity, stacking methods, and handling procedures, in which there is general technical and industrial concurrence (Fig. 12.2).

It is perhaps appropriate, at this point, to refer briefly to methods of determining product temperatures in frozen foods in the pipelines of distribution. Scientists of the U.S. Department of Agriculture (18)

Fig. 12.1. Refrigerated warehouse in production area—Hillsboro, Oregon. First frozen food processing installation on the Pacific Coast. Courtesy of Terminal Ice and Cold Storage Co., Portland, Oregon.

gave considerable study to this matter and finally devised a procedure (19) for taking frozen food product temperatures without damage to packages. This study was encouraged by industrial groups as a basis for industrial efforts to improve temperature control in frozen food handling. The procedure, simply described and illustrated in a widely distributed small folder, received almost universal industrial acceptance and has been satisfactory.

Through the years, a number of product-temperature indicating devices have been suggested, developed, and patented. Among these were the thaw-indicating type, which recorded by change of color of some chemical sensitive to temperature or to contact with liquid water the highest temperature experienced during the product exposure, but without any reference to the time during which such high temperatures prevailed. Even ice cubes in plastic bags were suggested, the melting of the cubes giving a warning that ice-thawing temperatures had been

Fig. 12.2. Refrigerated warehouse in production area—Nampa, Idaho. Courtesy of Terminal Ice and Cold Storage Co., Portland, Oregon.

reached in the product. From the discussions in previous chapters the reader will recognize that, while these devices were useful, they did have obvious limitations in sensing the influence of temperature upon product quality in a truly significant manner, other than to indicate that a certain higher or thawing temperature was reached at some time in the life of the product.

In this same context, there have been suggestions made that the individual packages be dated, so that the length of time since packing could be determined by the purchaser. Actually, such dating would mean very little, if anything, in truly and accurately predicting the quality potential or shelf- or use-life of a frozen product. This practice also might encourage unwarranted discounting of product value based only on a time-element factor without regard to prevailing temperatures in that time interval or any other quality criteria.

One small, flat, time–temperature indicating device which has shown considerable merit in actual use—experience is the Honeywell Time–Temperature Indicator (20). This indicator is so designed that it needs no external connections or power supply and can be placed in selected shipping cases or in packages in several locations within a rail car or truck-trailer loading in order to have a continuous record in selected time periods of the temperature conditions prevailing during the transit and distribution phases.

The Honeywell device is sensitive to environmental temperatures in the range wherein most frozen foods are transported and stored and over a period of time integrates temperature and time into one representative reading on a calibrated scale (see Chapter 13 of this volume). From this reading the observer, at any point in the distribution sequence, can interpret the time–temperature history of the frozen food with which the device was packed. Basically, this device could be very useful in monitoring transit, warehouse, and distribution temperature influences, if it were used consistently and intelligently. However, for various reasons, its general adoption has lagged, despite its acceptable accuracy and usefulness. The trade reaction has been that its readings could be misinterpreted and so result in wrong assumptions of presumed quality changes, with a consequent potential discounting of product value. These industry reactions must, of course, be considered in any commercial enterprise, even though the potential usefulness of such a device is recognized. The future may demonstrate that the usefulness outweighs any potential hazards of misinterpretation.

From operating and economic standpoints, one must weigh the added cost of providing unusually low storage temperatures against the commercial value of the added protection of frozen food quality. The power and other accessory costs increase rapidly as the storage room temperatures are lowered to the lower ranges, approaching −20°F. (−29°C.), which are below present industrial practice, as do the equipment maintenance, repair, and replacement costs.

Existing freezer storage capacity, perhaps not designed to maintain a temperature much lower than −5°F. (−20.5°C.) economically, will not usually be replaced or rebuilt simply to permit operation at lower temperature because of the high initial cost. However, in the design of new storage space consideration may well be given to providing greater refrigerating capability than has been customary, and using more effective thermal insulation, in view of the marked increase in production of types of frozen food whose stability is considerably better at −10 to −15°F. (−23 to −26°C.) than it is at 0°F. (−18°C.).

In commercial practice (Fig. 12.3) the operator of a refrigerated warehouse, and the packers, wholesalers, and distributors who utilize his facilities, are obliged to estimate, as best they can, the length of time a particular consignment of frozen food will remain in storage, and from their knowledge of the average storage stability of the product, and the quality requirements of the intended market, determine whether a storage temperature lower than the normal 0 to −5°F. (−18 to −20.5°C.) will be required. In general, an unusually low temperature will be maintained only if the expected market is thought

Fig. 12.3. Refrigerated warehouse market center. One of the first of its kind in the United States, located in Watertown, Massachusetts. Courtesy of Quincy Market Cold Storage and Warehouse Co., Boston, Massachusetts.

able to absorb the higher cost and dispose of the product at correspondingly higher prices.

Whatever freezing method is used, freezing should be completed to the stated product temperature before the individual packages are placed in shipping cartons or before bulk containers such as "tote boxes," are placed in the originating production warehouse.

When direct loading into refrigerated transport equipment follows soon after the processing sequence, it is even more important that only thoroughly frozen products be loaded. Transport equipment cannot be economically designed and operated so as to complete the freezing operation, while the product is in transit. In fact, it is always advantageous to retention of product quality to load frozen foods at a temperature somewhat lower than 0°F. (−18°C.). For this reason, progressive warehouse operators provide storage room temperatures of −5 to −10°F. (−20.5 to −23°C.) so that despite any slight product temperature rise during handling operations, the frozen foods still will not have warmed up to zero after loading (Fig. 12.4).

While the normal maximum marketing period for perishables such as frozen foods might be considered as one year before the next crop or

Fig. 12.4. Refrigerated warehouse market center and processing facility, located in Chicago, Illinois. Courtesy of Continental Freezers of Illinois, Chicago, Illinois.

supply materializes, yet of necessity considerable volumes of frozen items are carried over in storage from one production year to the next. This unavoidable fact also emphasizes the desirability of maintaining transport and warehouse room temperatures consistently as low as is economically feasible.

B. Storage Room Humidity

The many improvements in packaging materials and procedures in many areas of the American economy have materially benefitted frozen foods. They have also, to a large extent, lessened the need for precise humidity control in freezer storage rooms. Years ago, this problem was serious and was given much study. With the exception of frozen meat items in bulk, such as pork shoulders, bones, and bellies, and frozen fish in the round, which are protected by water glazing or plastic coverings, the bulk of the frozen foods for retail or institutional trade are packaged in waxed, plastic- or aluminum foil-laminated cartons, or are packed in plastic bags or other relatively water-vapor tight enclosures. All of these packages are, to some degree at least, resistant to the passage of moisture-vapor and possibly also other gaseous substances.

Through the years, Canadian scientists Huntsman (21), Young (22, 23), and Lentz (24, 25) have suggested methods of continuously humidifying freezer storage rooms or have advocated the so-called jacketed rooms, especially for fish products. The jacketed type of room consists essentially of a room within a room, so that the cold air circulating in the plenum between the two chambers remains nearly saturated with water vapor, and therefore does not have a dehydrating effect on the stored frozen items.

The basic reasons for such procedures were doubtless sound. However, in the rehumidification procedure continuous and unavoidable, even though slight, moisture vapor condensation on the refrigerating coils was again vaporized after intermittent or continuous defrosting of the coils and so the condensate was reintroduced into the chamber. Hence, there were some operating problems connected with this procedure. The unavoidably higher initial cost of jacketed rooms plus some operating problems have adversely influenced wider adoption of this type of storage room, although some of this type are in current successful operation in Canada.

Actually, when modern protective packaging materials are properly used for frozen foods, the product desiccation problem has become a somewhat academic matter provided that freezer storage temperatures are not allowed to fluctuate widely. If wide temperature fluctuations do persist, however, there is still some chance for in-package desiccation to occur, because temperature changes cause some of the product moisture to go into the vapor phase as temperature rises, later to be deposited on the inside of the package in the form of frost, when the temperature of the room falls again (26). This is the major reason for the frequently stated admonition that freezer storage room temperatures should be held relatively steady (see also Chapter 10, Sect. IV.D.1). Perhaps a brief reference to the cause of water-vapor loss from stored frozen products is appropriate. This results from the difference in temperature, and so also in water vapor pressure, between the stored product and the refrigerating surfaces or coils, of whatever type these may be. Rapid forced-air movement over the stored product may further intensify this moisture loss from the product.

The rate of moisture deposition as frost or ice on the refrigerated surfaces, even though they are continuously defrosted depends, then, on the amount of temperature difference between the stored product, the circulating air, and the surface of the refrigerating coils, whatever the refrigerant used. The greater the surface of the refrigerating coils, the greater can be the reduction in this temperature difference at the same total rate of heat transfer.

At times, especially in the past, competitive efforts to reduce initial construction costs have resulted in insufficient coil surface being provided. This was false economy, however, because product desiccation has thereby been increased and also the costs of operating the refrigerating compressor equipment.

When glazing of bulk meat or fish products has been practiced, this water glaze has to be renewed at frequent intervals, especially when high rates of water vapor transfer have prevailed. By adding chemicals, such as tri-sodium phosphate, to the water glaze, it has been claimed that a tougher, more desiccation-resistant glaze was created.

These bulk-frozen products, warehoused in the processing sequence, especially in the case of meat products, are really in a different product category from the retail and institutional frozen food items in packages, but they are referred to in order to illustrate the desiccation problem in its historical significance.

The inappropriate term "freezer burn" (which is not a burn in any sense of the word), caused only by the desiccation and oxidation which arises from moisture loss from the product surface, has fortunately gradually disappeared from trade language as a result also of the marked improvements in protective packaging.

Through the years, there have been two schools of thought regarding the comparative merits of direct-expansion coil installations on ceilings and walls or both (sometimes with portable fans for more positive air circulation) and of the forced air system. There was perhaps more reason for this engineering dialogue before the improvements in product packaging reduced the product desiccation potential. Obviously, the more air that is forcibly circulated about any refrigerated object, the more boundary conductance of sensible heat is effected and the greater is the potential water-vapor loss from the surface. Actually when frozen foods are thoroughly frozen to 0°F. (−18°C.) or colder before being warehoused or transported, there is only minimal internal heat still to be dissipated. The real function of air circulation then is to eliminate local warm spots in the room and handle heat leakage from outside the building; also, internal circulation reduces the temperature differential between the top and bottom of the storage room.

Thus, the use of floor racks, of at least 4-in. height, is strongly advised. Frozen foods should never be stacked directly on the refrigerated room floor. Stacking at least 6–8 in. from walls is also important, to permit air circulation beneath and about stacked frozen products. Pallets in themselves provide such air space beneath the stack. Wooden stripping or dunnage is use to tie the pallet or stock load together, 3/8- to 1/2-in. wooden strips being quite adequate.

When frozen merchandise is received at the warehouse at product temperatures above 10 or 15°F. (−12 or −9°C.), due to environmental temperature influence en route or because the product has not been sufficiently frozen, responsible warehouse operators take steps to spread the frozen foods in order to bring the product temperature quickly to an acceptable point, after notifying the owner of the goods; there is usually an extra charge made for this essential service (Fig. 12.5). Or, if the frozen food temperature permits, chimney stacking may be used, *i.e.*, the shipping cartons will be so placed as to leave space for air passage through the pallet load. If the merchandise is thoroughly frozen, this practice is not generally necessary. The sound answer to this awkward situation is to store only goods which have been thoroughly frozen before delivery to the warehouse.

The concensus of warehousing experience with frozen foods is that some forced-air circulation is desirable, especially in modern one-story refrigerated warehouses with substantial ceiling heights.

Except in an emergency, a freezer storage room, in which already frozen products are held, would usually not be used for freezing any considerable volume of unfrozen or only partially frozen products. The reasons should be obvious to the reader of this chapter. The

Fig. 12.5. Shipping and receiving covered truck platform depicting two methods of handling frozen foods. Operator on right is using an electric pallet jack to deliver a pallet load to the tailgate of a refrigerated truck. Operator on left is taking a pallet load into the freezer storage room by means of a reach-fork lift truck. Truck dock extension plates enable the refrigerated truck-trailers to back in before opening the trailer doors. Courtesy of Quincy Market Cold Storage and Warehouse Co., Boston, Massachusetts.

resulting significant rise in the temperature of other goods already in the room, would perhaps not be fully reversed for many days or even weeks, and would adversely affect the quality of the entire quantity stored. The refrigerating equipment for storage rooms is not normally designed to remove a large amount of residual latent and sensible heat from the room contents.

C. Handling at the Refrigerated Warehouse

The modern developments in palletizing and containerization have markedly reduced the time period during which frozen foods received on the dock of a warehouse are delayed before they are brought under proper freezer storage environment.

In addition, modern refrigerated warehouse plans now generally provide for enclosed loading docks, possibly with limited refrigeration, so that delivered pallet loads do not stand exposed to sun heat or higher atmospheric temperatures. There are also available so-called protective dock shelters—devices which seal off the atmospheric elements when a truck, trailer, or rail reefer is spotted at the dock door for loading or unloading. Such devices to a considerable extent prevent the in-leakage of warm air, and exposure to either wind or precipitation.

In recent years, progressive warehouse operators have provided a temperature-reading service for deliveries or selected shipping cases from significantly different parts of the truck-trailer or reefer load upon arrival, and permanent records have been kept. Some warehouses have also taken photographs of the load when the trailer doors are opened, using Polaroid cameras, in order to have a record of any damaged cases or improper stacking in the transport vehicle.

Any discussion of frozen food warehousing requires at least a brief reference to the types of such warehouses now being built. This development has paralleled the rapid progress in modern materials handling. Palletized and containerized handling of such goods has favored relatively high stacking patterns in order to utilize storage space most effectively. Hence, modern refrigerated warehouses are invariably one-story structures (construction details are not germane to this discussion) with relatively large storage rooms, a minimum number of door openings to semirefrigerated covered docks, and with ceiling heights up to 24 feet. Experience has indicated, however, that 16- or 18-ft. ceiling heights possibly result in better space utilization. An exception may be in-transit warehouses, wherein truck or carload lots are held temporarily nearer to markets, under special rail tariff arrangements, in order to expedite diversion in response to distribution demands (Fig. 12.6).

Fig. 12.6. A pallet load of frozen foods is placed on a 10-ft. "holding" conveyor on the upper level. The pallet conveyor carries the product to an "inclined" conveyor for transport to a 15-ft. conveyor on the lower level. The product is conveyed over this system at 35 ft./min., taken immediately from the end of the conveyor, and moved directly into the freezer storage room. Courtesy of Quincy Market Cold Storage and Warehouse Co., Boston, Massachusetts.

The refrigerated warehouse industry historically recognizes three distinct types of warehouse operation, although the functions sometimes overlap in the first and last named. These are (*1*) the production warehouse close to packing operations, (*2*) the in-transit house, and (*3*) the distribution warehouse, more recently developed into a so-called market center warehouse, from which mixed-lot distribution is made to retail establishments or institutions. The stacking of pallets or containerized frozen product, identified by lot numbers, such as the large, so-called "tote boxes" and 10- to 20-ton bins in the first two warehouse types, entails placement of relatively large masses of product, and then their removal for transport, in essentially the same magnitude. Incidentally, the "tote boxes," usually made of heavy paperboard, may contain as much as 1,000 lb. of frozen product, especially of those items that can be individually quick frozen (IQF) such as peas, cut snap beans, cut corn, lima beans, and the like (Fig. 12.7).

These "tote-boxes" are readily handled on pallets with materials handling equipment. This development is regarded favorably for

Fig. 12.7. Pallet load of frozen foods leaves the conveyor (Fig. 12.6) and is brought directly to the freezer room door. The rollers in the conveyor are automatic—motion and stop are controlled by electric motor. Courtesy of Quincy Market Cold Storage and Warehouse Co., Boston, Massachusetts.

better utilization of storage space. For the packers, there is an advantage because such bulk-stored frozen items can be packaged for retail or institutional sales as specific orders require, affording to the packer greater marketing flexibility in relation to production peaks. It may strongly influence his retail or institutional package inventories.

The constant effort of both public and private warehouse operators engaged in the distribution function has led to much study and discussion of the "warehouse of the future." This is not surprising when one considers the constantly developing new distribution patterns. The ultimate answers on the nature of future distribution center warehouses have not yet been written, but studies so far made indicate a different structure format and, of course, the use of automated procedures. Brief reference is made to these future potentials in refrigerated warehousing merely to alert the reader that present distribution practices are not final (27). This whole phase of the food economy is in a state of dynamic development.

Another striking change in the format, functions, and location of refrigerated warehouses, even in distribution areas, must be mentioned. Older warehouses in the past were located near or in relatively congested urban traffic areas, such as produce market centers, and were usually multi-story structures. Modern refrigerated warehouses are located generally in generously-sized undeveloped areas, purposely situated near superhighways for easy access by automotive transport equipment as well as rail transport. Instead of being simply warehouse operations, such modern warehouses are really refrigeration centers, with processing plants located about them. These plants may store their raw materials, as well as their manufactured products, in such warehouses.

As the prepared or precooked frozen items have grown in numbers and volume, their preparation has been done nearer to population centers rather than in farm-production areas, because of distribution advantages.

The multipurpose warehouses are thus related to production enterprises, but also, as distribution centers, have many accessory functions such as product freezing, rental of office space, inventory financing, breakup and distribution services, rental of processing space, frozen fish and meat glazing, inventory reports, in-transit storage, traffic and marketing information, packaging services, issuance of uniform warehouse receipts, and other clerical services for their customers.

At the present time, the public refrigerated warehouse industry comprises more than 800 plants, operating in excess of 640 million cubic feet of storage space, of which about 70% is devoted to freezer occupancy, maintained at 0°F. (−18°C.) or lower for frozen foods. The increase in freezer space has been more than 65% in the last ten years. Of the total number of such public warehouses, more than 90% handle frozen foods (Fig. 12.8).

In recent years, many private refrigerated warehouses have been constructed by processors, distributors, and large-volume retailers such as food chains, which for various policy reasons prefer to operate warehouse facilities in connection with their other food-production and sales activities. However, Powell (28) has stated that the user of public refrigerated warehouses has certain economic advantages such as maximum flexibility in use of storage space, without tying up capital in fixed facilities (with attendant problems of maintenance costs and efficient space and manpower utilization), widest choice of storage locations, low insurance rates, services of experienced personnel and advance knowledge of storage, handling, and distribution costs.

There has been, for some time, a consistent effort by progressive warehousemen to integrate their services with the requirements and procedures of the users of public warehouse facilities. This integrated function has been named "space-services" (Fig. 12.9).

One of the distribution services, performed by or emanating from the modern market center, is the assembly and transport of small lots

Fig. 12.8. Freezer storage room of a major frozen food distribution center (Watertown, Mass.). All essential refrigeration equipment is duplicated to prevent loss of refrigeration caused by mechanical failure. Single large refrigeration unit is continuously self-defrosting, eliminating temperature fluctuations caused by intermittent defrosting. Air from the refrigeration unit circulates across the roof, down the four walls, and across the floor, blanketing the storage area and its contents with continuously circulating refrigerated air. Fluorescent lighting is used. Courtesy of Quincy Market Cold Storage and Warehouse Co., Boston, Massachusetts.

of mixed frozen foods in less-than-carload (LCL) or less-than-truckload (LTL) lots received at the warehouse, as ordered out by the individual retail establishment. The "breakup" procedure is generally performed at night by special crews of product selectors so that the mixed product lots can then move to the stores early the following morning.

The special requirements of breakup activities, especially when mechanized order selection is installed, has caused the development of metal pallet racks for more orderly and better utilization of warehouse space.

One of the space problems of distribution warehousing is the so-called "honeycombing" of pallet loads, as a result of withdrawals of items to make up mixed lots for delivery and the necessity for maintaining pallet load identity of product. There is no general formula

Fig. 12.9. Fork-lift operating in a refrigerated warehouse. Courtesy of Clark Equipment Co., Battle Creek, Michigan.

for solving this problem but alert warehouse control, involving ingenuity and planning, can lessen this inventory problem. In warehouses with very high ceilings, there is a natural tendency for personnel not to utilize adequately the higher stacking spaces, but alert supervision can also diminish this problem.

Stacking problems, especially those caused by shipping case fatigue and failure have caused the use of metal side-irons, rope ties, and metal racks to give greater support to the pallet loads as they are stacked. Tilting or falling stacks create not only an industrial hazard to operating personnel but also cause product damage. Some operators believe that a three-pallet stack is most economical and safe (and if adopted, this would mean lower ceiling heights) but the high-ceiling storage rooms permit stacking of four or more pallets. Such high stacking may then require the use of support irons or racks, since the superimposed weight may be greater than the cartons on the lower pallets can bear, especially if the shipping case contents do not have an internal weight-bearing capability, such as the metal containers of heat-processed, "canned" foods do.

A competition-inspired tendency to use the least expensive shipping cartons has for some years aggravated this frozen food warehousing problem. Its solution would not be easy without an industry-wide adoption of shipping case standards.

A standard carton for frozen foods should have both protective and warehousability characteristics (30, 31). The carton is essentially a microclimate, insofar as protection of the frozen food is concerned (32), a concept first proposed years ago by Dr. C. W. Thornthwaite, now deceased, Director of the Climatology Laboratory, then located at Seabrook, N.J.

However, emphasis at present is on the ability of the shipping cases to withstand warehouse handling and stacking stresses under the expected air temperature and air moisture conditions. In conformity with traditional industry practice, the cases are tested and rated for bursting strength in relation to impact hazards during shipping. However, the pallet stacking problem is influenced by package fatigue and failure under compression. Efforts of technologists to inaugurate a standard compression test and rating procedure, however desirable, have not so far been possible because of the complexity of the matter; it would involve changes or modifications of package specifications, already established and recognized by such agencies as freight bureaus.

In recent years, there have been many developments in warehouse equipment, useful in palletized operations to expedite handling and permit comparatively high stacking. Progress continues to be made. Space limitation prevents detailed reference to the many forms of equipment now available for specific warehousing needs. Power-driven, materials-handling equipment used in refrigerated warehouse rooms must be electrically activated because the combustion vapors of fuel-driven vehicles cannot be tolerated for reasons of personnel safety and product protection. So far no method of absorbing combustion gases has proved to be sufficiently effective or economical. Hence, every modern warehouse has electric battery recharging equipment.

While there have been many improvements in construction procedures and materials in the last quarter century, notably in insulation materials and vapor seals, these also must be left to other sources of information and reference, such as the ASHRAE Guide and Data Books (33).

The use of pallets and materials-handling equipment has necessitated changes in storage door opening dimensions and has markedly influenced storage door fabrication and operation. Light-weight insulated doors, electrically operated, have significantly expedited the in- and out movement of frozen foods. Obviously such handling improvements all tend to lessen also the exposure of frozen merchandise on loading docks to higher ambient air temperatures. Door openings protected by forced air curtains have not only expedited the in-and-out movement of warehouse transport equipment but have also reduced the condensation on the walls and ceilings (and refrigerating pipe coils, if present) of storage rooms.

Where such air-curtains are not used, many warehouses have installed vestibules or air-locks equipped with flexible door curtains to reduce temperature fluctuation in storage room air, especially close to the door openings. It is customary also to locate the stacking pattern in the storage room so that there is a distance of 6–10 ft. between the door opening and the nearest stack.

The most satisfactory results in frozen food handling at the warehouse, in or out, have been obtained to mutual advantage by consistent communication between the transport and the warehouse personnel, so that orderly scheduling of receipts and deliveries and most effective use of dock personnel and handling equipment can be achieved. Such cooperation can significantly minimize delays in receiving or delivering frozen foods and thus help to protect the product.

IV. REFRIGERATED TRANSPORT OF FROZEN FOODS

Refrigerated transportation, by rail-car or truck-trailer, is an indispensible facet of frozen food marketing because many production areas in the United States are distant from the potential markets. In fact, even local transport to retailers from distribution or market-center warehouses, the so-called Less-than-Carlot (LCL) or Less-than-Truckload (LTL) transport, is required to place the desired frozen products before the consumer-buyer, and for all frozen foods, such transport must provide adequate low-temperature environment.

Basically, one must recognize that no refrigerated transport equipment can be economically designed and operated as a freezing chamber. Its only proper function is to protect the frozen lading, presumed to be already adequately frozen to a temperature at least as low as 0°F. (−18°C.), against heat and moisture leakage from the outside air by means of adequate insulation in the ceiling, walls, and floor, by proper stowing of the load, and by some form of added refrigeration. The latter is now increasingly furnished by mechanical equipment rather than by the traditional ice–salt method; this is now deemed generally inadequate to give the desired temperature protection to the frozen food load.

A protective shield of constantly circulated low-temperature air must entirely encompass the lading. If the desired low temperature is to be maintained economically in the transit period, the product must be loaded with a minimum of delays at a temperature somewhat lower than 0°F. (−18°C.). This will allow for a possibly unavoidable temporary product temperature rise during transit to the range of 0–10°F. (−18 to −12°C.) for so-called over-the-road transport, by means of truck-trailers. Some warehouse operators cooperate by maintaining storage room temperatures not higher than −5°F. (−20.5°C.) so that there will be some margin of safety in temperature control in the transport phase.

A. Refrigerated Rail Transport of Frozen Foods

Historically, the rail transport of frozen foods has benefited from the long experience of refrigerator car companies in the transport of nonfrozen perishable food stuffs. Through many years car insulation and construction have been improved. In order to obtain adequate air circulation about the lading, even before mechanically refrigerated special frozen food cars, or "reefers" were developed, 4- to 6-in. floor racks and side racks and appropriate bulkheading became accepted practice, replacing the small-dimension wooden or metal strips that were prevalent in the early years of reefer service (Fig. 12.10).

When frozen foods had to be transported by rail, the necessity for free air circulation from the car's salted-ice bunkers, which then supplied the refrigeration, was demonstrated by car tests in actual transit operation. Many of these tests were conducted cooperatively by the U.S. Department of Agriculture with the refrigerated car companies and railroads, and the technical information obtained was widely applied. Therefore, the rail car transport of frozen products did not

Fig. 12.10. Compartmentalized, refrigerated truck-trailer for LTL operations. Trailer has standard shoulder side door as well as an additional side door on the road side up front. Courtesy of Frozen Food Express, Dallas, Texas.

entail major changes in rail transport concepts but merely required the modifications necessary to provide much lower inside air temperatures (Fig. 12.11).

The economic necessity for meeting established minimum load requirements in order to take advantage of lower tariff rates resulted in the creation of the now well-known longer and more heavily insulated frozen food refrigerated rail cars. The mechanically refrigerated frozen food reefer has now come into very wide usage as improvements in refrigerating equipment, car insulation, and air distribution have provided much better protection for the ladings.

In the course of these frozen food developments, many experimental reefers were tested, among them some which were refrigerated by the sublimation of solid carbon dioxide, either directly or with a secondary liquid refrigerant circulating in coils or plates within the rail car interior. Operating problems and relatively high refrigerant costs have gradually eliminated such cars, except for meat service, from common usage (Fig. 12.12).

Tests giving the numbers and other pertinent information about reefers for frozen foods are available to users of this type of rail trans-

Fig. 12.11. Interior view of compartmentalized, refrigerated truck-trailer. Movable bulkhead is in place and with an entry door, permitting access to the forward freezer compartment, if necessary. Forward compartment has side door also. One unit and two evaporators provide refrigeration, the evaporators operated separately temperature-wise to provide 0°F. (−18°C.) temperature in the forward compartment and 40°F. (4°C.) in the rear space for unfrozen merchandise. Courtesy of Frozen Food Express, Dallas, Texas.

Fig. 12.12. Standard fleet refrigerated trailer with a temporary bulkhead. An inflated "rubber raft," which can be placed in a trailer to provide for separation of commodities at different temperatures in transit. Courtesy of Frozen Food Express, Dallas, Texas.

port services from several sources, including the Association of American Railroads (Transportation Building, Washington, D.C.) and the National Association of Frozen Food Packers (34) (Fig. 12.13).

The so-called stop-off LCL reefer has caused some problems from the standpoints of consistent, adequate temperature maintenance and occasional damage to shipping cases and contents if these have not been properly loaded and braced between stop-off delivery points. Reefers sectioned by movable sturdy partitions have been tried as aids to better protection of loads both for temperature control and for

Fig. 12.13. Over-the-road refrigerated truck-trailer for frozen food transport. Courtesy of Alterman Transport Lines, Inc., Miami, Florida.

Fig. 12.14. Over-the-road refrigerated truck-trailer for frozen foods, operated by a refrigerator rail car company. This series of trailers has the latest-design, heavy-duty, piggy-back kingpin and upper coupler assembly. Courtesy of Fruit Growers Express Co., Washington, D.C.

prevention of lading damage (Fig. 12.14). But again, these modifications increase costs and are not without operational problems. However, all these efforts to improve LCL rail transport are constructive and should be continued as new technical and engineering knowledge becomes available.

In the main, modern rail transport of frozen foods has been quite satisfactory for frozen foods when these have been properly loaded in frozen food reefers at properly low product temperature and with adequate bracing.

B. Refrigerated Truck Transport

A large quantity of frozen foods produced in the United States finds its way to markets by truck transport in mechanically refrigerated insulated truck-trailers.

In 1966, Tunnel Publications, Inc. of Houston, Texas, publishers of *Refrigerated Transporter,* issued an interesting booklet (35) regarding the size of the refrigerated truck transport industry. Of approximately 900,000 truck-trailers in highway use in 1965 in the U.S., about 85,000 were refrigerated. Total refrigerated truck equipment numbered 287,230; of those there were 84,670 refrigerated semitrailers (72,000 of them equipped with mechanical blower refrigeration units), and 202,560 refrigerated van and delivery vehicles. Each month approximately 2500 new refrigerated vehicles go into service. For such truck equipment, the use of mechanical refrigeration units doubled in the years 1960–1965, the 1965 production being nearly 28,000 units.

Of the 200,000 refrigerated trucks of all types on the road, over 100,000 had, in 1965, eutectic refrigeration plates. Production of liquid-nitrogen refrigeration systems for transport vehicles has doubled nearly every year since 1961, the 1966 production being about 2500 units.

To the credit of the trucking industry, it can be said that the majority of responsible refrigerated "over-the-road" truck operators long ago recognized their responsibility to provide adequate protection of frozen foods in transit. It is true that, at times, in the past, some of the so-called "gypsy operators," for various reasons did not employ equipment capable of giving adequate protection to the frozen foods which they transported. However, through constructive cooperation, involving the U.S. Department of Agriculture and the National Bureau of Standards of the U.S. Department of Commerce, together with the American Trucking Associations and the Truck-trailer Manufacturers Association, as well as a number of manufacturers of refrigerating units and truck components, much progress has been made in recent years in the improvement of the refrigerated truck transport of frozen foods.

A broad research program had to be initiated because previously there had been no recognized standard procedure for determining accurately the performance of an insulated trailer body or for selecting an adequate refrigerating unit capable of consistently maintaining the desired air temperatures within the trailer body.

From these several researches, a number of significant reports have resulted. A very important one (36, 37), deals with a standard method for rating refrigerated trailer bodies by the heat-sink principle. The standard test conditions involve the maintenance of 0°F. (−18°C.) air temperature within the trailer against an ambient temperature of 100°F. (38°C.) and 50% relative humidity in the test chamber. Canadian (38, 39) and U.S. investigators have investigated and tested trailer design in relation to air circulation in the trailer body when it is hauling loads of frozen foods (Fig. 12.15).

With respect to the effect of ambient conditions on refrigerated trailer performance, new information indicated that the pressure of air on the head of the trailer when it was operated at an average road speed of 50 mph., resulted, with some types of insulation and vapor seals, in considerable leakage of outside warmer and moister air into the insulation and trailer body; and if floor drainage vents were not plugged, this air infiltration was further increased. It was determined that this air leakage caused an increase of 10–27% in cooling load, above the load found during stationary tests in the test laboratory.

Also, the underside of the trailer was sometimes heated as much as 15 degrees above ambient air temperatures during road trials, prin-

Fig. 12.15. Full rear view with open doors of a piggy-back refrigerated truck-trailer. A welded aluminum T-slot floor and cross-ventilation slots go from each wall to refrigerant line trough. Duct air distribution system insures protection of lading from exterior heat leakage. Courtesy of Fruit Growers Express Co., Washington, D.C.

cipally because of waste heat from the tractor engine. Reflected heat from the pavement also is a factor affecting floor heat leakage. In some tests, frost accumulation in trailer walls caused an increase of 0.52–0.98 lb./hr. in nonrevenue-producing weight. Flexible, rubberlike coatings have been applied to insulation as an air and vapor seal and provide a practical and satisfactory remedy. However, the more permanent solution of the problem appears to have been found in the "foamed-in-place" polyurethane insulation (40) which presents an unbroken surface to impinging air streams. This form of insulation has a theoretical k value of 0.14 (BTU)/(sq. ft.) (hr.) (°F./in.) and a design value of 0.20 is suggested.

Before this problem found a practical solution, the infiltrated warmer and moister air imposed an unduly great strain on the mechanical refrigerating unit in its attempt to maintain a desired low air temperature within the trailer body. Furthermore, since the moisture in the infiltrated air is eventually deposited, for the most part, as ice in the insulation, undesirable nonrevenue-producing dead weight was added to the trailer. There are, of course, certain weight limits prescribed for such trailers by highway regulations and it is obviously uneconomical to transport this dead weight, to say nothing of the adverse effect of moisture accumulation on the efficiency of the insulation.

In the industry, it has been customary until recently to state that 6 in. of insulation is necessary for refrigerated trailers and it has seemed to some that this figure has been given undue emphasis in the light of improvements in modern insulation types for truck-trailers. The choice of insulation thickness should be made on the basis of demonstrated performance data in relation to the expected ambient environment conditions, and to the rated performance of mechanical or other refrigeration systems used in trailers (Fig. 12.16).

The Air-Conditioning and Refrigeration Institute has published up-to-date directories of certified refrigeration units for specific purposes based on controlled testing, engaged in by the ARI's Mobile Refrigeration Section in cooperation with five manufacturers of such equipment (41). The program, based on ARI Standard No. 1110-64 for speed-governed transport refrigeration units employing forced air circulation coolers, deals with such units for use in trucks, trailers, and large refrigerated chambers or boxes, including both those designed for temperatures of 25°F. (−4°C.) and above and for low-temperature service below 25°F. (Fig. 12.17).

For refrigerated truck-trailers, as for rail reefer cars, circulation of cold air about the loading is of paramount importance. Hence, the installation of side racks, to hold the loading away from walls, and the enlarging and perforating of the extruded aluminum floor strips have followed. Case loading patterns intended to insure more continuous circulation have also been introduced. If the frozen foods are frozen to 0°F. (−18°C.) or lower throughout, however, it is preferable to load in a tight stacking pattern, because the loading tends to protect itself from temperature changes, through its own thermal inertia.

Although considerable interest has been shown in so-called cold-wall trailers (42), their adoption has been slow. The basic principle involved in this trailer design is the creation of space within the external trailer structure itself so that cold air can be constantly circulated to capture heat leakage from the outside, before it can affect the internal air temperature of the trailer body. Such trailer construction is worthy

Fig. 12.16. Modern high-cube mechanical refrigerator rail car for frozen foods transport. Courtesy of Pacific Fruit Express Co., San Francisco, California.

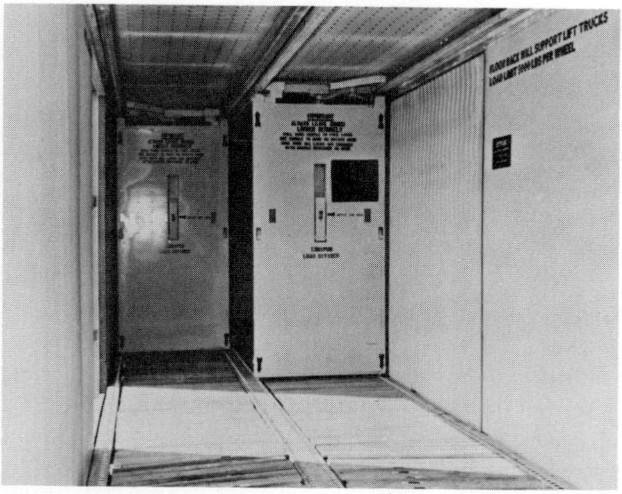

Fig. 12.17. Interior view of a modern high-cube refrigerator rail car, shows the load divider gates, the floor racks, the convoluted fiber-glass wall lining, and perforated cold-air distribution ducts. Courtesy of Pacific Fruit Express Co., San Francisco, California.

of study. U.S. Department of Agriculture workers have now developed considerable information that is available to the public.

The use of liquid nitrogen for refrigeration of truck-trailers is currently of considerable interest. It has proved its value in supplying supplementary cooling in LTL delivery of frozen foods. Its adoption for "over-the-road" truck transport has been considered, but at present there are unsolved economic and operating problems which will probably delay for some time the adoption of this refrigeration system in this area of truck-transport activity. Proponents of this so-called "cryogenic" refrigeration system claim several advantages for this method, such as the avoidance of maintenance problems incident to the use of mechanical refrigeration sources. Proponents of the latter method, of course, present advantages for it (43).

Initially the cost of liquid nitrogen and its perhaps limited availability in replenishing stations may have deferred its adoption, but the alleged disadvantages have now been largely overcome. Liquid nitrogen refrigeration has so far found its most useful application in LTL, or local delivery service, because in that service the rather frequent necessary opening of truck doors will cause some fluctuation of air temperature, with a possible continuing rise of product temperature.

A study made by the U.S. Department of Agriculture (44), cooperating with a manufacturer of liquid nitrogen and a frozen food distributor, revealed that the time required to restore the air temperature surrounding the lading in an insulated delivery truck to its desired low initial level was much less if the liquid nitrogen system was used than that achieved with conventional refrigerating units. The warm external air which enters the trailer when doors are opened to make deliveries was very rapidly cooled again to the air temperature prevailing before the doors were opened. This repetitive interior rise in air temperature, which cannot be avoided in making deliveries, has been a most difficult problem to solve. The USDA studies, while not conclusive or complete, showed that liquid nitrogen refrigeration for this type of trucking is definitely useful. As the pressure on the liquid is released by a thermostatic device, it begins to boil almost instantly and expands about 700 times in volume. The cold nitrogen gas, distributed over the lading by appropriate duct or pipe dispersal, displaces the warm air that has entered the truck body. The increase in gas pressure causes a dissipation of the warm air and thus is an effective refrigerating medium.

In this LTL use, stations for replenishing the liquid nitrogen supply can, of course, be available at the distribution warehouse. This may not be equally true for over-the-road truck transport, and this factor may be one of the important ones which will ultimately determine whether liquid nitrogen refrigeration is economic in the latter type of transport.

Since about half the trucks in local delivery are refrigerated by the so-called cold plate method, brief reference here is appropriate. These plates, attached to the ceiling and walls inside the truck body, are filled with a eutectic chemical salt solution which is refrozen after each day's use. As it melts, while the delivery truck is in operation, it absorbs the heat leakage and maintains the truck body air temperature at an acceptably low level; however, its response is not as rapid as that of the liquid nitrogen system.

Eutectic solutions can be obtained for different melting points and consequent temperature-reducing effect. However, for local delivery truck use, the solution is generally composed of a salt and water, melting at a temperature range of −12 to −14°F. (−24.4 to −25.5°C.). The solution thus melts and refreezes at a predetermined temperature. The refrigeration capacity depends on the weight of the eutectic solution used in the plates.

An adequately insulated truck body, properly equipped with cold plates, can normally maintain an internal air temperature with 8–

10°F. (4–5°C.) of the eutectic solution temperature. The degree to which the desired low air temperature can be sustained during delivery operation is, of course, markedly influenced by the product temperature at the initial loading time. This should preferably be −5°F. (−20.5°C.) or lower.

A significant report (45), issued in 1966, makes a statement which applies to all handling of frozen foods: "The solution of most field operating difficulties lies in the proper education of responsible personnel." Responsible refrigerated truck and truck-trailer companies, engaged in the transport and local delivery of frozen foods, do have training courses and operational manuals for their employees and exercise operational supervision. Some companies have regional service training schools to instruct operating personnel in mechanical equipment maintenance and repair. Some begin with a two-week basic instruction in refrigerating equipment and refrigerant performance, combined thereafter with practical equipment operation experience under supervision of qualified instructors.

Motion picture instruction has also been used for this purpose. The American Trucking Associations some years ago produced an excellent motion picture film entitled "Watch That Temperature" for briefing truck operators on important factors in frozen food temperature protection (46).

The local LTL delivery operation poses special problems. It must be studied in the light of its purpose, problems, and function (47). The main difficulty is the consistent maintenance of proper low air temperatures in the truck body, as already discussed, despite the necessary repeated opening of truck doors to make store deliveries.

At times, these delivery trucks are called upon to make mixed product deliveries—frozen foods, nonfrozen but refrigerated items, and so-called dry grocery products—in order to achieve delivery economies. Such trucks can be equipped with moveable insulated partitions, which in the time period required for a day's deliveries can provide a reasonably temperature-segregated environment for each type of product (47, 48). In compartmentalized truck bodies of this kind, the freezer section air temperature may not be optimum for extended delivery, but in limited time periods and given careful operation and supervision, these vehicles do serve a very useful purpose, especially when the costs of making separate truck deliveries of each type of product are considered (Fig. 12.18).

In the never-ceasing search for improvements in frozen food transport in a dynamic food economy, considerable thought has been devoted to insulated, containerized transport and delivery on flat-bed

Fig. 12.18. Sea-Land trailer-ship loads units by crane, while trailer units on truck chassis await detachment and loading. Courtesy of Sea-Land Service Inc., Elizabeth, New Jersey, a subsidiary of McLean Industries, Inc.

rail cars and trucks or in uninsulated transport equipment. While this development has not achieved its full potential, reference to it is desirable in the context of new developments in the frozen food industry.

In a sense also, the "piggy-back" or "fishy-back" operations (*i.e.*, detached insulated and refrigerated trailer bodies, transported on rail flat cars or on ship deck) must be considered in this connection, since they are large containers on wheels.

Primarily, however, unit containerization refers to insulated but unrefrigerated boxes for store or institution delivery in which store orders of differing volumes of mixed products can be delivered. In order to be easily handled in stacking on delivery vehicles and because of various order volumes, it has been suggested that these boxes be constructed in module sizes. Also, since they must be returnable for reuse, considerable thought has been given to making use of "knock-down" construction, so that they can be conveniently returned to the producer in less bulk than when they contain frozen foods (49). Since the amount of dead weight to be handled is important, very efficient insulation materials have been recommended; these permit only a minimal thickness to be used to achieve the desired temperature protection.

The search is obviously for some adequately protective delivery service, where multiple-order lots go to a considerable number of drop-offs in any delivery period.

Some thought has been given to the possibility of direct delivery from producer to retailer or from market center to retailer, wherein the pricing of standard items, which vary only slightly or not at all, would be done by the producer or market center warehouse, thus permitting the retailer to put such items directly in his freezer display cabinets. It has been said that 80% of the retailer's cost of operations arises from labor and handling actions after merchandise is delivered to the retail store; the expediting of frozen food movement to the display cabinets or to back-room storage during this handling cannot be ignored as a good way to reduce the exposure of frozen products to high ambient air temperatures while the segregation and pricing are being done in the retail store.

This practice is admittedly more in the study or pilot stage at present than in actual practice, but it is mentioned here to illustrate again the fact that there is continuous search for better and less costly methods of bringing frozen products to the merchandiser and placing it in display positions for maximum customer response.

In closing this discussion, emphasis should be given to the joint responsibility of the transport operator and the warehouseman to integrate their services, so that frozen foods will have maximum quality protection at all times.

Consistent communication must prevail between the transport operator and the warehouseman in the scheduling of deliveries or receipts, with sufficient advance notice so that dock spots at the warehouse can be available to the trucker with minimum delay, and best use be made of warehouse personnel and equipment.

For instance, local deliveries are expedited from the warehouse if the dock foreman has been given prior notice so that he can not only provide loading dock spots promptly, but also initiate assembly of the ordered merchandise. Orders telephoned in on the day prior to the delivery date can be assembled without delay or confusion by night crews, so that early morning deliveries from the warehouse can be handled efficiently.

V. DISTRIBUTION OF FROZEN FOODS

Insofar as refrigerated transport is used in the distribution of frozen foods, the discussion of rail-car, truck-trailer, and truck performance has dealt with some technical aspects. However, there are some historical and policy factors which should also be considered. There are available two excellent publications, by Sawyer, Midura, and Vondell, dealing with frozen food distribution (50, 51).

Some years ago, there was much variation in the practices and equipment used for the distribution of frozen products. Some of the vehicles used, such as unrefrigerated and uninsulated station wagons, pick-up trucks, and even passenger cars, scarcely fulfilled even the most elementary requirements for the protection of frozen items from excessive heat absorption.

Ordinary practice was based on the supposition that since only relatively short time periods were involved in delivery of mixed lots to retail stores the adverse effect on product quality would be minor. This may have been true occasionally if ambient air temperatures were not high and time of exposure was short, but such casual operating practice failed to give sufficient weight to the cumulative, irreversible effect upon product quality of minor rises in product temperature.

One of the basic facts with which the frozen food industry has learned to live is that this new technology requires a quite unusual degree of conscientious carrying of the burden of careful compliance with rules, by one separate group of people after another in the whole "cold chain," and this extends all the way from the processing plant to the consumer's kitchen. Ownership or control of a lot of products may be vested in a number of different people en route; any one of them, through carelessness, laziness, or ignorance, can ruin the product that has passed through his hands—or may damage it significantly, without obviously ruining it. No one of them can relax his efforts in the belief that his minor participation could make no difference.

The distribution segment of the frozen food industry, probably more than the other segments, faces multiple hazards arising from accident, equipment breakdown, inexperienced operators, and the like, and this fact, a few years ago, led to official and industrial concern that the public interest required better control of practices at the packer level and in the entire pipeline of distribution, even though the public health record of frozen foods was singularly good, considering the billions of pounds of these products that have been consumed during the years since the industry became a significant part of the American food economy.

One result of this desire to keep the record good was the drawing up of model frozen food "codes" by committees of the Association of Food and Drug Officials of the U.S. (See Chapter 14), and by a voluntary group representing all segments of the frozen food industry (the All Industry Frozen Food Coordinating Committee). The draft codes which have resulted from the cooperation of these two groups since 1960 have served as partial models for regulatory statutes adopted by several of the states and by some foreign countries (52).

A. Distribution Systems

A number of modern industrial plans or systems of frozen food distribution are presently under study or in operation. These are designed to improve this important function and they contemplate better service and greater economies in distribution, but they are also based on sound procedures of product quality protection. One such plan, designated as the Total Distribution Plan, or TDP (53, 54), was begun by a refrigerated warehouse group in January, 1966. It proposes an essentially simple and practical action—the pooling of less-than-truckload (LTL) shipments into combined truck-trailer loads to major market centers.

The benefits of the procedure are evident—savings on transportation and warehousing costs by organization and simplification, minimizing warehouse inventories and increasing turn-over of products, centralization and potential automation, and consistently planned product protection with approved transport equipment and specific temperature control. Even in its initial stages of operation, TDP has already proved to be practical and economical. Extension of this service may well bring about even more values, as time passes.

Another plan, the March Plan (55), involves containerized frozen food delivery, enabling the processor to move frozen products in large volume to centrally located public refrigerated warehouses, where retailers can then pick up low-volume containerized shipments, adding nonfrozen and dry grocery items to the delivery truck for unloading at the individual retail establishments.

It is claimed that this system eliminates duplicate warehousing and LTL transport, minimizes small lot handing, reduces exposure of frozen products to high air temperatures, eliminates repetitive, low-volume deliveries, increases product turn-over, promotes pallet exchange, and permits full carlot or truck-load shipments from the processor's plant.

There are doubtless other plans and systems of similar character. These two are mentioned, without prejudice to other systems, as examples of the consistent trend to achieve transport and distribution economies and by lessening repetitive handling of frozen foods and reducing their exposure to undesirable environmental temperature.

One of the pioneers of the industry once remarked that the ideal way to distribute frozen foods would, no doubt, be to employ a long refrigerated tunnel, extending from the production plant to the consumer's kitchen. This ideal can, by constant effort, be approached as closely as may be permitted by realistic economy and practicability.

Frozen foods will then have the protection essential to the maintenance of the highest quality reasonably possible.

It may be well to repeat that commercial frozen foods have had, through the years, a singularly trouble-free record insofar as spoilage of a character dangerous to human health, such as botulism, is concerned. The reasons for this are easily understood but their discussion is not germane to the matters considered in this chapter (56–58).

B. Retail Handling of Frozen Foods

Every article of commerce must eventually justify its presence in the market place, and frozen foods are no exception. Quality, convenience, utility, and price are the primary criteria of their acceptance by the consuming public.

In a sense, the merchandising phase is part of distribution—delivery to a place where the potential consumer can see, select, and buy the product.

Whatever the processor has done to create a high-quality product and whatever other agents in transport, warehousing, and distribution have done to protect that initial quality as nearly as possible must be continued by the retailer in his establishment and in his display cabinets.

When deliveries of cased frozen foods are made to the retail store, these must be transferred as rapidly as feasible to a proper frozen food temperature environment, whether this be in a so-called back room storage chamber or through distribution of the retail packages to the retail freezer cabinet.

If the produce department of the store is responsible for the placement of frozen food items, its personnel must be instructed specifically that frozen foods must never be handled in the same manner as fresh produce. The latter is usually considered to be somewhat more tolerant of limited delays at store air temperatures, even if they are ultimately placed for sale under ice or in refrigerated display cabinets.

In view of all that has been published about frozen food temperature protection, it would seem to be unnecessary to emphasize anew this indispensable action. Experience has shown, however, that unless store employees are instructed to give first priority, under managerial supervision, to this matter, the many duties of these employees may sometimes cause neglect of this important action.

Some food chain companies which nationally merchandise almost three-quarters of the volume of frozen foods sold in this country, have special training programs and specific rules in order to emphasize the need for consistent and adequate frozen food protection.

For example, so-called dollies, holding frozen foods, should never be allowed to stand in the store work-room or in the aisles for extended periods of time, unprotected against store air temperatures, while other store tasks are performed.

C. The Maintenance of Frozen Food Retail Cabinets

Frequent, at least daily, inspection of the cabinet temperatures is another important protective action. Past experience has shown that accidental malfunctioning of cabinet refrigeration equipment, blocking of air circulation by ice formation in air ducts, or improper carton positioning may cause temperatures to rise; if this is sustained, because of inadequate supervision, it may contribute substantially to product quality degradation. In the older type of box cabinets, shoppers may disarrange the retail carton placement and cause some packages to be above the "frost line," exposed to higher air temperatures.

Frozen food packages must never be displayed in stacks outside the freezer cabinet, for instance during special sales.

At reasonable intervals, depending on the product turn-over, the contents of freezer cabinets must be checked and rearranged so that slow-moving items do not accumulate at the bottom and remain there unnoticed for undesirably long time periods. Actually, this is just good, basic merchandising practice.

D. Modern Frozen Food Retail Cabinets

As in the case of refrigerated transport equipment, striking improvements have been made in recent years in frozen food retailing equipment.

Initially, frozen food cabinets were of the box type, open at the top or with sliding top closure. The purchaser was compelled to lean over the edge to make the desired selection of items which were piled up inside the cabinet. This often resulted in disarrangement of the initial piling pattern and resulted in some items being covered by others. Hence, there was often little opportunity for visual examination, selection, and purchase, such as would be possible if all items were visible to the shopper in orderly array, as is true for other grocery items.

Subsequently, frozen food cabinets took the form of sloping and more shallow display arrangements and this was an improvement.

The rapid increase in number of frozen food items and brands, especially as prepared foods and bakery goods have gained wider public acceptance, created display-space problems. Only a limited number of items can be displayed for sale in cabinets of this type in relation to occupied floor space, which is an important factor in determining the store revenue derived from frozen food sales. Hence, it was inevitable that inventive genius would finally design upright frozen food cabinets, which, first of all, would permit the shopper to see clearly the individual packages, thus increasing potential incentive purchases. These cabinets also permitted more items to be displayed in a given occupied floor space (Fig. 12.19).

The first of the really significant upright cabinets was the Dual Jet type, in which the air flow was controlled by aerodynamic jet flow principles. One such jet at 0°F. (−18°C.) or lower is directed inward over the frozen products. The parallel guard-jet stream at store air temperature, forming a laminar interface without significant turbulence, continuously pushes away the warmer store air, thus providing a highly effective dual-stream air curtain. The products are clearly visible and the jet air flow at about 250 linear feet per minute does not interfere with the selection of the frozen food items by the shopper.

This pioneer cabinet type was one of the most important developments in the history of frozen food marketing. It had been suggested by a system of freezing in a designated area by creating a self-enveloping "air skin" of jet-directed refrigerated air invented by Edward Simons of Mill Valley, California, (59,60). It accomplished this without significant interference with air conditions in adjoining areas or creating

Fig. 12.19. Modern open-front upright frozen food merchandising cabinet. The frozen products are protected by a jet stream of air at 0°F. (−18°C.) or lower, while a parallel guard jet of air at ambient temperature prevents turbulence from exterior sources and cold air leakage to the store environment. Courtesy of Dual-Jet Refrigerating Co., Chicago, Illinois.

significant turbulence outside the space in which the freezing was to be done (Fig. 12.20).

From a protective standpoint, this upright cabinet provides consistent, quite uniform interior subzero temperature throughout the display area of the cabinet, thereby overcoming a problem of product temperature variation in cabinets of other types, as pointed out by Peterson, Gunderson, and others (26,61,62).

Upright cabinets refrigerated by gravity air flow principles are prone to spill cold air into the aisles, especially if exterior store air turbulence occurs. This turbulence is caused by movements of store air past or into the cabinet by accidental ventilation or movement of persons passing the cabinet. The cabinet function does not, of course, include air-conditioning of the aisles. Some gravity air flow cabinets have been modified by having glass doors which give better air leakage control and do permit considerable visual product exposure for the customer who is making selections for a purchase (63).

With the advent of the upright cabinets some study has been given to new display procedures and even to redesigning package labels so as to permit end-on display (64).

This chapter can well be ended with the assertion that few new industries have made so much technical progress in so short a time (approximately one generation) as has the frozen food industry. Its striking progress, so far, despite some unsolved economic and distribu-

Fig. 12.20. A modern Dual-Jet upright open-front cabinet installation for the sale of ice cream. Interior temperature is maintained consistently at −10°F. (−23°C.) throughout the cabinet. Courtesy of the Dual-Jet Refrigeration Co., Chicago, Illinois.

tion problems, has been achieved through unusual constructive cooperative action on the part of the several functional segments which comprise the industry, and private, educational, and governmental technological and engineering research groups.

REFERENCES

(1) W. B. Van Arsdel, *Food Technol., 11,* No. 1, 28 (1957).
(2) J. A. Peters and J. W. Slavin, *Bull. Inst. Intern. Froid,* Annexe *1962–1,* 553 (1962).
(3) J. P. Lane, *Food Technol., 18,* No. 7, 156 (1964).
(4) J. P. Lane, *Food Technol., 20,* No. 4, 197 (1966).
(5) Anon., *Refrigerated Transporter, 3,* No. 4, 45, (1966).
(6) G. C. Thomas, *Modern Materials Handling,* reprint, October, 1963.
(7) J. A. Eaton, *Modern Materials Handling,* reprint, November 1963.
(8) G. C. Thomas, *Modern Materials Handling,* reprints, March and April 1965.
(9) J. J. Strobel, *Better Management Services,* reprint, November 1965.
(10) Anon., *Quick Frozen Foods, 29,* No. 2, 106 (1966).
(11) J. J. Strobel, Summary of report prepared for Forest Products Laboratory, Northeast Forest Experimental Station, U.S. Forest Service, U.S. Department of Agriculture, September (1966).
(12) W. H. Sardo, Jr., *Soft Drink Industry,* reprint, October 25, 1966.
(13) P. L. Breakiron, *Refrigerated Transporter, 2,* No. 4, 27 (1965).
(14) L. L. Westling, *Maritime Reporter,* reprint, December 1, 1960.
(15) L. L. Westling, *Bull. Inst. Intern. Froid, Annexe 1965–5,* 219 (1965).
(16) L. L. Westling, *Bull. Inst. Int. Froid,* Annexe *1965–5* 227 (1965).
(17) D. K. Tressler and C. F. Evers, *The Freezing Preservation of Foods,* 3rd ed., Avi Publishing, Westport, Connecticut, 1957. p. 886.
(18) D. G. Guadagni and E. F. Jansen, *Food Eng., 29,* No. 7, 74 (1957).
(19) Anon., Agricultural Research Service, U.S. Dept. of Agric. ARS 74-14, September 1959.
(20) Anon., Honeywell, Inc., Minneapolis Minn. Descriptive brochure, 1961.
(21) A. G. Huntsman, *Bull. Biological Board (Canada), 20,* 44 (1931).
(22) O. C. Young, *Contrib. Can. Biol. Fisheries, 7,* No. 39, 407 (1933).
(23) O. C. Young, *Can. Refrig. J., 18,* No. 11, 21 (1952).
(24) C. P. Lentz, *Can. J. Technol., 33,* 265 (1955).
(25) C. P. Lentz, *Proc. Congr. Intern. Froid,* 10^{e}, Copenhagen *3,* 298 (1959).
(26) L. van den Berg, *Food in Canada,* May 1966, 37.
(27) Anon., *Frozen Food Age 15,* 28 February (1967).
(28) R. M. Powell, *Quick Frozen Foods 29,* No. 5, 227 (1966).
(29) R. M. Powell, *Food Industry Yearbook.* Unnumbered reprint (1966–1967).
(30) W. J. Hoover, *Food Eng. 28,* No. 6, 56 (1956).
(31) A.C. Beardsell, *Pulp Paper Mag. Canada, 61,* T589 (1960).
(32) H. C. Diehl, *Good Packaging 15,* No. 7, 99 (1954).
(33) American Society of Heating, Refrigerating, and Air Conditioning Engineers. *ASHRAE Guide and Data Book.* Fundamentals and Equipment, 1965 and 1966; Applications, 1966 and 1967. The Society, New York.

(34) Anon., NAFFPC Committee Approved Refrigerator Cars for Shipping Frozen Foods, Revised July 15, 1967. National Association of Frozen Food Packers, Washington, D.C.
(35) Anon., Refrigerated Transporter, reprint (1966).
(36) C. W. Phillips, W. F. Goddard, P. R. Achenbach, H. D. Johnson, and R. W. Penney, *Agricultural Marketing Service.* U.S. Dept. of Agric. Marketing Research Report *433*. September 1960.
(37) S. M. Stubbs, *Refrigerated Transporter, 3,* 24 (1966).
(38) C. P. Lentz and E A. Rooke, *Can. Food Indus.,* reprint, June 1966.
(39) D. W. Kuenzli, *Agricultural Marketing Service,* U.S. Dept. of Agric., AMS *446,* June 1961.
(40) Anon., *Refrigerated Transporter, 3,* 17 (1967).
(41) Anon., *Quick Frozen Foods, 29,* No. 3, 152 (1966).
(42) D. W. Kuenzli, *Agricultural Marketing Service,* U.S. Dept. of Agric., Marketing Research Report *540* (1962).
(43) R. K. Krishef, *Refrigerated Transporter, 3,* 25 (1966).
(44) J. J. Dougherty, Jr., *Agricultural Marketing Service,* U.S. Dept. of Agric., AMS *426.* December 1960.
(45) Anon., *Refrigerated Transporter, 3,* 33 (1966).
(46) R. F. Allyne, address delivered at All-Industries Conference on Frozen Foods, Cornell University, Ithaca, N.Y., April 10, 1962.
(47) M. D. Ratner, *Refrigerated Transporter, 3,* 168 (1966).
(48) Anon., *Refrigerated Transporter, 3,* 27 (1966).
(49) Anon., *Quick Frozen Foods, 29,* No. 3, 160 (1967).
(50) F. M. Sawyer, T. F. Midura, and R. M. Vondell, *Handling Frozen Food—A Guide to Distributors,* Univ. of Mass., Amherst, Cooperative Extension Service Publ. 350. January 1960.
(51) F. M. Sawyer, T. F. Midura and R. M. Vondell. *Handling and Merchandising Frozen Food.* Univ. of Mass., Amherst, Cooperative Extension Service, November 1960.
(52) Anon., *Conference on Frozen Food Quality,* Agricultural Research Service, U.S. Dept. of Agric. Albany, Calif. ARS 74–21. November 4–5, 1960.
(53) Continental Freezers of Illinois, *Total Distribution Plan—New Economies in Distribution for Frozen Food Packers,* Brochure, January 1966.
(54) Anon., *Quick Frozen Foods, 29,* No. 8, 112 (1967).
(55) Anon., *Quick Frozen Foods, 29,* No. 3, 160 (1966).
(56) M. F. Gunderson, *Quick Frozen Foods, 24,* No. 5, 50 (December 1961); *24,* No. 6, 89 (January 1962); N. 7, 131 (Feb. 1962).
(57) A. C. Peterson, K. E. Karlson, and M. F. Gunderson, *Cornell Hotel Restaurant Quart., 4,* No. 1, 25 (1963).
(58) H. K. Burr and R. P. Elliott, *J. Am. Med. Assoc., 174,* 1178 (1960).
(59) Edward Simons, *Inst. Intern. Froid.* Commissions III, IV, V. Annexe 1960–63, 543 (1960).
(60) Edward Simons, *Food Eng., 25,* No. 11, 57 (1953).
(61) M. F. Gunderson, *Quick Frozen Foods, 24,* No. 1, 41 (1961).
(62) A. C. Peterson, In *Low Temperature Microbiology Symposium,* Camden, N.J. (1961).
(63) Anon., *Quick Frozen Foods, 27,* No. 12, (1965).
(64) Anon., *Frozen Food Age, 12,* No. 10, 5 (1964).

Chapter 13

SURVEYS OF INDUSTRY OPERATING CONDITIONS AND FROZEN PRODUCT HISTORIES

C. H. Byrne and K. G. Dykstra

Birds Eye Division, General Foods Corp., White Plains, New York

I. INTRODUCTION

Time and temperature are two equally important factors in the retention of high quality in frozen foods during distribution to the consumer. The producers of frozen foods, recognizing the fact that they cannot readily control the time their products are held in the several phases of distribution systems, have urged the maintenance of zero temperature (0°F., −18°C.) or lower all the way. Where exposure to uncontrolled high temperatures is likely—as in the transfer from delivery trucks to store cabinets—much effort has been exerted to encourage frozen food handlers to keep the time of exposure to a minimum.

The frozen food industry is based on the premise that most of the products it produces will, if held at zero or below, retain good quality for at least one year. As Kertesz (1) concludes, only a very small percentage of the total volume of frozen foods marketed is exposed, through accident or equipment breakdown, to temperature and time conditions which can lead to a public health hazard, but there are exposures to temperatures that lower quality. When consumers complain that the product they purchase is not of good quality and the producer establishes that the lack of quality is attributable to exposure to high temperatures, there is a need to determine where and for how long the product was handled improperly. It is the purpose of this discussion to present some of the procedures that have been used to determine what time and temperature conditions prevail in the frozen food distribution system, and the results obtained from the studies.

II. TEMPERATURE DETERMINATION

The temperature of the frozen food in a package can accurately and readily be determined if a thermocouple is frozen into the product and the leads are attached to an appropriate instrument. This method has been used in laboratory studies but is not practical for field work. Munter, Byrne, and Dykstra (2) in their report of field surveys of frozen food distribution systems presented a plan for determining temperatures in all parts of the system. A summary of their recommended methods is presented in Table 13.1. They used dial-type thermometers throughout, but modified the manipulation to attain the most reliable reading in each location. Guadagni and Jansen (3) compared the thermometer and thermocouple methods of determining temperature of frozen foods and concluded that the temperatures determined by the two methods showed good agreement. After trying several procedures for determining the temperature of frozen foods in cases they concluded that the preferred procedure was as follows:

1. Open top of case.

2. Remove corner package.

3. Punch a hole through the case wall (from the inside) at the exact position desired.

4. Replace the package.

5. Close the top.

6. Insert the thermometer in the punctured hole for a distance of 3 in. They further suggest waiting 5 min. for the thermometer to reach equilibrium. The best location for the thermometer for different types of packages, according to these authors, is between top and bottom for products with air spaces, between sides of solid pack items and cans, and between top and bottom of metal-ended containers.

Frozen food producers and distributors not only have familiarized

Table 13.1. A Summary of the Methods of Determining Temperatures in the Several Areas of a Frozen Food Distributing System (2)[a]

Area	Phase measured	Procedure
Storage room	Air	Read thermometer after 10 min. at front, center, and rear of room.
Case	Air and product	Insert stem of thermometer through top flap opening so that tip is about 1 in. from bottom of case. Allow 10 min. before reading.
Delivery truck	Air	Place thermometer on top of case, close doors, and read after 10 min.
Product temperature	Product	Concentrates: Without removing can from case, puncture holes in side and center of lid. Insert thermometer in outside hole and allow to precool 30 sec. Then move thermometer to center area and read after 30 seconds.
		Fruits and vegetables: Drill holes 1 in. from end of narrow face and in center of narrow face. Precool thermometer in first hole 30 sec. Then move thermometer to center hole and read after 1 min.
Retail cabinet	Air	Place thermometer on top of packages 2 in. below top of load and 2 in. from bottom of cabinet. Read after 10 min.

[a] Equipment: all-metal, dial-type thermometer.

themselves with the proper and best methods of determining temperatures in frozen foods and in storage areas, but have made opportunities to communicate these methods to frozen food handlers and warehousemen responsible for seeing to it that shipment temperatures are properly maintained.

In a recent survey of temperatures in the top level of retail store frozen food cabinets, several investigators were employed in widely separated areas. To insure uniformity in the collection of data, specific

Table 13.2. Instructions for Taking Freezer Temperatures

1. Go to freezer cabinet containing frozen vegetables (if more than one contains frozen vegetables, use one that carries the regular line rather than combinations, boil-in-bag, or other specialty vegetables).
2. Remove one package from the top layer of packages in the display case. If product is at various levels, select the row which is piled the highest. If cabinet is uniformly filled, select a center row. Whichever row is selected, take the center item going back in freezer.
3. Place the thermometer on top of the package that was just below the one removed. Before setting down thermometer, bend outside frame back so that it can be read without having to touch it.
4. Leave the thermometer in place for 5 min., then read it *without* touching it at all. (Touching the thermometer in any place throws it off considerably.)[a]
5. Take the thermometer reading from the *center* of the needle. Even though the scale divisions are in divisions of five degrees, record readings in *one degree units.*

   Example: a little left of + 5 would be + 4; halfway between − 10 and − 5 would be − 8

   Be careful to note the distinction between temperatures above and below zero since it is not particularly obvious on the thermometer.
6. Go to the store's own thermometer in the same freezer cabinet and record its reading. This will generally be a verbal rather than a numerical reading. Please be as accurate as possible within the limitations of each thermometer.

[a] Note: It would be advisable to attach some sort of little sign to each thermometer which advises consumers to keep hands off.

instructions on how to secure a temperature reading were issued. These instructions are reproduced in Table 13.2.

III. TIME–TEMPERATURE INDICATING DEVICES

Honeywell, Inc., has developed a small, inexpensive, expendable device for indicating combined time–temperature exposure, the Time–Temperature Indicator or TTI. The manufacturer describes the device and its application as follows: The TTI mechanism is a tiny electrolytic battery that starts to operate as soon as it is activated by the user. Its action speeds up as the temperature increases within the range of 0–80°F., moving the red coloration band along its yellow scale face. The user notes the position of the red–yellow boundary line in reference to the scale reading. To tell the equivalent time–temperature

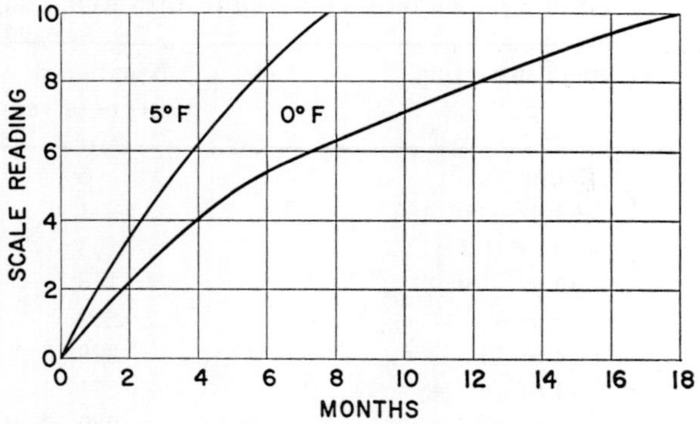

Fig. 13.1. Typical performance response curve for Time–Temperature Indicator.

history since activation, simply refer to the performance curves which are furnished with the device. These are similar to Fig. 13.1.

To activate the TTI, lay the device on a firm surface with its printed face down. Using any convenient, smooth, hard object such as a coin or the rounded end of another TTI, press firmly and roll along the centerline of the back of the case for about 1 in. A cracking sound will be heard when the vial containing the indicating fluid breaks; this activates the device. The indicator should immediately be placed into service, as the recording reaction starts within 1 min.

In one typical application a TTI device rated to show a scale reading of 8 after 12 months of exposure to a temperature of 0°F. showed a reading of 8 after only 5½ months of exposure. By reference to the performance response curve (Fig. 13.1) it can be seen that the curve which passes through a scale reading of 8 at a time of 5½ months is the curve for a temperature of +5°F. This, then, would be the "equivalent temperature" indicated by the device. Time–Temperature Indicators have been used in rather extensive surveys of food distribution facilities, as discussed later in this chapter. See also Guadagni (4).

IV. MATHEMATICAL TREATMENT OF SURVEY DATA

Temperature or time data collected in surveys usually can be expected to follow the normal distribution pattern. A convenient treatment of such data which permits simple estimation of percentages of temperatures above or below any chosen level is to plot the values on probability paper. The validity and application of this procedure is discussed by Velz (5) in his "Graphical Approach to Statistics." If the number of observations is large, the number of occurrences within convenient temperature intervals, such as 5°, will be tabulated, (see

Table 13.3. Top-Level Temperatures Observed in 1610 Retail Store Cabinets

Temperature group, °F.	Number of times observed
Below −25	6
−24 to −20	31
−19 to −15	53
−14 to −10	148
−9 to −5	269
−4 to 0	328
+1 to +5	274
+6 to +10	229
+11 to +15	130
+16 to +20	81
+21 to +25	32
+26 to +30	18
Over 31	11
	Total: 1610

Table 13.3). An estimate of the probability of encountering an instance of a temperature lower than some specified value can then be computed from the formula $P = 100\ (m/n+1)$, where m is the accumulated number of observations at the midpoint of the interval in question, and n is the total number of observations. (See Table 13.4, footnotes.)

In a recent survey of temperatures found at the top level of product in frozen food cabinets in stores, the data shown in Table 13.3 were collected. When these data are treated as described a work sheet as in Table 13.4 is developed. Finally, using arithmetic probability graph paper (probability scale one way, uniform scale the other), plot the probabilities calculated in the work sheet as in Fig. 13.2. (Individual points fall closely enough upon a straight line to indicate that the frequency distribution is nearly normal.) From such charts any desired level of reference can be readily extracted.

For example:

1. The mean (50% probability) temperature in the 1610 cabinets was 1.0°F.

2. Twenty percent of temperatures were above +9.5°F.

3. Three percent of temperatures were above +20°F.

4. Fifteen percent of temperatures were below −10°F.

5. The standard deviation of the temperature data is 10°F. (In Fig. 13.2 this is indicated by the difference in ordinates at points on the

Table 13.4. Work Sheet for Calculating Probability of Encountering Temperatures Reported in Table 13.3

Temperature group, °F.	Plotting temperature, °F.	No. times observed	Prior [a] accumulated total of observations	m no.[b]	P value
Below −25	−27	6	0	3	0.2
−24 to −20	−22	31	6	21	1.3
−19 to −15	−17	53	37	63	3.9
−14 to −10	−12	148	90	164	10.2
−9 to −5	−7	269	238	372	23.1
−4 to 0	−2	328	507	671	41.7
+1 to +5	+3	274	835	972	60.3
+6 to +10	+8	229	1109	1223	75.9
+11 to +15	+13	130	1338	1403	87.1
+16 to +20	+18	81	1468	1508	93.6
+21 to +25	+23	32	1549	1565	97.1
+26 to +30	+28	18	1581	1590	98.7
Over +31	+33	11	1599	1604	99.6
			$n=1610$		

[a] The prior accumulated total of observations column is developed for use in calculating the m number. Values in this column are the accumulated total of number of observations made prior to the group in question.

[b] m number, which is in fact the consecutive number of the "middle" value in the group, is determined by adding one-half the number of observations in the group in question to the prior accumulated total of observations.

straight line at cumulative percentages 34.1% above and 34.1% below the mean of all observations.

A particularly interesting comparison was made possible by superimposing two such probability charts resulting from surveys made in 1952 and 1966. The 1952 mean temperature was 7°F. higher than the 1966 mean, indicating that conditions have improved by a general lowering of temperatures this much during the elapsed 14 years. From the same two charts it can be seen that in 1952 40% of temperatures in the top levels of cabinets were above +10°F. while in 1966 only 20% were above +10°F. Thus if two observers looked at the two sets of results one might conclude: "Conditions are only 7°F. better than they were 14 years ago," while the other could say: "Things are twice as good as they were." If the probability charts are presented, both of these conclusions and many more can be readily drawn.

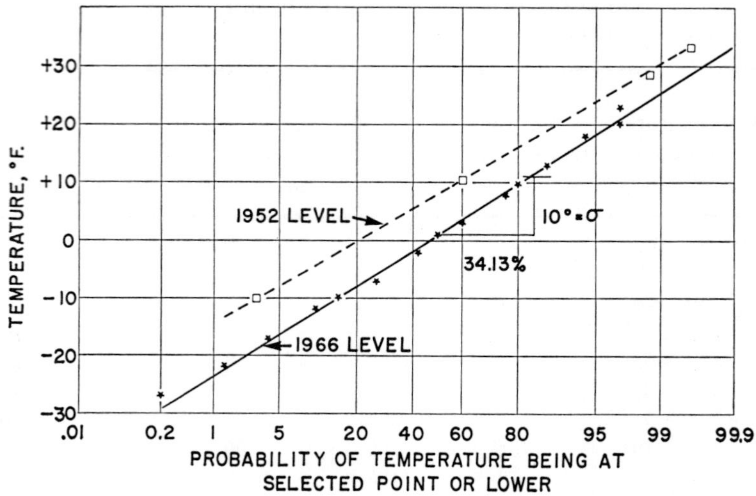

Fig. 13.2. Top temperatures in store cabinets.

V. TIME DETERMINATION

Time of exposure to a given temperature condition is as important as the temperature itself. Most reports of time found in the literature are estimates of "minimum to maximum," based on a number of observations.

Munter, Byrne, and Dykstra (1) made the following estimates:

Warehouse rooms	One day to one year
Assembly rooms	One day to two weeks
Delivery trucks	24 hr.
Cabinets	One day to two weeks

McColloch *et al.* (6), basing their time estimates on thermograph records, developed the following time pattern:

Producer's warehouse	1 month
Transportation	10 days
Terminal warehouse	6 months
Marketing	14 days
Home refrigeration	7 days

These times have a distribution curve, just as temperature does, and each phase has its characteristic probability curve. By placing customer-return questionnaires in packages of a group of selected frozen foods, and on their return recording the time elapsed from date of production to date of consumption, it has been possible to construct the time–probability curve for a total distribution cycle. During the period of

Table 13.5. Time Between Production and Use for Five Frozen Vegetables in Commercial Distribution System

Time between production and home use, months	Number of replies	Probability of product having been in distribution this length of time or less
1	2	0.3
2	29	5.2
3	36	15.3
4	72	32.2
5	35	48.9
6	49	62.0
7	30	74.4
8	18	81.9
9	25	91.7
10	7	93.6
11	8	95.9
12	8	98.4
	Total 319	

over one year during which the products were produced and distributed in normal channels, 319 returns were received and the time taken for the products to travel from producer to consumer was noted. The data shown in Table 13.5 were collected and the probabilities calculated. When plotted on probability paper as described above, the chart Fig. 13.3 was developed. The curved line indicates a skewed

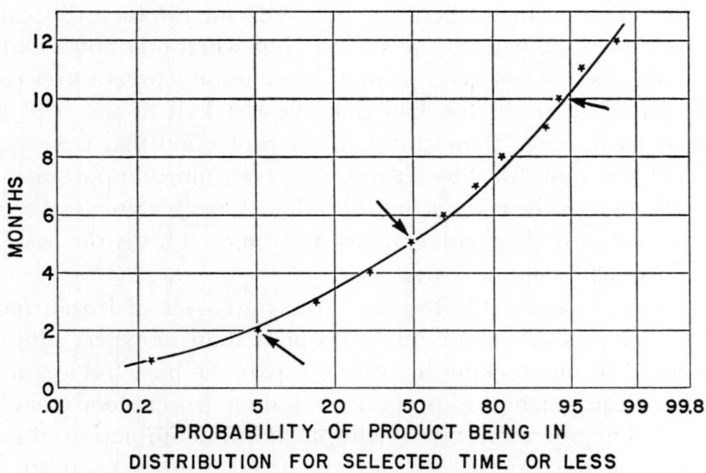

Fig. 13.3. Time frozen food is in distribution.

distribution of times, corresponding to the fact that it is physically impossible to move product from producer to consumer in the very short time periods. This curve tells us that:

5%	of	product	reaches	the	consumer	in 2	months	or	less
50%	"	"	"	"	"	" 5	"	"	"
94%	"	"	"	"	"	" 10	"	"	"

When the total distribution chain was broken into three phases the mean times for each of the phases were as follows:

In primary warehouses	1.95 months
In transportation and delivery to stores	1.3 months
In store and home	2.25 months

The sum of these three phases is 5.5 months, which is very close to the 5-month mean discovered for the total distribution cycle described above.

VI. RESULTS OF TIME AND TEMPERATURE SURVEYS

Having thus far considered the methods of determining temperature and time in frozen food distribution channels and in the various segments of the channels and how to treat the data, the obvious next question is, what picture develops when these methods are employed? Some of the results have already been revealed in the examples used in discussing procedures. When time–temperature devices were sent through a normal distribution channel and through separate segments of the channel, the mean times and temperatures found were those shown in Table 13.6. The mean temperature of +3°F. for the total distribution chain does not look bad on the surface, but when one notes that it is this low only because of good primary warehouse temperatures (in the −10°F. range), it seems too bad that the last half of the trip to the consumer should more than cancel all the protection that was supplied in the first few months. The second, and even more important, point of view which must be taken from this table of mean values is that there is a "top half' of higher temperatures and times. This is the area from which consumer complaints arise.

It is no more reasonable to expect the purchaser of frozen food to accept 5% of product whose quality is poor than to expect him to be unconcerned if one stocking in every ten pairs he buys has a run. The mechanical equipment for properly handling frozen foods has been developed. The operation of the equipment and elimination of misuse are matters of human error which must be taken more seriously.

Table 13.6. Mean Lengths of Time Frozen Foods are in the Three Basic Phases of Distribution, and "Equivalent" Temperature in Each Phase [a]

Distribution phase	Mean time products remain in phase, months	"Equivalent temperature" in phase, °F.
Primary warehouse	1.95	−10
Transportation and distribution to store	1.3	+3
In store and home	2.25	+8
Total distribution chain	5.5	+3

[a] "Equivalent" temperatures were determined by use of time–temperature integrating indicators.

VII. STORE CABINET TURNOVER

Another facet of frozen food distribution which requires that the "mean" time and temperature values not be thought of as maximums is the normally slower turnover rate of some products. Not all frozen food items have the relatively rapid turnover rate characteristic of the most popular items. To get some actual values for turnover rate, data on deliveries to stores over a period of time were collected and analyzed. Knowing the number of facings or stacks of product in a retail cabinet through which the individual products were sold and the number of units in each stack, it was possible to calculate the average number of hours a package of the particular product remained in the retail cabinet. The results of these calculations are shown in Table 13.7. These are values for a group of metropolitan high-volume stores. The implication to be derived therefrom is that these are the minimum lengths of time any of these packages of product can be expected to have to spend in a retail cabinet.

The time and temperature values in Table 13.6 were based on the high-volume, fast-moving products (peas, potatoes, French green beans, broccoli spears, and cut corn), which have turnover rates of 3.8–8.5 days. The slower-moving products in the 10- to 50-day category can be presumed to have proportionally longer average dwell times in the other phases of distribution. Their salvation probably lies in the fact that they are also held proportionately longer in the low-temperature primary warehouses.

Table 13.7. Store Cabinet Turnover Rates for 26 Frozen Food Items [a]

Product	Days dwell in cabinet [b]
Peas	3.8
Chopped spinach	5.2
Cut corn	5.3
Fried onion rings	5.3
Squash	5.3
Peas and carrots	6.1
Corn on the cob	6.1
French-cut green beans	6.1
Mixed vegetables	7.1
Broccoli spears	7.2
Cut green beans	7.6
Fish sticks	7.8
Cauliflower	7.9
Succotash	8.2
Asparagus spears	8.3
French-fried potatoes	8.5
Orange juice	8.7
Crinkle-cut, French-fried potatoes	8.8
Mixed fruit	9.8
Baby lima beans	10.4
Brussels sprouts	11.4
Fordhook lima beans	13.3
Artichoke hearts	13.5
Raspberries	15.4
Strawberries	22.2
Peaches	55.6

[a] Based on units sold through supermarket stores in a large metropolitan area. Data available: number of units of each item sold, facings or stacks of product through which each item was sold, time required for sale of item. Assumptions: stack height of ten cartons, and first-in-first-out movement. From C. H. Byrne, 1966.

[b] Part of dwell time is probably spent in store holding room.

VIII. HOME HANDLING AND HOME EQUIPMENT

Once a product is in the hands of the consumer no one but the consumer himself has any control over the way it is handled. The frozen food industry has tried to explain the need for holding frozen foods at zero temperatures or lower, particularly if they are held more than rather short times. The consumer may agree, but if facilities for maintaining zero storage are not readily available there is little he can do except to buy more frequently and store the product he buys only for short times. The temptation to do otherwise often presents itself; attractive sale prices which encourage "stocking up" are offered, and the convenience of having the product on hand is obvious.

McCracken and Churchill (7) have investigated the subject of frozen food storage in household refrigerators rather extensively. They found that the "actual frozen food load" was only about half of the amount computed by the standard formula for "manufacturer's rating." This could lead to overloading and misuse. Then in both gas and electric types of frozen food holding equipment these workers found, in 16 different pieces of equipment, with several different probable operating conditions, frozen food holding temperatures ranging from −12.7°F. to +25.9°F. The distribution of 148 observations made by these authors is shown in Table 13.8.

These are values found in new equipment; in used or improperly operating equipment, the results would most certainly be worse. Even so, 33% of the temperatures in household frozen food holding equipment were above +10°F.

Previously discussed data indicate that the phase of distribution between entering the retail store and final use in the home is one of the longest, and that the holding time in the store is usually not long for economic reasons; the home holding facilities would therefore appear to be a very fruitful area for improvement. The "equivalent temperature" of the "in store and home use" phase of distribution has been shown to average +8°F., while the mean temperature in store cabinets is only +1°F. Again, the added burden of misuse in this phase seems to be in the home area. This of course does not excuse the people in the earlier phases of distribution from properly handling the product. One improperly held shipment in the early stages of distribution could nullify all the best efforts of a thousand home users.

Table 13.8 Temperatures found in 16 Frozen Food Holding Compartments in Household-Type Freezers and Refrigerator–Freezer Combinations Under a Variety of Operating Conditions (7).

Temperature range in frozen food holding area	Number of times observed	per cent of observations
Below 0°F.	44	30
0 to +10°F.	55	37
+10 to +20	33	22
Over +20	16	11

IX. SUMMARY

The methods for determining the time frozen food spends in the various phases of distribution and for determining the temperatures which prevail in each phase and overall have been described and the results are presented and discussed. The most significant finding from all the studies is that both times and temperatures in all phases of distribution are spread over wide ranges and that conditions in the "top 10%" or "bad quarter" (or however one cuts off the undesirable end of the curve) are sufficiently severe to create deterioration in frozen food quality observable by the consumer. The consumer may complain, and the producer will then try to correct the problem, but too often the consumer may accept what he finds as "par for the course" and develop a resistance to frozen foods which is in the long run much worse for the producer.

There appears to be no way to place "limit controls" on the top portions of either the time or the temperature pattern, but a lowering of the temperature or of the total distribution time will decrease the exposure to damaging conditions. This is being done in practically all segments of distribution. The question is whether it is being done fast enough and far enough to produce the maximum acceptance of frozen foods in the shortest possible time at the least cost. It is inevitable that until the conditions are made widely known nothing can or will be done to correct them. The data already available should be studied, and where information is missing it should be sought out. Corrective action should then be undertaken along the lines proposed.

REFERENCES

(1) Z. I. Kertesz, in *Precooked Frozen Foods,* A Symposium. Advisory Board on Quartermaster Research and Development. Chicago, May 20, 1954. National Academy of Science, Committee on Foods, National Research Council, Washington, 1955, p. 68.
(2) A. M. Munter, C. H. Byrne, and K. G. Dykstra, *Food Technol., 7,* 356 (1953).
(3) D. G. Guadagni and E. F. Jansen, *Food Eng., 29,* No. 7, 74 (1957).
(4) D. G. Guadagni, *Frosted Food Field, 36,* No. 4, 42 (1963).
(5) C. J. Velz, *Water Sewage Works, 97,* 195, 309 (1950).
(6) R. J. McColloch, R. G. Rice, M. B. Bandurski, and B. Gentili, *Food Technol., 11,* 444 (1957).
(7) E. C. McCracken and F. Churchill, *ASHRAE J., 4,* No. 11, 27 (1962).

Chapter 14

VOLUNTARY AND OFFICIAL STANDARDS OF GOOD PRACTICE

F. M. Sawyer and K. M. Hayes

Department of Food Science and Technology, University of Massachusetts, Amherst, Massachusetts

I. INTRODUCTION

Most frozen foods reach consumers with a quality that is the pride of the industry. Occasionally, the excellent quality of frozen foods achieved by conscientious packers has been degraded by inappropriate handling in the postfreezing cold chain (*i.e.,* the succession of storages, transfers, and displays to which frozen foods are subjected in moving from the processing lines to the consumer and throughout which they must be protected by a suitable low temperature). Such mishandling may result from ignorance, carelessness, or even irresponsibility on the part of one or more persons responsible for frozen food care. It is conceivable that foods may become unsafe for consumption because of microbial growth while frozen products are inadvertently held at above thawing temperature. The extent of quality change and the limitations of microbial growth relative to frozen food temperature are explained in detail in other chapters of this book.

It is the responsibility of public health officers that frozen as well as other foods in the marketplace are safe, unadulterated, and equal to the representations of purveyors. These are also goals of the frozen food industry. In addition, the various components of industry desire that the quality of frozen food achieve standards of constancy and perfection that will assure an expanding and profitable continuation of its joint enterprise. To achieve these ends, regulatory agencies and industry groups have developed and offered official codes and statements of voluntary good practices to guide the proper care, handling, storage, and distribution of frozen foods.

In the late 1950's and early 1960's two associations bent their efforts toward developing statements of good practices for preserving frozen food quality. The Association of Food and Drug Officials of the United States (AFDOUS) and the Frozen Food All Industry Coordinating Committee (FFAICC) provided, respectively, the efforts of public health and industry groups to develop such statements.

Confusion existed at first between safety and quality aspects of frozen food temperature requirements. The distinctions between the two in terms of rate of quality degradation and proliferation of pathogenic microorganisms relative to temperature, which are presented in this book, were clarified in the minds of those most concerned at an historic meeting of the AFDOUS committee and the FFAICC with scientists of the U.S. Department of Agriculture at the Western Regional Research Laboratory in Albany, California in 1960.

In 1961, AFDOUS adopted a code (1) to serve as a guide for regulatory agencies that were considering the establishment of laws or regulations to govern the production and handling of frozen foods. The FFAICC, which was made up of representatives of trade associations and other groups covering all commercial aspects of food freezing, developed, at about the same time, recommended voluntary operating practices. Its activities were aimed both at protecting consumers and assuring commercial growth for food freezing. This coordinated effort by the industry represented a serious attempt at self-improvement and had an important influence on the course of many state and municipal movements toward promulgation of frozen food handling regulations.

At the state and municipal levels, AFDOUS-type regulations for temperature control have been established legally in a total of seven states at present and are currently under study in several others. A few municipalities across the United States have established local ordinances for temperature control. At the international level, numerous governments have implemented regulations governing production and handling of frozen foods, especially certain food commodities such as fish. Recently, important progress is being made toward establishment of international commodity standards for many frozen products.

Our objective in this chapter is to review briefly the circumstances related to development of voluntary and official standards of handling practice of frozen foods, primarily in the United States, and to summarize certain aspects of the more important codes and regulations.

II. BACKGROUND AND IDENTIFICATION OF PROBLEMS

A. Consumer's Viewpoint

Immediately after World War II, the American consumer was confronted with a relatively new form of food. As wartime controls relaxed, frozen foods were produced in larger quantities and in wider variety than ever before. Many new packers entered the market. sometimes with little previous food packing experience. Some products originally packed for the wartime trade became available. The consuming public, with money to spend, tried and tested the new frozen products eagerly.

The equipment and personnel of companies at all stages of the cold chain were overtaxed. Desperate competitive positions forced some relaxation of quality standards that had existed in the prewar industry. As the consumer bought these foods, many questions on quality, variety, packaging, and usage were asked (2–4). Quality and quality improvement were among the factors deemed most important. Technologists (5,6) echoed the importance of quality and quality control in the frozen food area, and pointed up the need for further research.

Not long after the lowering of price control barriers, and the entrance of many new firms into the field, price grades of frozen foods entered the market place (7). Although quality of these brands was of a questionable and varied level, they did serve to introduce frozen foods to many consumers who otherwise might not have tried them. Consumers used more frozen foods, became more knowledgeable and discriminating, and began to differentiate between price and quality. As demand rose, the need for better quality control became evident (8–10).

B. Problem Areas

In a 1946 industry survey (11), the areas requiring improvement were as follows: original product, processing, quality uniformity, new products, temperature protection, packaging, pricing, descriptive labeling, and merchandising.

From the time of the above survey in 1946 until the present, these areas have been of concern. To better identify these areas and their problems, they will be discussed by market function.

1. Production

Within the concept of production, raw material for processing has been and is the subject of intensive research by numerous investigators. Among the important factors relating to final product quality that have been studied are: varietal selection, maturity, yield, mechanical harvesting adaptability, disease resistance, color and flavor, and suitability for freezing (4–6,12). In the processing plant, equal emphasis has been placed on methodology, equipment design, materials handling and flow, quality control techniques, and quality maintenance (5,13,14).

Quality improvement in frozen foods has continued through adoption of improved processing equipment and methods, wider application of grades and standards, and increased emphasis on research and development programs (12,14,15). Product protection in the marketing chain through use of improved packaging materials, and application of packaging technology to food characteristics and properties resulted in lengthened shelf life and better protection (16). While new techniques in production and quality control improved the quality and stability of frozen foods, other areas of the market system were changing equally as rapidly.

2. *Storage and Distribution*

Immediately after the end of World War II, the frozen food industry realized that facilities for handling frozen foods from processing plant to retail outlet needed improvement (2). Problem areas in warehousing centered around the need for improved temperature control, materials handling, inventory control, and sanitation. The rise in production due to consumer demand increased the pressure for needed changes and improvements. Lack of temperature control in assembly areas, both at the warehouse level and at distribution points, caused loss of quality. Also, lack of adequate freezing facilities for lowering product temperature from above-zero levels caused loss of quality (4,8,17–19).

The movement of frozen foods from the processor to the retail outlet has been implicated throughout the years as an area where quality loss occurs. All segments of the transportation industry—rail, truck, and water—were concerned with temperature control and physical equipment. High-priority problems included improvement of equipment, temperature control, reduction in use of poorly insulated carriers, and educational efforts to emphasize the effects of temperature on frozen food quality (20,21).

3. *Retail*

Quality loss in frozen foods at the retail level was attributed to insufficiently trained personnel and inadequate physical facilities for handling, storage, and display. Efforts to change these areas resulted in many improvements, yet did not fully attain the level needed for quality protection. With the advent of voluntary and official standards of handling for frozen foods, the retail area of frozen foods was closely scrutinized and controlled in an effort to improve practices and prevent frozen food quality loss (4,22,23).

As frozen food consumption increased, along with the introduction of frozen convenience foods and especially their more extensive use in public food service, public health agencies showed more interest in studying this area of the food field (24). Of potential concern were foods such as cream pies, stuffed poultry, frozen dinners and entrees, and other foods that historically in the fresh form had proven troublesome to the public's health. Although minor infractions in processing, handling, and storage were found and corrected, no major hazard to human health from frozen foods was found. Surveys by numerous investigators indicated that sanitation in the frozen food field was as necessary as in other areas of food production (24–26). Both state and federal agencies, along with interested industry groups, sponsored educational programs to insure that safety of frozen foods remained at a high level (19).

While interest by public health agencies in frozen food safety was of deep concern, industry leaders were also concerned over the reaction of consumers to quality, and the possible effect on future acceptance and market growth. In 1948, government and industry joined together to investigate frozen food quality and related areas.

Spearheaded by the U.S. Department of Agriculture, Western Regional Research Laboratory, a massive study on all aspects of frozen food quality was conducted (26a). The FFAICC and the U.S. Department of Agriculture Research and Marketing Advisory Committees provided a first-hand industry knowledge that guided the research program and guaranteed early acceptance and use of its findings. Additional emphasis was gained in extending research findings by use of seminars, workshops, technical publications, exhibits, and oral presentations. The principal trade associations and scientific groups, including the Refrigeration Research Foundation, The National Association of Refrigerated Warehouses, The National Association of Frozen Food Packers, The National Frozen Food Association, The North-West Canners and Freezers Association, The California Freezers Association, and others became cooperating and supporting implements for extending the new knowledge. Use of the information to improve frozen food handling at the grocers' level was a project sponsored by the U.S. Department of Agriculture Federal Extension Service and conducted by the University of Massachusetts. Trade magazines and technical journals covered all the material developed to bring it to the immediate attention of industry. The Federal Extension Service prepared an illustrated lecture that was used in many places to help industry improve frozen food quality and safety.

Additional research work by many groups in the areas of transportation, warehousing, and retailing, combined with the results of the Albany laboratory, provided the basis for proposals of frozen food handling codes (22,23,27–32).

III. VOLUNTARY AND OFFICIAL STANDARDS FOR HANDLING

A. AFDOUS Code

The Association of Food and Drug Officials of the United States (AFDOUS) Frozen Food Handling Code had its beginnings in the early to middle 1950's when frozen foods in wide variety began to appear on the market. With the increase in frozen prepared foods, leading authorities discussed and studied the needs and requirements of frozen food processing, storage, and handling (10,14,16,17,21,22). Although the frozen food industry had sponsored and supported the time–temperature tolerance investigations beginning in 1948, several years passed before the industry and other groups began to emphasize the need for improvement (7,13,23,25).

At about the same time, AFDOUS invited the National Association of Frozen Food Packers to work with it on developing uniform handling rules (23,30,32). The Code was published by AFDOUS primarily as a guide for state and local legislation (1).

Publication of the Code brought immediate action in all segments of the industry (28,29) to support the strong movement to improve frozen food handling voluntarily (23,32–35).

An analysis of the AFDOUS Code reveals that it is concerned primarily with temperature control, and sanitation, without setting microbiological standards, because the methodology of the latter remains unsettled and no known public health hazard exists. A condensation of the AFDOUS Frozen Food Handling Code (1), and the pertinent points within the various sections follow:

1. Retail Section

Section I. Definitions

a. Defines terms used in code pertaining to retail section, including frozen foods.

b. Details method for measuring internal product temperature.

Section II. Equipment

a. Outlines requirements for back room storage facilities, method of storage, defrosting, and temperature measurement.

b. Explains display case design, construction, equipment (including a temperature measurement), product load line requirements, and display case location.

c. Specifies that both back room storage facilities and display cases must be capable of maintaining a product temperature of 0°F. (–18°C.) or lower for frozen foods when used for the appropriate purpose.

Section III. Handling Practices

a. Stipulates that delivery of frozen foods shall not be accepted by a retail outlet when internal product temperature exceeds 0°F. (–18°C.).

b. Offering of frozen food for sale whose internal temperature exceeds 0°F. (–18°C.) is forbidden.

c. Requires that frozen food be held, stored, and displayed under conditions to prevent any contamination of the product.

2. *Warehousing Section*

Section I. Definitions

a. Defines frozen foods, warehouse and associated areas including breakup rooms.

b. Describes methods of determining internal product temperature in trucks and rail cars for both full and partial loads.

Section II. Equipment

a. Requires that warehouses be capable of maintaining a product temperature of 0°F. (−18°C.) or lower under all temperature and load conditions. Associated warehouse area must maintain air temperatures of 0°F. (−18°C.), or lower.

b. Specifies methods and locations of temperature measuring devices, and maintenance of dated temperature records.

Section III. Handling Practices

a. Specifies that internal product temperature of frozen food shall not exceed 0°F. (−18°C.) during receipt, loading and unloading, and shipment.

b. Describes procedure for acceptance and handling of incoming frozen food lots with an internal product temperature in excess of 0°F. (−18°C.). Freezing of foods must be done in rooms, or areas distinctly separate from storage rooms.

c. Outlines coding of frozen food lots, method of storage, and sanitation procedures during storage and defrosting.

Section IV. Sanitary Provisions

a. Specifies sanitary conditions within and without warehouse including toilet, washroom, and dressing room facilities not regulated by state legislation. In states where such legislation is constitutional, warehouses must comply with state ordinances covering sanitary codes.

3. *Transportation Section*

Section I. Definitions

a. Defines terms and meanings of frozen foods, transportation, vehicle, carrier, device, route delivery, and delivery truck.

Section II. Equipment

a. Describes refrigeration methods or facilities to maintain an air and product temperature of 0°F. (−18°C.) or lower in vehicles of transportation loaded with any frozen foods, and a temperature–measuring device that measures inside temperature that can be read from the outside of the vehicle.

b. Specifies that route delivery trucks meet the above requirements, and be equipped with curtains, flaps, or port doors to maintain refrigeration during stops.

Section III. Handling Practices for Over-the-Road Transportation

a. Enumerates precooling requirements, sanitary packaging, loading techniques and air circulation, use of refrigeration during unloading and loading, and other previously stated practices for temperature maintenance. Internal product temperature will not exceed 0°F. (−18°C.) during any handling, or in transportation.

Section IV. Handling Practices for Route Delivery

a. Requires that in addition to the above specifications, each individual consignment lot will be refrigerated to a temperature of 0°F. (−18°C.) or lower, and that devices will be precooled to a temperature of 0°F. (−18°C.) or lower before being loaded with frozen foods.

b. Specifies closing of truck doors during cessation of loading or unloading.

Section V. Sanitary Provisions

a. Requires interior surfaces of vehicles and devices used for transporting frozen foods be clean and odor-free prior to being loaded.

4. Recommendations for Design and Construction of Frozen Food Processing Equipment

a. The recommendations detailed within this portion of the Code apply to the design, materials, construction, and installation of equipment used in the processing, holding, and packaging of ready-to-eat frozen foods and the processing and holding of gravies, batters, or other food components capable of supporting rapid bacterial growth.

b. Includes standards, regulations, definitions, and classifications of equipment and equipment standards that apply to the above foods and their processing.

c. Discusses general principles, materials, design and construction, and installation of equipment used with these foods from a sanitation and engineering aspect.

B. The FFAICC Voluntary Operating Practices

The Frozen Foods All-Industry Coordinating Committee (FFAICC), with Dr. H. C. Diehl as chairman, was established in 1960 after informal meetings and discussions by leading industry representatives over a period of more than five years. In addition to their long-time concern for maintaining high quality in frozen foods they had a particular concern that if regulatory codes were to be introduced they should be founded on scientific fact, so that neither consumers nor industry would be penalized by unnecessarily rigid requirements. They wished also to avoid the crippling effect on interstate commerce such as would occur if wide disparities in handling requirements were developed by individual states or local governments. They were concerned that immediate enforcement of rigid temperature requirements would mean that much of the industry's existing equipment (especially for transportation and retail display) would become arbitrarily obsolete. This, it was believed, would impose a staggering financial burden of replacement costs upon the industry, that would price frozen foods beyond the reach of many consumers.

Every segment of the frozen food industry participated to some extent in the activities of the committee through key representatives from the private sector, their trade-association counterparts, and other interested persons. On June 5, 1961, the Recommended Voluntary Operating Practices for the Handling of Consumer Packaged Frozen Foods was published and officially endorsed and subscribed to by the following organizations: American Trucking Associations, National Association of Food Chains, National Association of Frozen Food Packers, National Association of Refrigerated Warehouses, National Association of Retail Grocers of the United States, National Fisheries Institute, National Frozen Food Association, and National Prepared Frozen Food Processors Association.

The FFAICC Voluntary Operating Practices (VOP) recommended temperature control limitations which were initially more lenient than the AFDOUS Code. These tolerances were to be tightened progressively (from 1961 through 1964) so that its temperature requirements by 1965 would be at least as strict as those of AFDOUS in most respects. The intervening period would allow time for equipment replacement, and for education and training of personnel. The following is a summary of the salient features of the VOP (36):

1. *Foods for Freezing*

a. Deals with processor handling conditions.

b. Specifies processor warehouses temperature 0°F. (−18°C.) or lower.

c. Specifies limit of 10°F. (−12°C.) and transfer in an insulated refrigerated vehicle for off-premise completion of freezing, with an 8-hr. time limit for reduction of temperature to 0°F. (−18°C.) or lower.

2. *Warehouse Equipment*

a. Specifies maintenance of an air temperature of 0°F. (−18°C.) or lower, in all storage rooms.

b. Requires accurate temperature measuring devices in all storage rooms.

3. Warehouse Handling Practices

a. Specifies that when frozen foods are received with product temperatures of 15°F. (−9°C.) or higher, the warehouseman should recommend to the owner or consignee that such products be subjected to special handling procedures designed to reduce product temperature to 0°F. (−18°C.) or lower as rapidly as possible.

b. Describes other handling procedures, marking or coding, order assembly, defrosting, stacking, etc.

4. Transportation Equipment

a. Specifies that vehicles should be able to maintain a product temperature as follows:

1961 through 1962	0°F. (−18°C.), tolerance to 15°F. (−9°C.)
1963 through 1964	0°F. (−18°C.), tolerance to 10°F. (−12°C.)
by 1965	0°F. (−18°C.), tolerance to 5°F. (−15°C.)

(It should be noted that the AFDOUS Code allowed a tolerance of 10°F., or 5.5°C.)

b. Vehicles should be equipped with an appropriate temperature measuring device.

c. Route delivery trucks should be equipped with curtains, flaps in doorway, or with port doors and should be capable of maintaining a product temperature of 15°F. (−9°C.). (The AFDOUS Code specifies maintenance of a product temperature of 0°F., or −18°C.).

d. Over-the-road equipment should be capable of maintaining a product temperature of 5°F. (−15°C.).

5. Handling Practices for Over-the-Road Transportation

a. Specifies that any frozen food shipment should not be tendered to, or accepted by a carrier for transportation when product temperature exceeds 0°F. (−18°C.).

b. Specifies that vehicles should be precooled to an air temperature of 20°F. (−6.6°C.) or lower before loading; that products should be packaged or wrapped; that 2½ hr. should be allowed for loading, with a tolerance of 2°F. (1°C.) for each additional hour of loading, not to exceed a total tolerance of 10°F. (5.5° C.).

c. Sets a minimum product temperature rise during loading not to exceed 10°F. (5.5°C.) years 1961–62; 5°F. (3°C.) years 1963–64; 0°F. (0°C.) by 1965. Product temperature during movement should not exceed 15°F. (−9°C.) 1961–62; 10°F. (−12°C.) 1963–64; 5°F. (−15°C.) by 1965, with a 5°F. (3°C.) tolerance allowed if multiple pickup and/or delivery is performed.

6. Handling Practices for Route Delivery

a. Specifies product temperature be maintained as follows:

1961–62	15°F. (−9°C.) or lower, with a tolerance of 5°F. (3°C.) for a period not to exceed 6 hr.
1963–64	15°F. (−9°C.) or lower, with a tolerance of 5°F. (3°C.) for a period not to exceed 3 hr.
by 1965	10°F. (−12°C.) or lower, with a tolerance of 5°F. (3°C.) for a period not to exceed 3 hr.

b. Product should not be tendered to, nor accepted by a carrier when product temperature exceeds 0°F. (−18°C.).

c. Vehicles should be precooled to an air temperature of 20°F. (−6.6°C.), or lower, before loading and frozen foods should be packaged or wrapped.

7. Equipment for Retail Stores

a. Frozen Food Storage Facilities

1. Facilities should be capable of maintaining an air and product temperature of 10°F. (−12°C.), or lower, during 1961–62; 5°F. (−15°C.), or lower, during 1963–64; and 0°F. (−18°C.), or lower, by 1965.

2. Requirements for defrosting, temperature measurement, and air circulation for cabinet-type and walk-in type storage facilities are given.

b. Display Cases

1. Cases should be capable of maintaining 10°F. (−12°C.), or lower, during 1961–62; 5°F. (−15°C.), or lower, during 1963–64; and 0°F. (−18°C.), or lower, by 1965.

2. Specifies defrosting, temperature measurement, product load line, air circulation, etc.

8. Retailer Handling Practices

a. Specifies that frozen foods should not be accepted when product temperature exceeds 10°F. (−12°C.) during 1961–62; 5°F. (−15°C.) during 1963–64; and 0°F. (−18°C.) by 1965, with any permitted delivery tolerances allowed.

b. Procedures for inventory rotation, display cabinet stocking, receiving and storage are given.

9. Definitions

Product temperature is defined and the procedure recommended for its measurement is described.

The over-all impact of FFAICC's Voluntary Operating Practices has been to stimulate the industry to improve its equipment and handling procedures. At the state level, FFAICC has provided guidance and assistance to local industry committees, and has aided in preventing establishment of unrealistic official regulations by some local authorities for handling frozen foods. As of this writing, FFAICC is still functioning in this capacity, and maintains its headquarters in Washington, D.C.

C. State Regulations

Consideration has been given by many states in the U.S. to establishment of official regulations governing the handling of frozen foods. In only seven states, however, have such regulations been promulgated, and in each instance the requirements stated have been similar to those recommended in the AFDOUS Code. States with regulatory handling requirements are as follows (37):

1. Connecticut and Maryland require 0°F. (−18°C.) temperature with 10°F. (5.5°C.) tolerance for unavoidable variation.

2. Florida requires handling at 0°F. (−18°C.) and recommends transportation and storage at −10°F. (−23°C.).

3. Georgia requires 0°F. (−18°C.) for frozen foods in retail cabinets with a 10°F. (5.5°C.) tolerance for unavoidable variation.

4. Illinois and Pennsylvania require handling at 0°F. (−18°C.) with a 15°F. (8°C.) tolerance for unavoidable variation.

5. Massachusetts: as in *1* above, but also specifies bacteriological standards (in spite of the fact that these were deliberately omitted from the AFDOUS Code).

Other states are expected to develop similar legislation at some time in the future. Industry compliance with established standards, according to state regulatory officials, is said to be reasonably good, but probably in no state have the regulations been strictly and widely enforced.

D. Standards for Handling in Other Nations

1. *International*

With the rapid development in recent years of food freezing industries in many nations and the growth of international trade in frozen foods, the need for uniform "rules" became apparent.

A group of experts on standardization of "deep-frozen" products representing the Economic Commission for Europe (ECE) held its first meeting on frozen food standards in Geneva, Switzerland, September 14–17, 1964. Austria, Belgium, Federal Republic of Germany, Finland, France, Italy, Netherlands, Poland, Spain, Sweden, Switzerland, Union of Soviet Socialist Republics, and United Kingdom took part. An observer from the United States mission in Geneva was present.

The work schedule established included: (*a*) definitions of deep-frozen foods, including processing, handling, transporting, and storage in all stages of distribution; (*b*) text of product statements to serve as pilot standards; (*c*) packaging and labeling; and (*d*) framework of standards.

The group of experts decided that close working relations with the International Institute of Refrigeration (IIR) and the Organization for Economic Cooperation and Development (OECD) should be established. It was agreed that the IIR booklet, *Recommendations for Processing and Handling of Frozen Food* (38) constituted a basic document for the processing and handling of frozen food. The principles in the IIR booklet do not differ markedly from the AFDOUS Code. Temperature recommendations consist of the following: (*a*) storage at 0°F. (−18°C.) or lower; (*b*) loading and transport no higher than 0°F. (−18°C.), except for fruit juices, their concentrates, and fatty fish to be no higher than −4°F. (−20°C.) and ice cream no higher than −9°F. (−23°C.). No part of a cargo should rise above 5°F. (−15°C.) during transport and unloading after transport, except ice cream is limited to −4°F. (−20°C.); and (*c*) retail trade requirements are that product temperature should be constant at 0°F. (−18°C.), or lower and should not be higher than 5°F. (−15°C.) when received.

At the first meeting, "General Rules and Definitions" were adopted for processing and handling frozen food. They closely resembled those described in the IIR booklet.

These principles, designed as "rules," lacked the flexibility necessary for commercial freezing technology and handling practices, and this weakness led to broadening the scope of international standard formulation to include the FAO/WHO Codex Alimentarius Commission. The first meeting of the joint ECE/Codex Alimentarius Frozen Food

Standardization group was held in Geneva, Switzerland, September 6–10, 1965. Fourteen countries were represented.

At the first meeting the general rules and definitions for frozen fruits and vegetables were expanded to cover all quick frozen foods and they generally relaxed or eliminated restrictive requirements on rate of freezing. The requirements of the general standard relative to handling followed as guidelines the AFDOUS Code and the IIR booklet. The general standard was revised in 1966 at Geneva and again in 1967 at Rome.

The question of making the requirements of the General Standard mandatory was discussed at length by the expert committee with no general agreement. However, the Codex Alimentarius Commission at its meeting in Paris in October, 1967, ruled that the requirements should be accepted as a code of general principles. Specific requirements to be mandatory would have to be expressed in specific commodity and product codes.

The joint ECE/Codex Alimentarius meetings for quick (deep) frozen food in 1966 and 1967 developed draft standards for quick frozen strawberries, peas, and a number of other individual commodities. Discussions on quick frozen strawberries and peas were advanced to the point where drafts were distributed to the different countries for comment.

General requirements for labeling and hygienic practices are being developed by Codex Alimentarius committees for these topics. As with the general statement on frozen foods, specific statements must be incorporated into commodity or product standards to become mandatory.

Basic sampling procedures and acceptance criteria, along with test methods for taking temperature, net weight, thawing techniques, cooking procedure, enzymes, etc., are in draft form and are soon to be discussed by the expert committee.

The IIR jointly with OECD is currently engaged in the development of a code of practice for frozen fish (39). Individual international commodity standards are being prepared also for a number of different frozen products (40,41).

2. *Individual Nations*

It is beyond the scope of this chapter to attempt to describe voluntary and official handling standards in each nation in which they have been established. Those discussed are representative.

a. Denmark. Regulations have been established for frozen foods containing meat. These regulations are commonly regarded as applicable to other types of frozen foods. Storage temperature specified is −4°F. (−20°C.) and it is anticipated that eventually −13°F. (−25°C.) will be required. In transport, a minimum temperature of 0°F. (−18°C.) is required, but for less than 24-hr. duration, 5°F. (−15°C.) is allowed. At retail, for storage and display, 0°F. (−18°C.) is required. Retail packages must be marked with a shelf-life date, calculated for a steady temperature of 0°F. (−18°C.) (40–42).

b. Norway. Minimum temperature requirements have been established for frozen fish products. Storage and transport for more than 24 hr. must be at −4°F. (−20°C.); at retail display, 0°F. (−18°C.) or lower must be maintained. Maximum storage time at −4°F. is 4 months, and for 9 months storage −18°F. (−28°C.) is required (40–42).

c. Other Nations. Regulations or voluntary codes for temperature control of frozen foods, especially fish products, have been established in England, France, West Germany, Switzerland, Sweden, Portugal, and Canada (39–43). In general, these specify storage at 0°F. (−18°C.) or lower; in transport, a rise in product temperature up to 5°F. (−15°C.) is often allowed; retail display must not exceed 0°F. (−18°C.). Temperature requirements for fish products are usually more strict than for other frozen products, *e.g.*, in England, a voluntary code specifies storage at a steady −20°F. (−29°C.).

Representatives of several nations are currently studying the advisability of modifying regulations to take account of the fact that the keeping characteristics or high-quality life of certain types of frozen food products may vary significantly at a given storage temperature, such as 5°F. or 0°F. (−15° or −18°C.).

E. Internal Standards in Industry

1. Processors

Processors of frozen foods have instituted quality control programs of their own to insure proper handling. The basis for these programs has been the AFDOUS Code, the FFAICC Code, or combinations of the two. An outstanding example is a manual of handling and shipping "Zero-Frozen Foods Most Important Figure," compiled by Gorton's of Gloucester (33). The manual covers six areas: *(1)* "Instructions for Handling Damaged or Defrosted Merchandise"; *(2)* "Instructions to Trucking Companies"; *(3)* "Instructions to Brokers"; *(4)* "Instructions to Warehouses"; *(5)* "Handling Instructions"; *(6)* "Protecting Frozen Food from Temperature Damage." Temperatures specified are 0°F. (−18°C.) or lower, except that an allowance of +10°F. (5.5°C.) is permitted upon arrival at destination. Product temperatures above this limit require the consignee to notify Gorton's Traffic Department, and to refuse the shipment. Suggestions for loading, stacking, and temperature measurement are also given. Incoming frozen products must meet a 0°F. (−18°C.) or below requirement, whether transported by truck or ship.

Another example of product care is the system used by the Kitchens of Sara Lee to insure low storage temperatures prior to shipment. The holding freezer, and the assembly area are held at −10°F. (−23°C.) with all products being palletized (44). The combination of palletizing and low storage temperatures insures maintenance of low temperatures during loading and in transport.

2. *Warehousing*

Within the warehouse industry, maintenance of temperature is not an acute problem, yet does require constant supervision. Quincy Market Cold Storage and Warehouse Company (45) sets −5°F. (−20.5°C.) or lower as the limit for all frozen food storage except fish, which is held at −10°F. (−23°C.) or lower. Constant temperature checks are made to insure adherance to these temperature limits.

3. *Transportation*

A set of "Uniform Highway Loading Requirements for Frozen Foods" was established in 1966 by the Refrigerated Division, Common Carrier Conference and The National Association of Frozen Food Packers (46). The requirements are for voluntary acceptance by shippers and carriers of frozen foods. The major requirements are that trailers be capable of maintaining product at 0°F. (−18°C.) or lower throughout the load in all movements; that all units be precooled to at least 20°F. (−6.6°C.) before loading; that product temperature be no higher than 0°F. (−18°C.) when tendered for loading and that it be in saleable condition; that drivers be instructed in temperature measurement methods, and be provided with testing thermometers, and that carriers protect the product until ready for unloading by the consignee.

4. *Retail*

In the retail area, numerous chains, voluntary groups, and independent stores have initiated programs designed to control the receipt, handling, storage, and display of frozen foods so as to protect their quality. Among the first to institute a chain-wide program was First National Stores (31). Operating in an eight-state area, First National in cooperation with the University of Massachusetts surveyed the practices being used within the chain. As a result of the survey, and to meet the requirements of the Massachusetts law, the following program was used to train store personnel in the care and handling of frozen foods. The course was divided into nine subject areas: *1.* Introduction, *2.* Product information including history, time–temperature tolerance, and the Massachusetts law, *3.* Work methods improvement, *4.* Preparation for ordering and receiving, *5.* Use of the order guide, *6.* Receiving, *7.* Price marking, *8.* Principles and methods of display, and *9.* Merchandising.

In all training, it was emphasized that the chain would meet the requirements of the Massachusetts law, *i.e.,* 0°F. (−18°C.), or lower, back room storage and display at 0°F. (−18°C.) or lower, and receiving at 10°F. (−12°C.) or lower.

Another supermarket chain, Stop and Shop, Inc. has set 0°F. (−18°C.) as the maximum temperature allowed in any phase of the frozen food chain. In addition, it requires that warehouse storage be at −10°F. (−23°C.) and that incoming loads meet the 0°F. (−18°C.) limit. At the store level, a 0°F. limit is imposed and enforced to ensure quality maintenance (35).

F. U.S. Military Regulations

Regulations in effect for the handling of frozen foods are specified in TM–743–200 by the Department of the Army (47). Temperatures required for storage are 0°F. (−18°C.) or lower for all commodities unless stated otherwise for a specific food. Products covered by individual specifications are dairy products, eggs and egg products, and baked goods. The length of storage permitted for various foods is also given.

IV. CONCLUSION

The extensive dissemination of the information which is summarized in this book has led to better compliance with the AFDOUS Code and the FFAICC statement by the frozen food industry and has created a technologically sound base for continued growth. Since 1960, efforts and capital expended to achieve this goal have resulted in improved product quality and higher consumer acceptance. The frozen food industry has benefited from this experience, and in the future it can be expected that stringent handling practices will continue to be observed.

REFERENCES

(1) Anon., *Assoc. Food and Drug Officials U.S., Quart. Bull. 23,* No. 4, 191 (1959).
(2) Anon., *Quick Frozen Foods, 8,* No. 6, 66 (1946).

(3) Anon., *Quick Frozen Foods, 8,* No. 6, 70 (1946).
(4) Anon., *Quick Frozen Foods, 13,* No. 9, 39 (1951).
(5) H. C. Diehl, *Frozen Food Industry, 1,* No. 2, 16 (1945).
(6) M. A. Joslyn, *Quick Frozen Foods, 8,* No. 8, 71 (1946).
(7) Anon., *Quick Frozen Foods, 13,* No. 12, 51 (1951).
(8) M. J. Copley, *Quick Frozen Foods, 19,* No. 8, 177 (1957).
(9) L. W. Hammack, *Quick Frozen Foods, 8,* No. 9, 101 (1946).
(10) G. Klomp, *Quick Frozen Foods, 13,* No. 8, 112 (1951).
(11) Anon., *Quick Frozen Foods, 8,* No. 6, 57 (1946).
(12) D. K. Tressler and C. F. Evers, *The Freezing Preservation of Foods,* 3rd ed., Vol. 1, Avi Publishing, Westport, Connecticut, 1957.
(13) D. O. Haynes, *Quick Frozen Foods, 13,* No. 8, 93 (1951).
(14) A. E. Symons, *Quick Frozen Foods, 13,* No. 5, 44 (1950).
(15) G. Matthews and W. G. Young, *Can. Food Ind., 28,* No. 3, 15 (1957).
(16) T. W. Koch, *Quick Frozen Foods, 13,* No. 11, 62 (1951).
(17) T. H. Allegri, *Refrig. Eng., 64,* No. 9, 60 (1956).
(18) K. G. Dykstra, *Refrig. Eng., 61,* No. 9, 54 (1956).
(19) R. P. Straka and J. L. Stokes, *Quick Frozen Foods, 18,* No. 7, 182 (1956).
(20) D. G. Hickey, *Quick Frozen Foods, 13,* No. 3, 53 (1950).
(21) W. Teagle, *Quick Frozen Foods, 13,* No. 3, 55 (1950).
(22) Anon., *Quick Frozen Foods, 18,* No. 7, 111 (1956).
(23) Anon., *Food Processing, 21,* No. 8, 36 (1960).
(24) Committee on Frozen Food Sanitation, International Association of Milk and Food Sanitarians, *J. Milk Food Technol., 20,* 145 (1957).
(25) R. P. Elliott and H. D. Michener, *Appl. Microbiol., 9,* 452 (1961).
(26) A. T. Miller, H. M. El-Bisi, and F. M. Sawyer, *Exptl. Sta. Bull.,* No 548, Univ. of Mass., Amherst, 1964.
(26a) W. B. Van Arsdel, *Food Technol., 11,* 28 (1957).
(27) Anon., *Frosted Food Field, 31,* No. 1, 8 (1960).
(28) Anon., Fleet Owner Special Report, 1960.
(29) Anon., *Frosted Food Field, 30,* No. 3, 18 (1960).
(30) Anon., *Canner and Packer, 130* No. 3, 31 (1961).
(31) Anon., Frozen Foods Training Program, Federal Extension Service, U.S. Dept. of Agriculture, Washington, D. C. 1962.
(32) H. P. Milleville, *Food Process., 21,* No. 7, 29 (1960).
(33) Anon., *A Manual of Handling and Shipping,* Gorton's of Gloucester, Gloucester, Mass. 1960.
(34) Anon., *Conference on Frozen Food Quality,* ARS–74–21, Agric. Res. Serv., U.S. Dept. of Agric., Washington, D. C., 1960.
(35) T. F. Midura, R. M. Vondell, F. M. Sawyer, K. M. Hayes, T. W. Leed, and E. L. Hayes, *Quick Frozen Foods, 23,* No. 2, 130 (1960).
(36) Anon., *Quick Frozen Foods, 23,* No. 11, 50 (1961).
(37) National Association of Frozen Food Packers, *Tech. Sci. Serv. Bull. No. 7,* 1965.
(38) International Institute of Refrigeration, *Bull. Inst. Intern. Froid,* Annexe, Paris, 1964 *Recommendations for the processing and handling of frozen foods.*
(39) J. Kuprianoff, personal communication, Federal Research Centre for Food Preservation, Karlsruhe, West Germany, 1967.
(40) H. Somm, *Annexe 1962–1 Bull. Inst. Intern. Froid,* 385.

(41) H. Somm, personal communication, ROCO Food Packing Co., Rorschach, Switzerland, 1966.
(42) J. E. Pedersen, personal communication, The Danish Meat Research Institute, Roskilde, Denmark, 1966.
(43) Anon., *Quick Frozen Foods, 29,* No. 10, 48 (1967).
(44) Anon., *Frosted Food Field 37,* No. 12, 5 (1964).
(45) P. D. Burrill, personal communication, Quincy Market Cold Storage and Warehouse Co., 1967.
(46) Anon., Report *B–152,* Bureau of Commercial Fisheries, *Market News Service,* Boston, Mass. 1966.
(47) Anon., *Storage of Special Commodities, TM–743–200,* Dept. of the Army, 1966.

AUTHOR INDEX

Numbers in parentheses are reference numbers and show that an author's work is referred to although his name is not mentioned in the text. Numbers in *italics* indicate the pages on which the complete references are listed.

K

L

M

N

SUBJECT INDEX

Q

R

T

U

V

W